TOURISM TODAY:

A Geographical Analysis

Second Edition

TOURISM TODAY:

A Geographical Analysis

Second Edition

Douglas Pearce
Department of Geography
University of Canterbury
Christchurch
New Zealand

LONGMAN

Addison Wesley Longman Limited
Edinburgh Gate
Harlow
Essex CM20 2JE, England
and Associated Companies throughout the world

First published 1987
Second edition 1995
Reprinted 1996, 1997

British Library Cataloguing in Publication Data
A catalogue entry for this is available from the British Library.

ISBN 0-582-22822-0

Library of Congress Cataloging-in-Publication Data
A catalog entry for this title is available from the Library of Congress.

Set by 3 in 9/10^1/2 Plantin
Produced by Longman Singapore Publishers (Pte) Ltd.
Printed in Singapore

For Chantal and Rémi

Contents

List of Figures

List of Tables

List of Plates

Acknowledgements

Preparation of the second edition of this book has involved the compilation of a large data base of recent statistics and new material. I am indebted to the many national, regional and local tourist organizations who responded generously to my requests for information. I would also like to acknowledge the many friends and colleagues in different parts of the world who provided advice, feedback, references and reprints or with whom I have spent time in the field since the first edition was written. My thanks also go to Paula Wilson for her patience, skill and care in assembling much of the new data, to Alastair Dyer, Anita Pluck and John Thyne for cartographic assistance, to Linda Harrison and Anna Moloney for typing the manuscript and to Longman Group for editorial advice and assistance. Research support was provided by a grant from the Geography Department, University of Canterbury. I am especially grateful to Chantal and Rémi for their continuing support and encouragement.

Doug Pearce
Christchurch, February 1994

We are indebted to the following for permission to reproduce copyright material:
The author, Dr G. A. Ashworth for Fig. 9.5 from Fig. 2 (Ashworth, 1985); British Tourist Authority for Tables 2.5 from Table 64 (*Digest of Tourist Statistics No. 16*, 1992) and 4.3 from Tables 20, 30 (*Digest of Tourist Statistics No. 9*, 1981; *Digest of Tourist Statistics No. 16*, 1992); the author, Dr S. G. Britton for Fig. 1.8 from Fig. 3 (Britton, 1980a); the Canadian Association of Geographers for Fig. 1.10 from Fig. 1 (Butler, 1980); Centre des Hautes Études Touristiques for Figs 1.11 from Fig. 3 (Gormsen, 1981), 4.4 from Fig. 3.3 (Kissling, Pearce *et al.*, 1990), 6.2 from Fig. 12A and Table 6.2 from p165 (Rognant, 1990); the editor, *Cornell HRA Quarterly* for Table 1.1 from pp13–16 (Plog, 1973); Elsevier Science Ltd for Figs 1.13 (Reprinted from *Annals of Tourism*, 20, 3, Oppermann, Tourism space in developing countries, Fig. 2, Copyright (1993)) and 4.7 (Reprinted from *Annals of Tourism*, 15, 1, Pearce, Tourist Time Budgets, Figs 1 and 2, Copyright (1988)), both reprinted with kind permission from Elsevier Science Ltd, The Boulevard, Langford Lane, Kidlington OX5 1BG; European Communities for Fig. 2.2 and Table 2.2 from Fig. p10 and Table p19 (Commission of the European Communities, 1986), the author, Dr P. C. Forer for Table 4.4 from Table 111 (Forer and Pearce, 1984); Florida Department of Commerce for Figs 5.5 and 5.6 from Figs p5, 7, 16, 18 (Coggins, 1990); the author, H. Peter Gray for Table 2.1 from Table p14 (Gray, 1970); the editor, *Journal of Tourism Studies* for Table 3.1 from Table 2 (Pearce, 1990); the editor, *Journal of Travel Research* for Fig. 5.8 from Figs 3, 6 (McHugh, 1992) and Table 4.5 from Table 1 (Pearce and Elliott, 1983); Kluwer Academic Publishers for Figs 5.1, 5.2, 5.3 from Figs 1–5 (Pearce, 1993), 4.8 from Fig. 2 (Jenkins and Walmsley, 1993) and 9.3a from Fig. 3 (Jeans, 1990); the author, N. Leiper and the editor, *Contemporary Issues in Australian Tourism* for Table 2.3 from Table 6 (Leiper, 1982b); London Tourist Board and Convention Bureau for Tables 4.6 from Tables p33, p35 (*London Tourist Statistics*, 1987); Dr M. Oppermann for Fig. 4.5 from Fig. 11 (Oppermann, 1993); Pergamon Press Inc. and the authors for Figs 1.4 from Fig. 1 (Pearce, 1979; after Miossec, 1976) and 2.1 from Fig. 1 (Iso-Ahola, 1982); Pergamon Press Ltd for Fig. 9.1 from Figs 6, 7 (Pigram, 1977); the editor, *Revue de Tourism* for Figs 1.6 redrawn from Fig. 1 (Lundgren, 1982) and 9.8 redrawn from Fig. 2 (Lundgren, 1974); State of Alaska Division of Tourism for Fig. 5.1 from Table 111-B-10 (Pearce, 1993a); State of Texas Department of Commerce for Fig. 5.7 from Figs p44, p45 (Texas Department of Commerce, 1993); the author, Dr G. Wall for Fig. 1.3 from Fig. 30 (Greer and Wall, 1979).
We are also grateful to the National Geographical Institute – 1050 Brussels – Belgium for permission No. A584 to reproduce Plate 9 Aerial photography S179-117 *De Panne-Wenduine*.
Whilst every effort has been made to trace the owners of copyright material, in a few cases this has proved impossible and we take this opportunity to offer our apologies to any copyright holders whose rights we may have unwittingly infringed.

Introduction

Tourism is essentially about people and places, the places that one group of people leave, visit and pass through, the other groups who make their trip possible and those they encounter along the way. In a more technical sense, tourism may be thought of as the relationships and phenomena arising out of the journeys and temporary stays of people travelling primarily for leisure or recreational purposes. While writers differ on the degree to which other forms of travel (e.g. for business, for health or educational purposes) should be included under tourism, there is a growing recognition that tourism constitutes one end of a broad leisure spectrum. In a geographical sense, a basic distinction between tourism and other forms of leisure, such as that practised in the home (e.g. watching television) or within the urban area (e.g. going to the local swimming pool), is the travel component. Some writers employ a minimum trip distance criterion but generally tourism is taken to include at least a one-night stay away from the place of permanent residence. These travel and stay attributes of tourism in turn give rise to various service demands which may be provided by different sectors of the tourist industry so that in an economic and commercial sense tourism might also be distinguished from other types of leisure activities.

The spatial interaction arising out of the tourist's movement from origin to destination has not been examined explicitly in much of the geographical literature on tourism. The majority of geographical, and other, studies have been concerned with only one part of the system, usually with the destination, as typified by the many ideographic studies which have appeared since the 1960s. A sense of this interaction and a more general spatial structure does, however, emerge from some of the earlier studies. In his seminal work on second homes in Ontario, for instance, Wolfe (1951: 28) observes: 'In ecological terms we have added segments of two zones to the city of Toronto: a buffer zone 50 miles wide, a recreation bridge to cross . . . and a summer dormitory zone seventy miles wide'. In a similar vein Defert (1966), one of the first French geographers to make a substantial contribution to the geography of tourism, writes of an *espace distance* which separates the permanent residence from the seasonal one and of an *espace milieu* where tourists enjoy their holiday. A decade later, Miossec (1976) speaks of *l'espace parcouru* and as an *espace occupé* involving a *lieu de déplacement* and a *lieu de séjour*. Miossec notes, however, that few studies have explicitly attempted to bring together the generating regions (origins), receiving regions (destinations) and the associated linkages.

The aim of this book is to analyse in a systematic and comprehensive manner the geographical dimensions of tourism, not only to increase our understanding of this important and growing industry but also to show how a geographical perspective can contribute to its planning, development and management. The basic framework used is an origin-linkage-destination system of the type outlined above as this provides an effective integrative device for investigating what is inherently a very geographical phenomenon. Geographic scale is a second major organizational feature, with the focus changing from the international to the national, regional and local scales. The links between these scales are also emphasized.

While this book stresses geographical methods of analysis and draws predominantly, but by no means exclusively, on the works of geographers, it is addressed to a much wider audience. An appreciation of the geographical dimensions of tourism and the adoption of a spatial perspective can provide valuable insights into this phenomenon for researchers in allied disciplines – economics, sociology, anthropology, resource management, business administration – as well as for those involved in the planning, development and management of the tourist industry at different levels.

Chapter 1 provides a theoretical base and reviews a variety of different concepts and models which give weight to different aspects of the origin-linkage-destination system. Chapter 2 concentrates on demand and motivation and considers what underlies people's desire to leave their home area to visit other places. Selected tourist flows at various scales – international, intra-national and domestic – are examined in Chapters 3 to 5. The focus then shifts to destination areas, with Chapter 6 being devoted to a review of ways of measuring spatial variations in tourism. Subsequent chapters then consider the spatial structure of tourism at various scales, at a national and regional level in Chapter 7, on islands in Chapter 8 and in coastal resorts and urban areas in Chapter 9. Finally, Chapter 10 reviews three major and recurring themes – concentration, spatial interaction and scale and considers the significance of these. Examples are also given of how geographical techniques and a spatial perspective can be applied in planning, development, marketing and the assessment of the impact of tourism. Conclusions are then drawn.

Each of these topics is examined systematically, with the emphasis being on identifying general patterns and processes and on distinguishing the general from the specific. In order to understand better the processes involved, an attempt is made to examine the evolution of flow patterns and the changing distribution of tourists and facilities over time. Discussion of patterns and processes, and particularly their evolution, cannot, however, be separated from questions of data and methodology. The weight given to these matters varies from topic to topic depending on the extent and nature of the related literature. As several reviews have pointed out, the coverage of topics is far from uniform and the literature on the geography of tourism is still rather fragmented (Pearce 1979a; Barbier and Pearce 1984; Duffield 1984; Lundgren 1984; Mitchell 1984; Pearce and Mings 1984; Mitchell and Murphy 1991). In some cases, as with domestic tourist studies at the local level (Chapter 5) or the morphology of coastal resorts (Chapter 9), it is possible to review a wide range of related research results. In other instances, more attention must be given to the availability and reliability of the data, as with international tourist flows (Chapter 3) or methodological issues, for instance measuring intranational tourist flows (Chapter 4) or analysing domestic tourist travel at the national and regional levels (Chapter 5). Given the diversity of data and techniques used in measuring spatial variations in tourism, it is appropriate to review and evaluate these in one chapter (Chapter 6) so as to allow a more fluid discussion of patterns in subsequent chapters. In topic areas which have hitherto attracted little attention, existing studies are complemented by original material or new treatment, such as the discussions of national and regional patterns of tourism (Chapter 7), tourism on islands (Chapter 8) and tourism in urban areas (Chapter 9). Avenues for future research are also identified.

Given the nature of tourism, particularly international tourism, as well as the aim of identifying general patterns and processes, it is appropriate to draw on examples from many parts of the globe. National reviews indicate that the coverage of topics varies considerably from country to country (Barbier and Pearce 1984; Benthien 1984; Duffield 1984; Lichtenberger 1984; Lundgren 1984; Mitchell 1984; Pearce and Mings 1984; Takeuchi 1984; A. M. Williams and Shaw 1988). Factors accounting for this uneven treatment include differences in the type of tourism practised, variations in available data, broader disciplinary traditions and emphases (regional studies in France and quantitative analysis in North America) and the interests of individual researchers. Drawing on a geographically diverse range of examples allows the generality of ideas and patterns to be examined in those topic areas which have attracted relatively more attention and enables the evaluation of a wider range of techniques and data sources with reference to those problems which are only just being explored. Inevitably the material selected has been biased by the author's own experiences, contacts and limited access to certain foreign language material. Any generalizations made have therefore to be seen in the light of the examples used. It is hoped nevertheless that the systematic approach adopted will readily enable other researchers to put their own results alongside the patterns and trends identified in this book and to compare and evaluate their techniques and methods with those discussed here.

CHAPTER 1

Tourism Models

The geography of tourism continues to lack a strong conceptual and theoretical base. A number of models dealing with various aspects of the spatial structure of tourism emerged in the late 1960s and 1970s, most of which were developed independently of one another, with little or no recognition of or attempt to build on previous efforts. Few new models have been presented since then, nor, with a couple of notable exceptions, have the earlier ones been either broadly adopted or subjected to much critical appraisal or empirical testing. In this respect geography does not differ from many of the other disciplinary approaches to tourism (Dann, Nash and Pearce 1988; Pearce and Butler 1993). Indeed, with the possible exception of planning (Getz 1986), geography, though limited in application, is perhaps better endowed than most other disciplines. Although some changes may be occurring, the general persistence of this situation is to be lamented. Models, as Getz underlines, have a crucial role to play in enabling us to describe and comprehend the complexities of the real world, to acquire, order and interpret information and to explain, understand and ultimately predict phenomena and the relationships between them. This chapter reviews and evaluates the spatial models which have been developed with the aim of providing a theoretical and conceptual base and general frameworks for examining the spatial dynamics of tourism and analysing and interpreting material in subsequent chapters.

As noted in the Introduction, a few early writers such as Wolfe (1951) and Defert (1966) outlined fundamental aspects of the patterns and processes of spatial interaction inherent in all forms of tourism. Later researchers have attempted to express these relationships more explicitly and to derive increasingly complex models of tourist space. The basis of most of these models remains an origin-linkage-destination system, with various writers giving different emphasis to these three elements and expressing them in different terms. Four basic groups of models might be identified: those emphasizing the travel or linkage component, origin-destination models, structural models and evolutionary models.

Models of Tourist Travel

The emphasis in the early explicit models of tourist systems tends to be on the linkage or travel component. Mariot (1969; cited by Matley 1976), for example, proposes three different routes which may link a place of permanent residence (origin) to a tourist centre (destination) – an access route, a return route, and a recreational route (Fig. 1.1). The access and return routes, which in some cases may be one and the same, essentially provide a direct link between the two places. Those travelling the recreational route will make use of various tourist facilities along the way, even if the intervening area does not constitute the main goal of the journey. Alternatively, the tourist may use the recreational route for only part of the journey, entering or leaving it at some stage en route between the origin and destination.

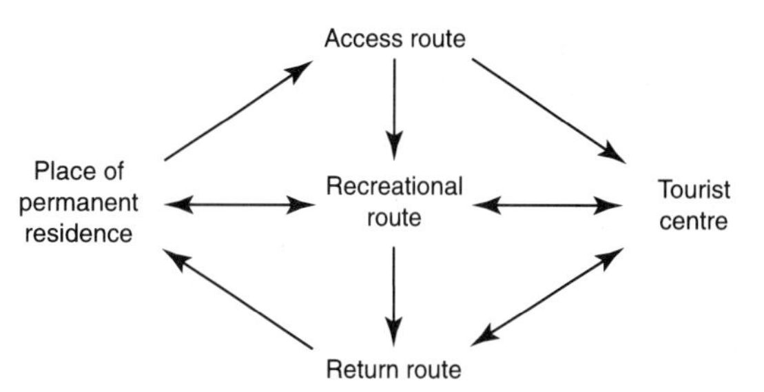

Figure 1.1 Mariot's model of tourist flows between two locations.
Source: Redrawn from Matley (1976) and Mariot (1969).

Implicit in Mariot's recreational route is the idea of touring, that is in visiting several places on one trip, rather than just a single destination. This notion is developed in Campbell's (1967) model which portrays different patterns of movement away from an urban centre (Fig. 1.2). Campbell distinguishes between various groups on the basis of the relative importance of the travel and stay components of their trip. For the 'recreationist' the recreational activity itself is the main element while for the 'vacationist' the journey as such constitutes the main activity of the trip, with a number of stopovers being made on a round trip away from the city. An intermediate group, the 'recreational vacationist', is shown to make side trips from some regional base. According to the model, 'recreational' travel is scattered radially from the city whereas 'vacation' travel is essentially linear and highway oriented, with 'recreational vacational' travel involving elements of both. Campbell's work is also supported by Rajotte's research on movement patterns in Quebec (Rajotte 1975). While the concept of these different types of travel is useful for analysing tourist flow patterns there are clearly semantic problems in restricting the term 'tourist' or 'vacationist' to someone whose primary interest is 'invariably in sightseeing and travelling'.

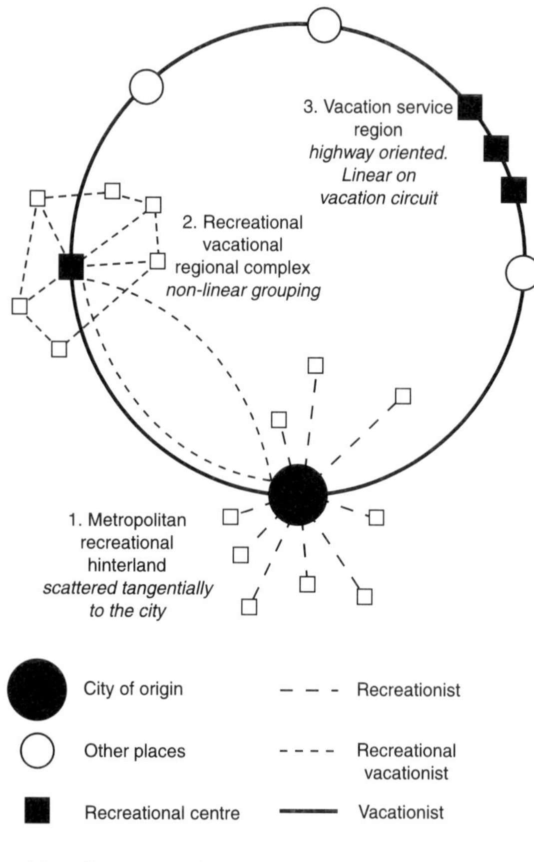

Figure 1.2 Campbell's model of recreational and vacational travel.
Source: Redrawn from Campbell (1967).

Other writers have been concerned not so much with travel routes and itineraries but with changes in the volume of tourist travel. Both domestically and internationally the volume of traffic is generally held to decrease away from the generating centre as travel costs in time, money and effort increase. Domestic travel is typically seen in terms of concentric zones surrounding the city as defined on the basis of blocks of available leisure time: a day trip zone, a weekend zone (which often corresponds with a second home belt) and a holiday or vacation zone (Mercer 1970; Rajotte 1975; Ruppert 1978). As Greer and Wall (1979: 230) point out, however, although demand may vary inversely with an increase in distance from the city, the potential supply of recreational and vacational opportunities will increase geometrically 'as each successive unit of distance gives access to increasingly larger areas of land'. This interaction of supply and demand, they argue, would theoretically produce a 'cone of visitation' peaking at some distance from the generating centre with the exact form of the cone depending on the nature of the activity and its sensitivity to distance (Fig. 1.3). The concept of successive, though overlapping, zones is thus retained in Greer and Wall's model, but the notion of a simple distance-decay function within each zone is rejected.

Theoretically, international demand and supply might be expected to interact in a similar manner to produce larger-scale cones of visitation. However, the models of international tourist space which have been proposed (Yokeno 1974; Miossec 1976, 1977) have concentrated on incorporating various modifications to hypothesized regular concentric zones. In Miossec's model (Fig. 1.4), the origin or core is surrounded by four major belts or zones and travel motives, means and costs change as well (Sector 1). In the real world these theoretical regular concentric zones are subject to modification by 'positive deformations' (low cost of living, favourable climate, historic links) which extend the belts and the 'negative' ones (essentially political) which compress them (Sectors 2 and 3). These positive and negative deformations are not independent. Puerto Rico, for example, benefited from the Cuban blockade, and the development of the Mediterranean was in part due to political barriers in Eastern Europe. Moreover, in reality a series of cores exists giving rise to concurrent spatial demands (Sector 4). Miossec also attempts to incorporate perception of this space in his model, although the schematic representation of this is not particularly clear. In general, knowledge of destinations declines with distance but there may be certain points of reference or evocative names so that the individual's mental map of the tourist space has both concentric and sectoral constraints. Miossec also suggests that the quality of the image will depend on the socio-political-linguistic environment of the points of departure and arrival. Finally, the model incorporates the idea of a hierarchy of resorts, an idea developed in a second model (Fig. 1.12).

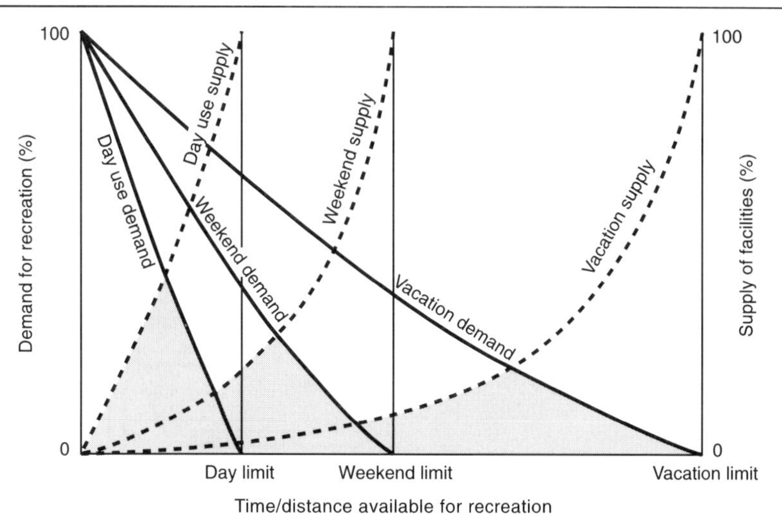

Figure 1.3 Distribution of recreational uses.
Source: Greer and Wall (1979).

Similar notions to these are embodied in Yokeno's model of international travel, a development of his earlier work applying the concepts of von Thunen and Weber to the tourist industry (Yokeno 1968). According to Yokeno, deformations of the hypothetically concentric zones may result from capital city tourism, major transport links and tourism price levels (an intervening country may be bypassed in favour of a more distant one where local costs are lower).

Certain travel models then emphasize the routes taken while others stress the volume, nature and direction of the tourist traffic. A distance decay function characterizes this latter group, whether dealing with domestic or international tourist travel.

Origin-destination Models

One of the basic features not considered in the models discussed so far is that most places are, in varying degrees, both origins and destinations. That is to say, as well as sending tourists to some destination or a series of destinations, a particular place may also receive visitors from those same destinations. Likewise, the routes and structures linking one place to another may convey travellers from each place to the other and back again or on to some third place. This double generating/receiving function and this reciprocal interaction are portrayed in the models of Thurot (1980), Lundgren (1982) and Pearce (1981a) as depicted in Figs 1.5 to 1.7 respectively.

Thurot's model concerns tourism at a national and international level. In each of the three national systems A, B and C depicted in Fig. 1.5, Thurot distinguishes between supply and demand and between domestic and international tourism. Part of the demand for tourism generated in system or country B, probably the larger part, will be fulfilled by that country's tourist facilities with the remainder being distributed to countries A and C. At the same time, part of the demand from country A will be channelled to country B (and to country C), which thereby becomes an international destination as well as a source of international travellers. In contrast, no international demand is shown to emanate from country C, although it may generate domestic tourism and receive tourists from countries A and B. Country C is said to represent certain Third World countries where the standards of living may generally be insufficient to generate international tourism (although often a small elite may indulge in a large amount of such travel) and the former Soviet bloc countries where restrictions on international travel existed. The different national systems, which need not be contiguous, are separated by an *espace de transit*, a transit zone comparable to those of Wolfe (1951) and Defert (1966) noted earlier.

Thurot's model was originally conceived as an aid to analysing carrying capacity (see Chapter 10) but it is equally useful as a means of conceptualizing the different levels of tourist flows and spatial structures. Clearly the model could be made more complex and sophisticated by increasing the number of national systems in the model and by compartmentalizing or classifying in different ways (geographic, qualitative, etc.) the supply and demand components in each system. One important question, which will be considered in Chapter 7, is the extent to which domestic and international tourists share the tourist facilities in any given country or region and the extent to which there is a marked separation in the resources and services used by each group.

Lundgren (1982) focuses on the role of different

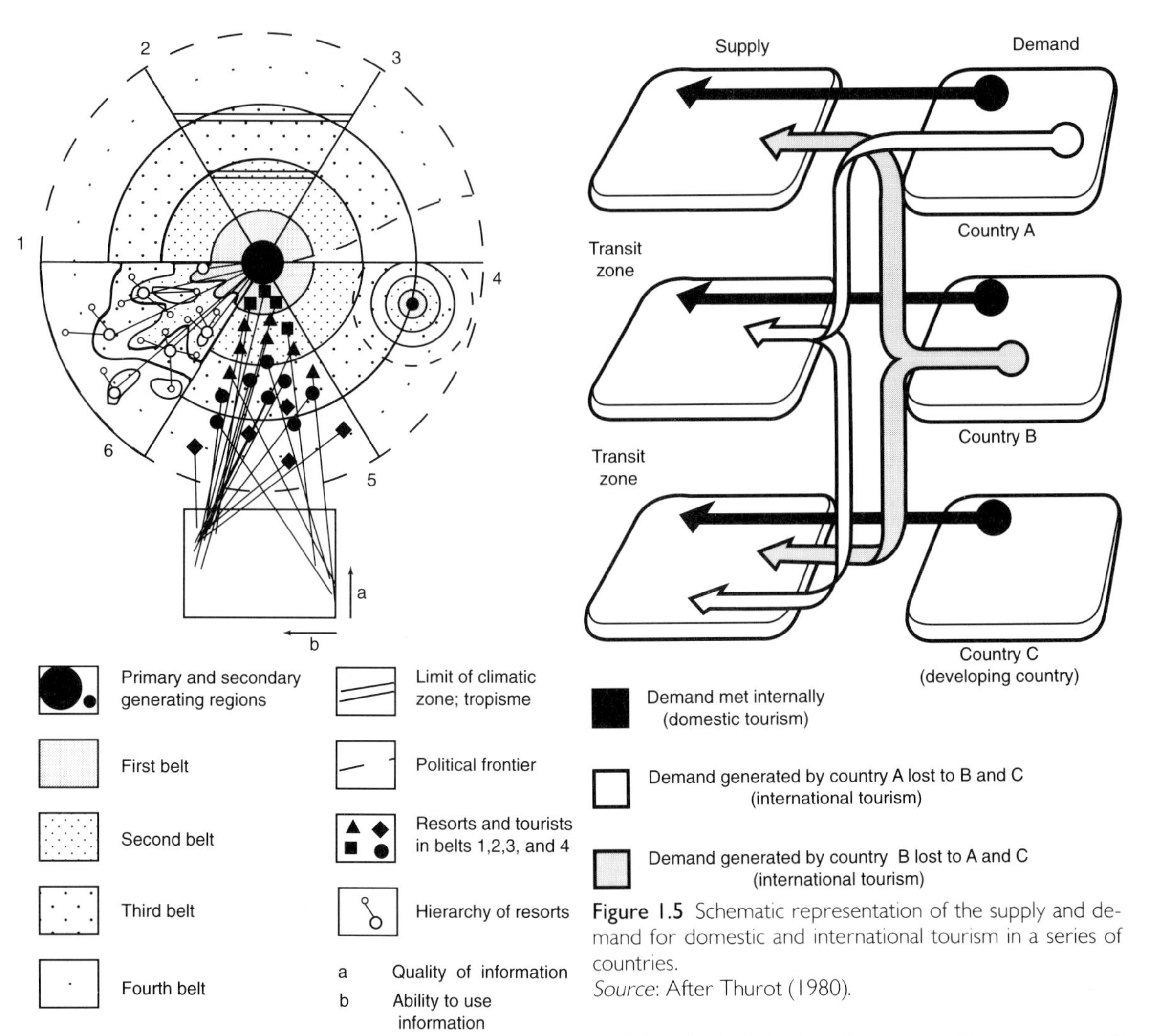

Figure 1.4 Miossec's model of tourist space.
Source: After Miossec (1976).

Figure 1.5 Schematic representation of the supply and demand for domestic and international tourism in a series of countries.
Source: After Thurot (1980).

places rather than countries. While he sees places essentially in terms of destinations, their 'degree of mutual travel attraction' (generation vs inflow of tourists) is one of the defining characteristics he uses in positioning destinations within what he calls the 'travel circulation hierarchy' (Fig. 1.6). The other characteristics used are relative geographical centrality, geographic place attributes (attractions) and the ability of places to supply tourist-demanded services from within their own local or regional economy. Four broad tourist destination types are identified by Lundgren in this way:

1. Centrally located metropolitan destinations which have a high volume of reciprocal traffic and function both as a generating area and a major destination. These include high-order metropolitan centres well integrated into the international and transcontinental transport networks.
2. Peripheral urban destinations, which have smaller populations, a less important central place function and which tend to have a net inflow of tourists.
3. Peripheral rural destinations, which are less nodal in character, depending upon a geographically more extensive environment which draws visitors through a combination of landscape characteristics. As the population of such areas is often small and dispersed, a strong net inflow usually results.
4. Natural environment destinations which are usually located at long distances from the generating areas, very sparsely populated and often subject to strict management policies, as in the case of national and regional parks and other reserves. Moreover, Lundgren (1982: 11) suggests, 'as the indigenous

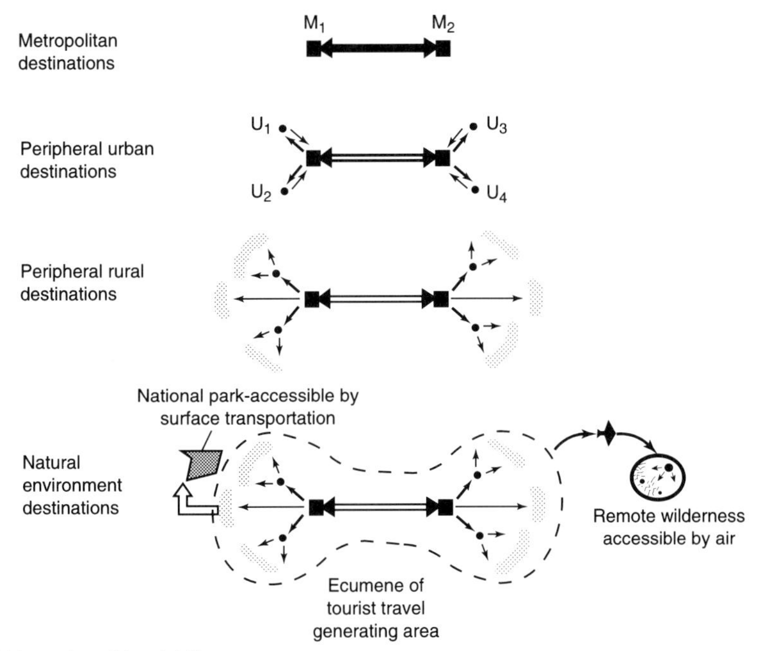

Figure 1.6 A spatial hierarchy of tourist flows.
Source: Redrawn from Lundgren (1982).

economic system for all intents and purposes is non-existent, these destinations can only function through importation into the region of various tourist services. This makes the destination completely dependent upon the tourist generating areas.'

Lundgren's spatial hierarchy is potentially very useful for identifying the functions of a particular place and its associated flows. Unfortunately he provides a detailed example only of the fourth type, the national environment destination, which is illustrated by reference to Nouveau Quebec-Labrador. Some of the features of the other categories require further explanation. Lundgren suggests (1982: 11), for example, that some of the peripheral countryside destinations are 'often explored by tourists via some urban centre acting as a staging point', but he does not elaborate on this function. It is also unclear whether or not the flows generated in one metropolitan area are redistributed via a second metropolitan area to lower-order destinations or whether visitors to such destinations essentially originate in the second metropolitan area.

Lundgren (1987) subsequently develops the basic notions of Fig. 1.6 at a national scale, presenting a Canadian application of what he terms 'a conceptual model of tourist destination space'. This emphasizes a high degree of horizontal interaction among the metropolitan centres of southern Canada and vertical linkages from this zone into the Canadian interior. Movement into the interior is accompanied by changes in the type of tourism found – nature progressively outweighs services and culture in terms of attractions, accessibility changes, seasonality becomes more pronounced and the volume of tourist traffic declines. The model also depicts outbound international flows from the metropolitan zone to settled temperate or tropical areas of the USA, Europe, Asia, Latin America and the Pacific.

The joint generating/receiving functions of urban areas and their associated flows have been integrated in the model proposed by this author (Pearce 1981a) as shown in Fig. 1.7. This model suggests that the city, especially the large city, not only can act as a source of tourists but also can play several different or complementary roles as an international and national destination. Each of these functions gives rise to specific types of flows. The most frequently studied function of the city is its role as a generator of tourist traffic (Mercer 1970; Rajotte 1975; Ruppert 1978). As noted earlier, the flows of city residents might be classified by duration of trip and distance travelled so that a series of concentric zones surrounding the city might be identified. Yokeno (1968) suggests, however, that it is not just a question of these flows spreading out radially for tourists will concentrate their activities in certain favourable localities, such as near a body of water, along a coastline or in an upland area. In any case, the movement is essentially centrifugal.

This may be complemented by a centripetal move-

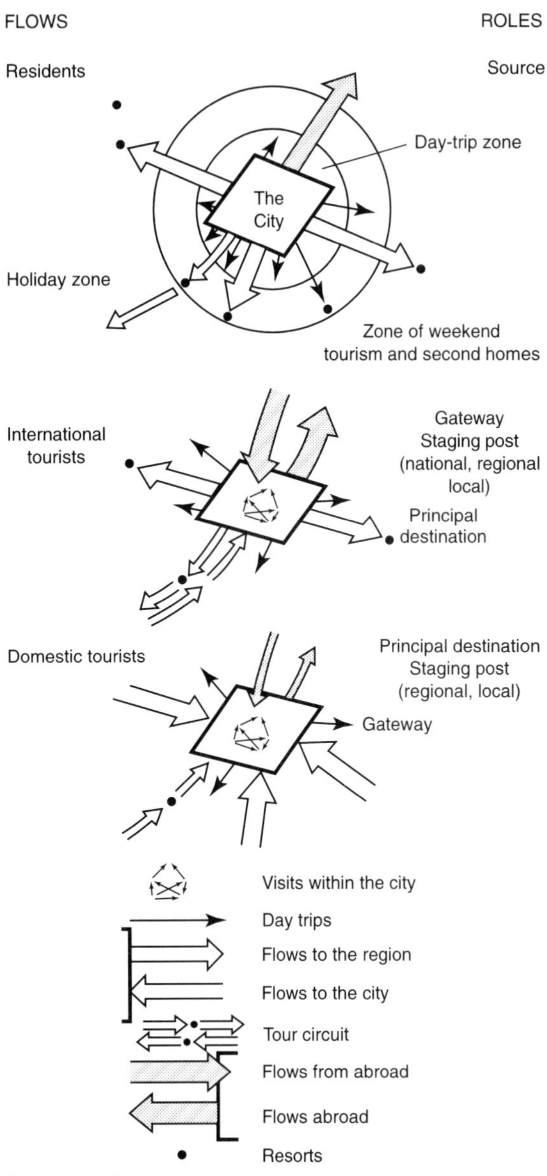

Figure 1.7 Schematic representation of tourist flows to and from major urban areas.
Source: After Pearce (1981a).

ment of other domestic visitors and international tourists. For international tourists, the large city may constitute a gateway, a point of entry into or out of a national territory. The city might also play a regional staging post role, sending the visitors on to other centres or resorts. In the case of circuit tourism, the city may be just one stop among several on a given tour; it may also provide a base from which day trips are made to surrounding areas. Of course, the large city may also constitute a destination in its own right, in which case tourists will travel around within the city, especially in the city centre where many of the attractions are frequently found. Similar functions may also be performed for domestic tourists from other areas, although the relative importance of each of these may differ significantly from the demands being made by international visitors. Indeed, the roles a city plays and the relative importance of each may vary not only from one group of visitors to another but also from city to city depending on such factors as city size, other functions, the nature and degree of development of the surrounding region and so on.

Structural Models

Writers examining the impact of international tourism in Third World countries have emphasized the structural relationships between origins and destinations (Lundgren 1972, 1975; IUOTO 1975; Hills and Lundgren 1977; S. G. Britton 1980a, 1982; Cazes 1980). Hills and Lundgren and Britton express these relationships in core-periphery terms with the former focusing on the Caribbean and the latter on the Pacific. The models they propose share many common features and only that of Britton is presented here (Fig. 1.8). In each case the market is concentrated upwards through the local-regional-national hierarchy, with the international transfer occurring between the national urban centres in the generating and receiving countries. Dispersal within the peripheral destination is more restricted, with the tourists moving from their point of arrival out to some resort enclave. Movement may occur between such enclaves but only limited travel to other areas occurs. Cazes (1980) presents a more general origin-linkage-destination model which does not include concentration and dispersion within generating and receiving countries.

These writers generally agree that the basic pattern shown in Fig. 1.8 largely arises out of the control exerted by metropolitan-based multinational corporations over the international tourist industry. Lundgren (1972: 86–7) suggests these relationships 'are basically a function of the technological and economic superiority of the travel-generating, metropolitan core areas as such and the willingness of the destination areas to adopt metropolitan values and solutions in order to meet the various demands of metropolitan travellers'. In particular, he stresses the dominant role of the metropolitan countries as air carriers who can effectively and selectively control the international links between the market and the destination. In this respect the metropolitan-based countries are further advantaged by their direct contact with the market (IUOTO 1975). Although Cazes, like Britton, stresses the role of the

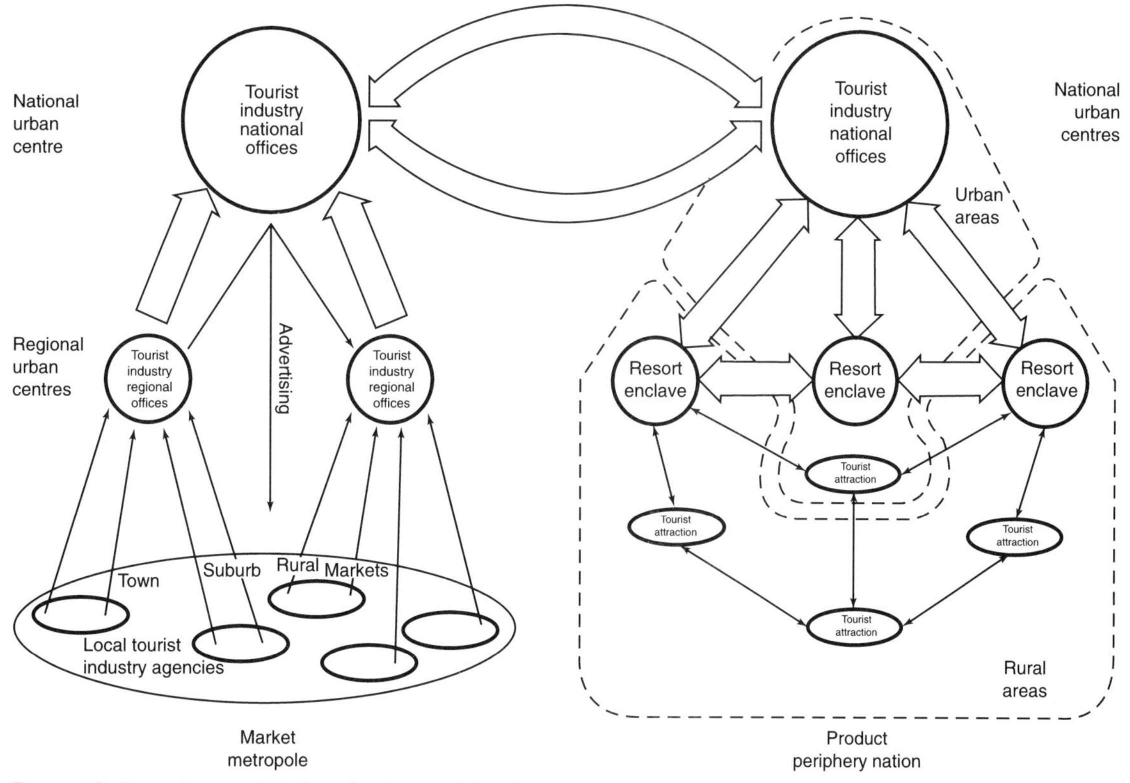

Figure 1.8 An enclave model of tourism in a peripheral economy.
Source: Redrawn from S. G. Britton (1980a).

'multinational commercial system' in fostering the tourist traffic to developing countries he also includes reference in his model to the demand created spontaneously by individual 'pioneer' tourists; unfortunately he does not elaborate on this point. Cazes also discusses how various international organizations – the WTO, World Bank, ILO, OECD etc. – have encouraged the expansion of tourism in developing countries and have acted as a means of channelling funds and expertise to them from the industrialized countries. Such encouragement has often confirmed the developing country's belief in international tourism as a means of increasing their foreign exchange earnings, creating new job opportunities, reducing regional imbalances and promoting their national image. But as Cazes, Britton and others have pointed out, including some international organizations (IUOTO 1975), these benefits are frequently illusory and over-stated. Britton emphasizes the structural weaknesses inherited from colonial times which enable the multinational corporations to impose their system on dependent destinations. As well as dominating the transport sector, these corporations may also provide many of the facilities within the destinations in the form of purpose-built resort enclaves which fulfil most of the package tourist's wants and remove the need for further interaction with the local society and environment.

While the models of Hills and Lundgren and of Britton must be seen within the context in which they were written, that is in terms of international tourism in Third World countries, they do highlight several useful general points which may be applicable to other forms of international tourism as well. Particularly useful is the idea of concentration up and subsequently dispersion down the hierarchy. That is to say, international tourist movements commonly involve not only flows between pairs of countries but also some prior movement within the generating country and a corresponding distribution within the destination, limited though that may be in the examples given. There is also the overriding emphasis on the factors giving rise to these structures. In other words, we should not limit ourselves to identifying patterns of movement or spatial relationships but should try to account for them.

Evolutionary Models

Models which stress change, whether the evolution of international tourist movements or the development of tourism structures, are also important in drawing attention to explanatory factors and underlying processes. These models again appear to have been developed in isolation with a variety of approaches being adopted and different explanations being proposed.

A concept allied to the zones and belts depicted in the models of Miossec and Yokeno is that of the 'pleasure periphery', a term coined by Turner and Ash (1975: 11–12) with reference to the distribution of international tourism. They write

> This periphery has a number of dimensions, but is best conceived geographically as the tourist belt which surrounds the great industrialized zones of the world. Normally it lies some two to four hours flying distance from the big urban centres; sometimes to the west and east, but generally toward the equator and the sun.

Two major examples are used by the authors to illustrate this concept, the Caribbean *vis-à-vis* North America, and the Mediterranean, the sun belt for North and West Europeans. Turner and Ash (1975: 12) then suggest: 'These Pleasure Peripheries are never static, possessing a dynamism of their own, which depends on the extension of the range of places and the increase of leisure and affluence in general. The pioneer tourists are ever moving outward looking for new destinations which have not yet been sampled by mass tourism.'

Different views exist on just who constitutes the pioneer tourists and what underlies the process of expansion. Thurot (1973) sees the evolution of tourism in the Caribbean in terms of class succession. He proposes a model based on an analysis of the evolution of airline routes in which the different destinations pass through three successive phases:

- *Phase 1*: Discovery by rich tourists and construction of an international class hotel.
- *Phase 2*: Development of 'Upper-Middle-Class' hotels (and expansion of the tourist traffic).
- *Phase 3*: Loss of original value to new destinations and arrival of 'Middle-Class' and mass tourists.

According to Thurot, the length of this process will depend on the time it takes for the upper-middle-class tourists to arrive and the speed with which the 'traditionally leisured classes' open up new destinations. In this way the Caribbean pleasure periphery is extended geographically as succeeding waves of tourism spread progressively and selectively outwards from the well-established resorts of Florida. In the Bahamas, for example, the tourist front moves from Nassau out on to the smaller cays and islands. According to Thurot, there was a hierarchy of development in the Commonwealth Caribbean, with the development of tourism in Jamaica preceding that of Trinidad and Barbados, which occurred before that of the smaller Leeward and Windward Islands.

Table 1.1 Travel characteristics of psychographic types

Psychocentrics	Allocentrics
• Prefer the familiar in travel destinations	• Prefer non-touristy areas
• Like commonplace activities at travel destinations	• Enjoy sense of discovery and delight in new experiences, before others have visited the area
• Prefer sun 'n' fun spots, including considerable relaxation	• Prefer novel and different destinations
• Low activity level	• High activity level
• Prefer destinations they can drive to	• Prefer flying to destinations
• Prefer heavy tourist accommodations, such as heavy hotel development, family type restaurants and tourist shops	• Tour accommodations should include adequate-to-good hotels and food, not necessarily modern or chain-type hotels, and few 'tourist' type attractions
• Prefer familiar atmosphere (hamburger stands, familiar type entertainment, absence of foreign atmosphere)	• Enjoy meeting and dealing with people from a strange or foreign culture
• Complete tour packaging, appropriate with heavy scheduling of activities	• Tour arrangements should include basics (transportation and hotels) and allow considerable freedom and flexibility

Source: Plog (1973: 15).

Plog (1973), on the other hand, emphasizes not class but the personalities of different types of travellers. From a series of motivational studies, initially of flyers and non-flyers, Plog suggests that travellers are distributed normally along a continuum from psychocentrism to allocentrism. At the one extreme are the 'psychocentrics', who tend to be anxious, self-inhibited, non-adventuresome and concerned with the little problems in life. In contrast, the 'allocentrics' are self-confident, curious, adventurous and outgoing; travel, according to Plog, is a way for them to express their inquisitiveness and curiosity. The travel characteristics of the two groups differ (Table 1.1) so that different types of travellers will visit different destinations.

Figure 1.9a represents the psychographic positions of destinations visited by US travellers (presumably New Yorkers) in 1972. Plog suggests that the market for a given destination evolves and that the destination appeals to different groups at different times. The destination will be 'discovered' by 'allocentrics', but as it

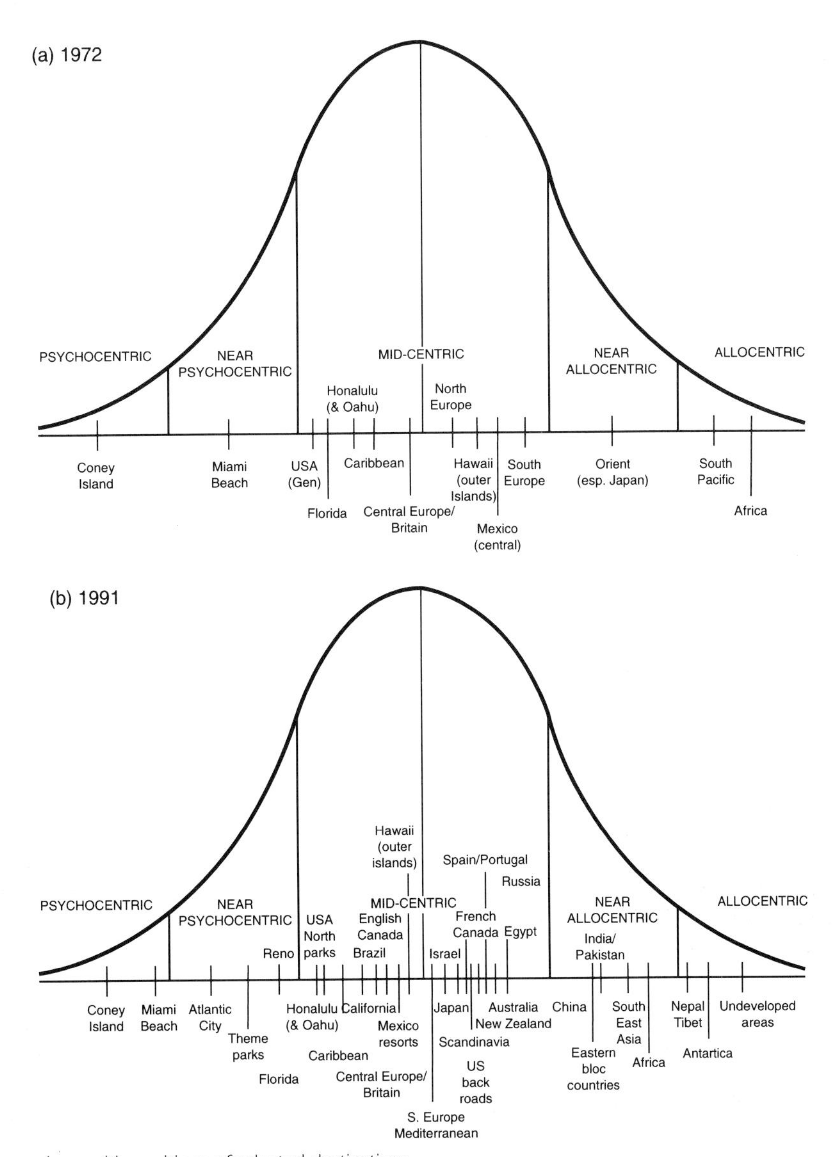

Figure 1.9 The psychographic positions of selected destinations.
Source: Redrawn from Plog (1991a). Copyright © 1991 John Wiley & Sons, Inc. Reprinted by permission of John Wiley & Sons, Inc.

becomes more well known, develops and attracts more visitors, for example the 'mid-centrics', it will lose its appeal and the 'allocentrics' will move on. As the population is said to be normally distributed, this means that an area will receive the largest number of visitors when it is attracting the 'mid-centrics', that is at a stage when it is neither too exotic nor too familiar. But from this point on, the implication is that the market will decline. According to Plog (1973: 16), 'we can visualize a destination moving across the spectrum, however gradually or slowly, but far too often inexorably towards the potential of its own demise. Destination areas carry with them the potential seeds of their own destruction, as they allow themselves to become more commercialized and lose their qualities which originally attracted tourists.'

This evolution is depicted in Plog's representation of the situation in 1991 (Fig. 1.9b) based on research carried out in the intervening period and also 'the conceptual "filling" of destinations on the curve' (Plog 1991a). Comparison of the two figures shows 'the sure but steady movement of most destinations toward more psychocentric characteristics, and the audiences they attract'. However, Plog continues, 'This process need not happen, but, without concerted effort executing a preconceived plan, it will'.

Widely cited, Plog's early model has only recently attracted more critical comment. P. L. Pearce (1993), in a comparison of motivation theories, raises questions of measurement (i.e. the measures used are not widely available), the dominance of a single trait approach, failure to distinguish between extrinsic and intrinsic motivations and lack of a dynamic perspective for individuals (compared with destinations). The issue of measurement was also central to an earlier exchange between S. L. J. Smith and Plog (S. L. J. Smith 1990, 1990b; Plog 1990, 1991b). Smith argued that his findings from an analysis of destination preferences for a sample of long-haul travellers from seven nations failed to confirm an association between personality types and destination preferences. Plog rejected Smith's criticism on the grounds that he had failed to replicate adequately the original study, using the wrong variables, classification system and sample. A study by Nickerson and Ellis (1991) using Plog's scale and other measures provided not only empirical support for the concept of allocentrism/psychocentrism but also, through the incorporation of activation theory into the analysis and interpretation, further theoretical explanation for travel behaviour.

A more widely tested evolutionary model is Butler's (1980) tourist area life-cycle concept (Fig. 1.10). Butler (p. 5) argues 'there can be little doubt that tourist areas are dynamic, that they evolve and change over time'. After citing the work of Plog (1973) and other earlier writers such as Stansfield (1978) and Noronha (1976), Butler draws on the product life-cycle concept to produce a six-stage hypothetical evolutionary sequence: exploration, involvement, development, consolidation, stagnation and rejuvenation or decline. Growth in the number of visitors is said to be accompanied by changes in their composition following the pattern from allocentrism to psychocentrism outlined by Plog and from explorers to institutionalized mass tourists suggested by Cohen (1972). Each stage is also accompanied by changes in the nature and extent of facilities provided and in the local/non-local provision of these. No specific facilities for tourists exist in the first stage, those in the involvement stage are provided primarily by locals then local involvement and control declines rapidly in the development phase as larger, more modern and elaborate facilities are provided by external developers and regional and national authorities assume responsibility for planning. Local involvement only increases again in the decline stage 'as employees are able to purchase facilities at significantly lower prices as the market declines' (Butler 1980: 9).

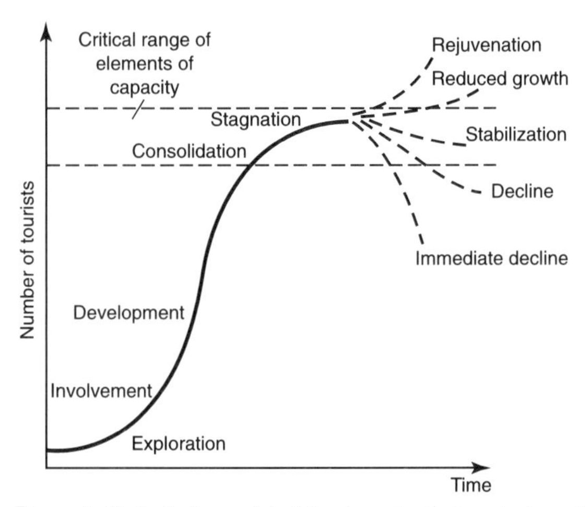

Figure 1.10 Butler's model of the hypothetical evolution of a tourist area.
Source: Redrawn from Butler (1980).

General empirical support for Butler's model has come from the dozen or more studies in which it has been applied, with different authors emphasizing particular aspects of the process or reporting variations in the timing and nature of the stages identified (e.g. Meyer-Arendt 1985; Strapp 1988; C. Cooper and Jackson 1989; Debbage 1990; Foster and Murphy 1991; Kermath and Thomas 1992). Choy (1992), however, finds little evidence for the S-shaped curve in his analysis of growth patterns in South Pacific destinations.

Most of these studies, however, as Foster and Murphy (1991: 564) note, are 'oriented toward the supply-side characteristics of the market whereas the model itself is predicated on the demand-side characteristics of tourism volumes'. Accommodation statistics are often used as the measure of tourism growth but even where figures for the tourist traffic itself are provided there is rarely any attempt to disaggregate these or provide anything more than anecdotal evidence of the 'type' of tourist. If measuring the nature of allocentrism/psychocentrism today presents challenges, then reconstructing past visitor types in Plog's terms is even more so. Other conceptual and measurement issues are raised by Haywood (1986) in terms of the unit of analysis (resort, region, island, etc.), relevant markets (whether or not there are distinct sub-segments), the patterns and stages of the life-cycle, the unit of measurement (number of tourists, length of stay, etc.), and the relevant unit of time to be considered. Operationalizing the tourist area life-cycle in these terms is essential, according to Haywood, if it is to be used as a management or planning tool.

Haywood sees potential for the application of the

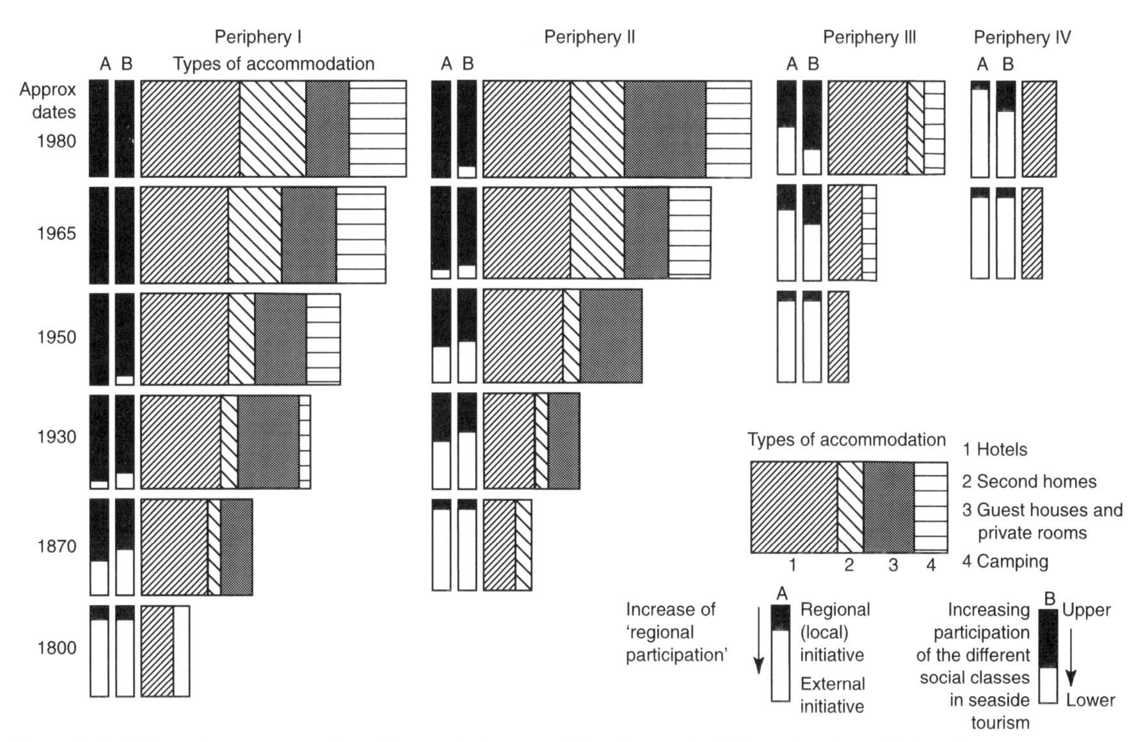

Figure 1.11 Schematic representation of the spatio-temporal development of international seaside tourism. *Source*: Redrawn from Gormsen (1981).

model in this way but others have debated its predictive value. Butler himself appears to have seen the life-cycle in descriptive terms although he does explore some of the implications of the evolutionary sequence postulated. C. Cooper and Jackson (1989) claim that Butler's model 'provides an enlightening descriptive tool for understanding how destinations and their markets' evolve but suggest it is inadequate as a prescriptive tool because the cycle simply reflects policy decisions and is 'destination specific, with stages and turning points only evident with hindsight'. Choy (1992: 31) argues 'Rather than using Butler's model to describe the evolution of a tourist destination, it is better to treat each destination individually as a unique entity and not be preconditioned in conceptualizing alternative future growth patterns'.

In his model of the spatio-temporal development of international seaside tourism, Gormsen (1981) attempts to incorporate not only the ideas of spatial and temporal evolution, but also corresponding changes in the degree of local or regional participation in the development process, in the social structure of the tourist traffic and in the quantity and range of accommodation available (Fig. 1.11). Gormsen's model is based on a study of the historical development of coastal tourism, essentially from a European perspective. Thus the first periphery refers to resorts on both sides of the Channel as well as those of the Baltic; the second incorporates the coasts of southern Europe; the third includes the North African littoral and the Balearic and Canary Islands; while the fourth periphery embraces more distant destinations in West Africa, the Caribbean, the Pacific and Indian Oceans, South East Asia and South America. Gormsen also suggests that comparable peripheries can be identified for the USA. In Fig. 1.11 approximate dates are given for the various stages in the development of each periphery.

According to this model, the initiative in the early stages comes from external developers but with time there is a growing local participation in the development process (column A). The extent of local participation shown for the third and fourth peripheries broadly corresponds with the work of Hills and Lundgren (1977) and S. G. Britton (1980a, 1982). These latter writers have not generalized their models as Gormsen has done, but the implication of their work appears to be that the structural characteristics they emphasize would lead to the continued dominance of external developers, at least in the immediate future. Gormsen's model, however, involves a comparatively long time-scale and covers a period of far-reaching social and economic changes. There is also general agreement between the changing social structures suggested by Gormsen (column B) and the processes discussed by Thurot (1973) in the Caribbean (Fig. 1.5). It should be noted, however, that a completely black column B (as in the first periphery since the 1960s) means that tourists from all social classes are participating in seaside holidays and not that the lower classes have entirely replaced the upper classes. With changes in the social structure of the tourist traffic, which in part results from growing accessibility, also come changes in the types of accommodation demanded and provided, notably an increase in second homes and later in camping facilities.

Aspects of the development process and changes in tourist behaviour were also emphasized by Miossec (1976, 1977) in his second model which depicts the structural evolution of tourist regions through time and space (Fig. 1.12). In particular, Miossec stresses changes in the provision of facilities (resorts and transport networks) and in the behaviour and attitudes of the tourists and the local decision-makers and host populations. In the early phases (0 and 1) the region is isolated, there is little or no development, tourists have only a vague idea about the destination while the local residents tend to have a polarized view of what tourism may bring. The success of the pioneer resorts leads to further development (phase 3). As the tourist industry expands, an increasingly complex hierarchical system of resorts and transport networks evolves while changes in local attitudes may lead to the complete acceptance of tourism, the adoption of planning controls or even the rejection of tourism (phases 4 and 5). Meanwhile the tourists have become more aware of what the region as a whole has to offer, with some spatial specialization occurring. With further development, Miossec suggests it is tourism itself rather than the original attractions which are now drawing visitors to the area. This change of character induces some tourists to move on to other areas in a manner similar to that suggested by Plog (Fig. 1.9).

In the most recent of the evolutionary models, Oppermann (1992a, 1993a) combines elements of the broader spatial structure of Miossec's second model with considerations of the role and behaviour of different groups of tourists. Oppermann (Fig. 1.13) focuses specifically on the tourist space of developing countries, recognizes (Phase 0) the influence of pre-tourism structures (i.e. unlike Miossec's and other models discussed here, tourism does not develop in a vacuum) and highlights the differing spatial patterns of the informal and formal sectors. The capital city exercises a dominant role, a role that is magnified by the presence close by of the sole or main international airport. The informal sector, characterized by drifters and adventurers with their less demanding needs in terms of services and facilities, is seen to open up the country progressively extending their visits and impacts to more and more locations. The formal sector is implanted first in the capital, later extending to coastal areas and selected nearby sites of special interest.

Discussion and Conclusions

During the 1970s and 1980s a variety of models dealing with particular aspects of tourist flows and the structure of tourist space have been proposed. In most cases these appear to have been developed independently of one another although references to earlier work in the same field are occasionally made and the influence of ideas from other areas of geography, notably the work of von Thunen, can also be detected. Miossec (1976), for example, acknowledges the earlier work of Yokeno (1968) and both writers make explicit reference to von Thunen. Butler (1980) draws directly on Plog (1973) and Oppermann (1992a) refers to a range of earlier works as well as to dependency and diffusionist theory. In general, however, there is little evidence of a concerted effort to build up a cohesive body of theory on tourism.

Nevertheless, some general themes do emerge from these disparate studies. The notion of spatial interaction underlies virtually all these models, with a potentially very complex tourist system operating if the various features stressed in individual models were combined. At this stage, however, it is perhaps more appropriate to identify basic concepts and ideas.

The basic geographic concept of distance decay, whereby the volume of tourist traffic decreases with distance away from the generating area, is embodied in several of the models, whether concerned primarily with international tourism or domestic tourism. It is recognized, however, that regular distance decay curves may be deformed in the real world and attention is directed to a variety of factors which may bring about these deformations.

Reciprocity is another important feature, with most areas having, in varying degrees, both generating and receiving functions. In most cases a two-way traffic will exist between pairs of places, whether two countries or two linked cities, although the flows may be stronger in one direction than the other. As most places are linked to a range of other places, a complex system of reciprocal flows may occur, with individual places having a variety of generating and receiving functions. In particular, several of the models draw attention to the receiving functions of urban areas which have often been neglected in the past in favour of more distinct tourist destinations.

Adding to this complexity is the range of links which may exist between origins and destinations, with some

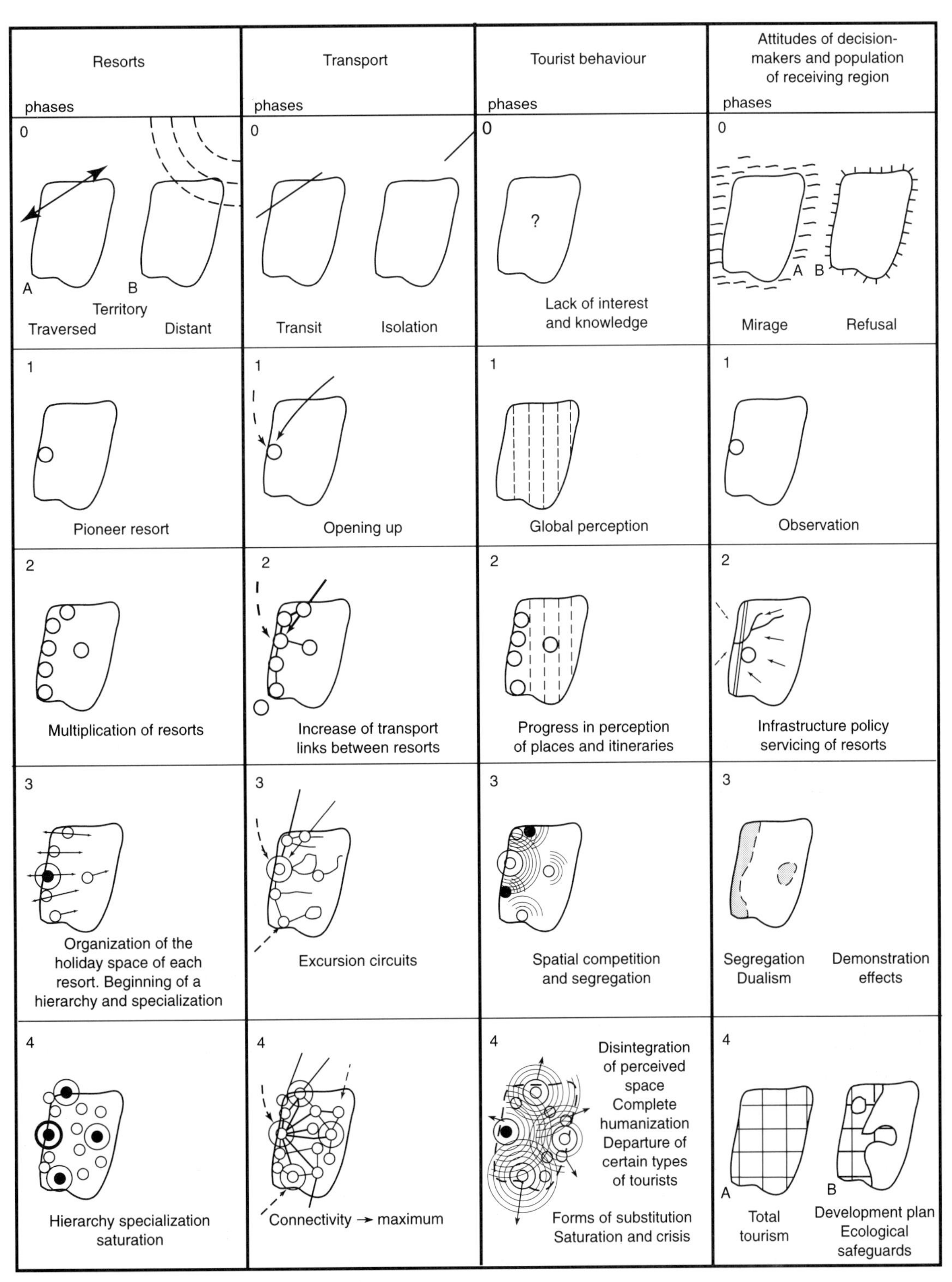

Figure 1.12 Miossec's model of tourist development.
Source: After Miossec (1976).

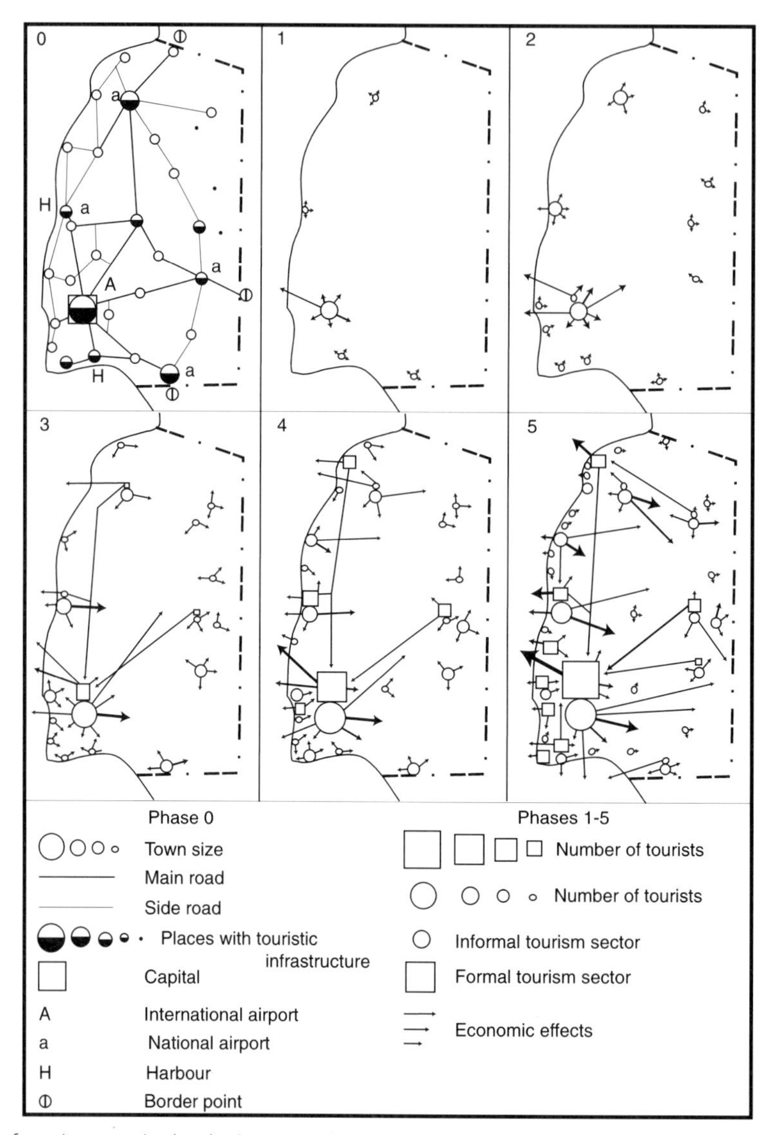

Figure 1.13 Model of tourist space in developing countries.
Source: Oppermann (1993a).

places being linked directly to each other, while others may form part of a circuit.

The notion of a spatial hierarchy is also central to many of these models, whether in terms of flows from an international to a national, regional and local level, or the distribution of tourist supply and demand at these different levels. How one flow relates to another and how one level is linked to another are particularly relevant questions. The relationships between domestic and international tourism are especially critical and should not be overlooked.

Spatial hierarchies may change over time, and the idea of evolving flows and structures is central to our understanding of why these developments may come about and what factors influence the nature of distributions and structures at a particular time. Different ideas have been advanced to account for changes in the nature of demand while other writers have stressed variations in the development of destinations. Clearly the changes in demand, flows and the development of destinations are closely interrelated and a variety of different factors acting in different ways and at different scales may play a role.

Although the majority of these models were first published in the 1970s and early 1980s, few have gained general acceptance or been subject to empirical testing. This does not necessarily mean that they are ill-conceived or are of little value. Rather it reflects more general characteristics of tourism research over this period, with Dann, Nash and Pearce (1988) noting a general lack of methodological and theoretical sophistication. Of the models discussed here only two, those by Plog and Butler (Figs 1.9 and 1.10), have been widely cited, and only the life-cycle model has been empirically tested by a number of different researchers. S. L. J. Smith (1990a: 40) suggests the 'widespread citation and use of Plog's model testifies to its strong, intuitive appeal and perhaps to its conformance to stereotypes' while P. L. Pearce (1993: 188) attributes its popularity 'partly [to] reasons of timeliness and an absence of competitors'. These factors may have been important but they perhaps do an injustice to the concept itself. Haywood (1986) attributes the acceptance of the life-cycle concept to its 'overall simplicity . . . and life-to-death analogy' but its widespread use may also reflect its application in one of the most common types of tourism research – the local or regional case study. That is, the popularity of the life-cycle model may not so much reflect its value as a descriptive or predictive tool, as the particular domain to which it has been applied. In contrast, more systematic studies involving the interaction of more than one place remain less common with correspondingly fewer calls being made on origin-destination or structural models. The more critical approach to both Plog's and Butler's models and to associated issues of measurement noted above also reflects a broader trend which began to appear among tourism researchers in the late 1980s of questioning accepted concepts and approaches and developing more sophisticated techniques and methods (Pearce and Butler 1993).

These appraisals of specific models have also been accompanied by more general critiques. After having summarized the discussion of the models presented in the first edition of this book, Stabler (1991: 20), an economist, suggests that geographic models lack two major elements:

> First they are insufficiently integrated because they fail to incorporate the tourist and tourism industry adequately into the spatial context. Second they lack an economic and management-marketing base as a means of explaining the actions of the tourism industry in both origin and destination areas.

As a more 'general analytical framework', Stabler (1991: 21) proposes the concept of opportunity sets. These are

> the holiday opportunities which exist at any given point in time, and a total opportunity set comprises all possible holidays, both commercial and non-commercial, which are available . . .
>
> Within the 'total opportunity' set can be embodied the 'consumer' or 'tourist', 'industry' and 'destination' sets.

A similar point regarding the neglect of tourism's supply side is made by S. G. Britton (1991), although from a completely different theoretical perspective. Britton claims (p. 451) that

> Geographers working in the field [of tourism] have been reluctant to recognize explicitly the capitalistic nature of the phenomenon.

and contends (p. 475):

> we need a theorization that explicitly recognizes, and unveils, tourism as a predominantly capitalistically organized activity driven by the inherent and defining social dynamics of that system, with its attendant production, social and ideological relations.

While Britton provides some examples of these issues with reference, for example, to multinationals' behaviour in the hotel industry and the commodification of places, his untimely death has meant that these have not been developed further.

The organization of subsequent chapters has been influenced by some of the concepts and ideas outlined above. In other instances, particular topics are examined with reference to these models, which may provide either a framework for analyis or a basis for explaining and interpreting observed patterns. It should be recalled, however, that very few of the studies reviewed in the following chapters have been originally based on the theories and concepts discussed here.

CHAPTER 2

The Demand for Tourist Travel

Why do tourists travel? What induces them to leave their home area to visit other areas? What factors condition their travel behaviour, influencing their choice of destination, itineraries followed and activities undertaken. These are basic questions which underlie the spatial interaction of the models depicted in Chapter 1 and the travel patterns analysed in subsequent chapters. However, it is only comparatively recently that researchers have started to address these questions systematically. Geographers have not been to the forefront of this research which has been led by psychologists, sociologists, marketeers and economists. Some of these researchers have touched on such issues as the potential significance of variations in motivation on destination choice but the geographical implications of differences in demand factors have yet to be fully explored.

Demand here is seen in terms of the relationship between individuals' motivation to travel and their ability to do so. After the literature relating to these two areas has been reviewed the two sets of factors are brought together in a discussion of the growth of international and domestic tourism.

Travel Motivation

Moutinho (1987: 16), in a useful review of consumer behaviour in tourism, defines motivation as referring to 'a state of need, a condition that exerts a "push" on the individual towards certain types of action that are seen as likely to bring satisfaction'. With tourism, the concern is with understanding why tourists travel, what conditions predispose them towards selecting specific destinations and engaging in particular activities. This is one of the more complex areas of tourism research and one which continues to offer many challenges (Mansfeld 1992; Witt and Wright 1992; P. L. Pearce 1993).

Pearce draws attention to a number of fundamental issues by outlining a 'blueprint for tourist motivation'. An adequate and insightful tourist motivation theory, he contends, must address the following factors:

- the conceptual place of tourist motivation
- its task in the specialism of tourism
- its ownership and users
- its ease of communication
- pragmatic measurement concerns
- the development of multi-motive perspectives
- adopting a dynamic approach
- resolving and clarifying intrinsic and extrinsic motivation approaches.

Few theories or studies to date measure up fully on all these criteria and many questions relating to motivation remain. Considerable progress has been made, nevertheless, in identifying and clarifying key issues.

Many of the basic psychological factors underlying vacation motivation and behaviour were put forward in the pioneering paper by Grinstein (1955). Grinstein identified a need to 'get away from it all', a need to escape from the demands of everyday life. Indulging 'one's pleasure principle to the maximum', he concluded, could best be achieved by a change of place. Later writers, while not drawing directly on Grinstein's work, have extended some of his ideas and developed some of the issues raised.

Gray (1970) saw two basic reasons for pleasure travel – 'wanderlust' and 'sunlust' (Plates 1 and 2). Wanderlust is defined as 'that basic trait in human nature which causes some individuals to want to leave things with which they are familiar and to go and see at first hand different exciting cultures and places. . . . The desire to travel may not be a permanent one, merely a desire to exchange temporarily the known workaday things of home for something which is exotic.' Sunlust 'depends upon the existence elsewhere of different or better amenities for a specific purpose than are available locally'. One expression of this, as the term suggests, is literally a 'hunt for the sun' as typified by tourist flows to the Caribbean or the Mediterranean.

Wanderlust might be thought of essentially as a 'push' factor whereas sunlust is largely a response to 'pull' factors elsewhere. Mansfeld (1992: 405) argues: 'This approach represents a confusion between person-specific motivations and resort-specific attributes'. While development studies and promotional efforts in destination areas have often continued to concentrate on the 'pull' factors with respect to the product they offer, recent motivational research has tended to emphasize the 'push' factor, the need to break from routine and to 'get away from it all'.

Leiper (1984a) points out that 'all leisure involves a temporary escape of some kind' but 'tourism is unique in that it involves real physical escape reflected in travelling to one or more destination regions where the leisure experiences transpire'. He continues: 'A holiday trip allows changes that are multi-dimensional: place, pace, faces, lifestyles, behaviour, attitude. It allows a person temporary withdrawal from many of the environments affecting day to day existence'. As such, Leiper argues, tourism enhances leisure opportunities, particularly for rest and relaxation.

Reporting on a motivational study of 39 individuals, J. L. Crompton (1979: 415) notes: 'The essence of "break from routine" was, in most cases, either locating in a different place, or changing the dominant social context from the work milieu, usually to that of the family group or doing both of these things.' Having established a 'break from routine' as the basic motivation for tourist travel, Crompton then suggests that it is possible to identify more specific directive motives which serve (p. 415) to 'guide the tourist toward the selection of a particular type of vacation or destination in preference to all the alternatives of which the tourist is aware. In most decisions more than one motive is operative.' Crompton goes on to suggest that his respondents' different motives can be conceptualized (p. 415) as being 'located along a cultural-socio-psychological disequilibrium continuum'. The socio-psychological motives were often not expressed explicitly by the respondents but Crompton identified seven of these:

1. escape from a perceived mundane environment
2. exploration and evaluation of self
3. relaxation
4. prestige
5. regression (less constrained behaviour)
6. enhancement of kinship relationships
7. facilitation of social interaction.

These broadly correspond to Dann's (1977) basic motivations of a reaction to anomie (1, 6, 7), ego-enhancement (4) and fantasy (4, 5). Two primary cultural motives were expressed by Crompton's respondents, novelty and education, though he suggests in many cases that these motives were more apparent than real. They would also appear to be closely associated with some of the socio-psychological motives. The search for novelty, for example, might well be a complement to the escape from the mundane environment.

Similar ideas are expressed by Leiper (1984a), but from a different perspective. Leiper distinguishes between recreational leisure which restores and creative leisure which produces something new. He sees the three functions of recreation as being:

1. rest (which provides recovery from physical or mental fatigue)
2. relaxation (recovery from tension)
3. entertainment (recovery from boredom).

These three functions broadly correspond to Crompton's socio-psychological motives while creativity covers aspects of the cultural motives.

Many of these factors are brought together by Iso-Ahola (1982) who proposes a more theoretical motivational model in which the escaping element is complemented or compounded by a seeking component (Fig. 2.1). One set of motivational forces derives from individuals' desire to escape their personal environment (i.e. personal troubles, problems, difficulties and failures) and/or their interpersonal environment (i.e. co-workers, family members, friends and neighbours). Another set of forces results from the desire to obtain certain psychological or intrinsic rewards, either personal or interpersonal, by travelling to a different environment. In general, Iso-Ahola's examples of personal rewards – rest and relaxation, ego-enhancement, learning about other cultures, etc. – and interpersonal rewards (essentially greater social interaction) correspond fairly closely with Crompton's specific motivational forces noted above. The point that Iso-Ahola is making here, and which is conceptualized in his model, is that tourism 'provides an outlet for avoiding something *and* for simultaneously seeking something'. Recognition that elements of both sets of forces, whose relative importance may vary from case to case, may be satisfied at the same time is particularly useful in clarifying some of the issues that arise in motivational research. Iso-Ahola also suggests that in terms of dominant motives, it is theoretically possible to locate any tourist or group in any cell at a given time but notes that this cell may not only differ from one individual or

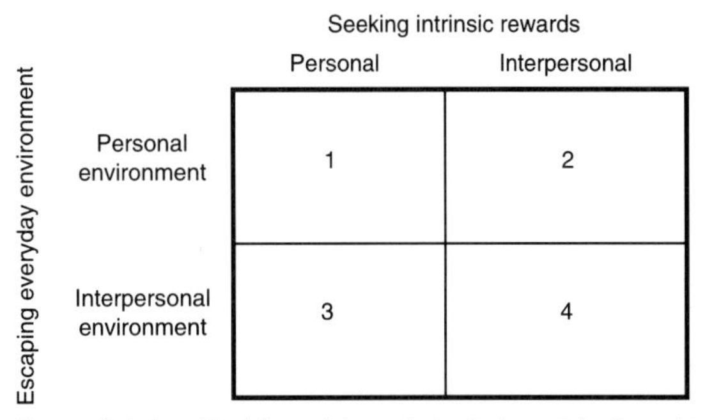

Figure 2.1 Iso-Ahola's social psychological model of tourist motivation.
Source: Iso-Ahola (1982).

Plate 1 Wanderlust tourism: a group of package tourists record their visit to Pulguksa Temple, Kyongju, South Korea.

group to another, but for any particular tourist or set of tourists it may change during the course of a trip or from one trip to another.

P. L. Pearce (1993: 125) also emphasizes change but over a longer period, arguing that people have a career in their tourist behaviour:

> Like a career at work, people may start at different levels, they are likely to change levels during their life-cycle and they can be prevented from moving by money, health and other people. They may also retire from their travel career or not take holidays at all and therefore not be a part of the system.

The levels to which Pearce refers are steps on a 'leisure ladder', a concept drawing on Maslow's hierarchical system of needs which has five motivational levels, ranging from bodily needs through stimulation, relationship development, self-esteem and development to fulfilment.

This theoretical work on tourist motivation has been complemented by a wide range of empirical studies, usually survey based. Assessing actual motives in this way is not without its limitations, but the results do yield some insights into the issues raised above.

'Change of environment' was found to be the single most important factor accounting for the desire to travel abroad for vacation in a 1980 survey of over 600 past and potential long-haul travellers in the United States (Opinion Research Corporation 1980). Plog (1991a: 73) concluded from a survey of 1200 potential long-distance travellers that

> The motives that sustain and support travel, in general, are not deep seated nor highly psychological. Ego support and a sense of self-discovery garner the fewest. . . . Rather, those items that suggest the need for a change of pace or that people get tired of daily routines in their lives, receive the most votes – 92 percent of all selections (life is too short, add interest to one's life, and a need to unwind and relax).

Other studies, however, give greater weight to the enhancement of kinship relationships. 'To meet with family, acquaintances, friends' headed the list of motivations given in a survey of Swiss tourists in 1982 (Schmidhauser 1989) while 'Being together as a family' was the top-ranked motivation of a large-scale survey of Canadian tourists in 1983 (Eagles 1992). In the latter case the scores between the 15 motivations given the highest rankings were not wide-ranging (2.76 to 2.30 on a 1 to 4 scale). Eagles (1992: 5) concluded:

> These data show Canadian travellers concentrate on travelling with family and friends to warm climates with predictable weather. They want to be safe and

Plate 2 Sunlust tourism: sunbathers shelter from the breeze at de Haan on the Belgian coast, their deck chairs oriented to the sun and away from the sea.

> in familiar surroundings. They like to see lakes, streams and beaches. They want to have fun and visit parks.

While some perspective on the relative importance of each motive is provided, these four examples illustrate the difficulties in drawing general conclusions from the disparate studies presented in the literature. The two studies in which 'change of environment' predominated involved long-distance travellers while the family-oriented motives were more important in samples of all holidaymakers. Are these results conflicting, do they simply reflect different survey designs or do they indeed indicate that different cultures or different types of tourists (e.g. long- or short-haul) have varying motivations?

Attempts have been made to address some of these issues by more systematic analysis of cross-cultural differences of market segments. Yuan and McDonald (1990), for example, examined the motivations of overseas travellers from four countries: Japan, France, West Germany and the United Kingdom. Similar rankings were found for all four countries with 'novelty' generally being followed by 'escape', 'prestige', 'enhancement of kinship relationships' and 'relaxation/hobbies'. Less consistency was found in the pull factors associated with the decision to visit overseas destinations – 'budget' was the top-ranked factor and 'hunting' the lowest in all four cases but differences occurred in the five intervening factors.

More specific empirical studies have sought to establish links between motivations and travel behaviour. Wahlers and Etzel (1985) examined the relationship between vacation activity preferences and individual stimulation needs. They hypothesized that individuals would select a holiday that corresponded with their optimal arousal level such that those experiencing stimulation deficiency in their everyday lives ('seekers') would prefer stimulating vacations while those subject to stimulation overload ('avoiders') would look for a more tranquil holiday. Stimulation deficiency or overload is relative to the individual's needs, the optimal level being 'a balance between the need for stability and the need for variety which is individually determined and changes over a person's life-span'. 'Small but detectable preferences' were identified between the two groups with the 'seekers' expressing preferences characterized as invigorating and/or innovative whereas the preferred holidays of 'avoiders' were described as structured and/or enriching. Fisher and Price (1991) found a negative relationship between the escape motive and intercultural interaction, which in turn raises questions for the promotion of exotic destinations for this group.

What becomes particularly important for the geography of tourism is the ways in which and extent to which different motivations influence destination choice and generate different travel patterns. As Mansfield (1992: 406) notes:

> A stimulus to travel can lead either to a motivation to travel to a very specific place or to a general motivation that places no specific preference on any particular destination.

Much of the motivational research is concerned with this latter type while more market-oriented studies may focus on motives for travel to particular places or, at least, examine the profiles of travellers to specific destinations.

Table 2.1 The attributes of wanderlust and sunlust travel

Sunlust	Wanderlust
• Resort vacation business	• Tourist business
• One country visited	• Probably multi-country
• Travellers seek domestic amenities and accommodations	• Travellers seek different culture, institutions and cuisine
• Special natural attributes a necessity (especially climate)	• Special physical attributes likely to be manmade: climate less important
• Travel a minor consideration after arrival at destination	• Travel an important ingredient throughout visit
• Either relaxing and restful or very active	• Neither restful nor *sportive*: ostensibly educational
• Relatively more domestic travel	• Relatively more international travel

Source: Gray (1970: 14).

Gray (1970) suggested that wanderlust and sunlust generate two distinctive forms of travel, the main attributes of which are listed in Table 2.1. From a spatial perspective, one of the crucial differences between the two is the extent to which they are likely to generate international rather than domestic tourism. Wanderlust tourism is more likely to be manifested in international travel than sunlust tourism, which in many cases can be realized elsewhere in an individual's country. The extent to which this is true, however, will depend, among other factors, on the size and geographical and cultural diversity of the country in question. Gray also recognizes here that the difference between the two kinds of travel is one of degree and cites as an example a beach resort in a foreign country offering some appeal to wanderlust. Although Gray did not elaborate on it or provide any empirical data, the suggested difference between the degree of travel associated with wanderlust and sunlust is particularly important to an analysis of travel patterns. Plog's (1973) work on the psychographic segmentation of tourists discussed in Chapter 1 suggests that tourists with different personalities will seek different travel experiences, selecting particular forms of travel and types of destination (Fig. 1.9; Table 1.1).

Writers emphasizing the need to escape, to break from routine and change of place, attach importance in the first instance to the characteristics and conditions of the origin rather than the destination. J. L. Crompton (1979: 422) notes:

> The escape from a mundane environment, exploration of self, and regression motives, require only a destination which is physically and socially different from the residential environment. Literally thousands of destinations could meet these criteria and thus serve as direct substitutes.

Similarly, Graburn (1983: 22–3), who portrays tourist travel as the spatial inversion of normal routines and situations, argues:

> the choice of tourist style stems from the culture and social structure of the home situation. . . . In the selection of changes that people wish to encounter in their tourism they choose those particular factors that they are *not* able to change in their home lives, within the constraints of opportunities offered by income and self-confidence.

The growing Singapore short-break market to neighbouring ASEAN (Association of South-East Asian Nations) destinations is a good example of this phenomenon. 'Island claustrophobia' is a term sometimes applied to the condition produced by the highly developed, densely populated nature of modern Singapore, a condition reflected in an increasing desire to take short two or three night breaks away as frequently as possible by those who can afford to do so. Readily accessible destinations offering good accommodation, food and entertainment at competitive prices have been the main beneficiaries of this demand, notably coastal resorts such as Penang and Phuket and hill resorts like Malaysia's Genting and Cameron Highlands.

While examples such as this can be found which correspond broadly with the outcomes suggested by the motivational studies discussed, the links between motivation and travel to particular destinations remain poorly developed. Oum and Lemire (1991: 295) point out that 'most survey data on tourist destination choice do not contain the choice maker's subjective evaluation of the attributes of alternative destinations' and note (p. 303) in their own analysis of Japanese international travel that their model 'explains the profile of people going to a specific destination, it does not explain why they go there'. Oum and Lemire report pronounced differences in individual socio-economic characteristics across destinations – the probability of Japanese travellers going to Hawaii, Europe and Asia, for example, decreases as income increases – but without further perceptual data were unable to ascertain why particular choices were made.

Motivation, of course, is not the only factor which comes into play in deciding to go on holiday and in selecting a particular destination. Rather, these processes are seen to involve a hierarchy of decisions (Hodgson 1983), successive stages (van Raaij and Francken 1984) and a complex of visitor characteristics, destination attributes and marketing variables (Moutinho 1987; Woodside and Lysonski 1989; Witt and Wright 1992). In particular, this and other work on destination sets (Woodside and Sherrell 1977; Um and Crompton 1990; J. Crompton 1992) indicates that potential tourists lack perfect knowledge in their decision-making and will not be aware of all possible destinations which might be able to fulfil their motivations or satisfy their needs. On the contrary, many tourists are selecting from a relatively small number of destinations which they are likely to visit in any given year.

In the case of non-tourist travel, the motivations, attitudes and policies of decision-makers other than the actual travellers themselves may be needed. O'Brien (1991: 94) argues: 'In the case of the corporate travel market it is necessary to conduct research amongst travel budget decision makers in order to understand more clearly the company perspective'. O'Brien suggests that the business travel sector has been poorly researched, available information is expensive to obtain and that 'Demand for business travel services is strongly correlated with economic output'. Similarly, research on conference travel must focus not only on the motivations of conference goers but also on the decision-making processes of all the organizations involved, both in terms of site selection and attendance policies. Further work is also needed on incentive travel, particularly in understanding how and why travel rewards appears to act as an effective motivation for generating increased sales. Hampton (1987: 15) reports that 'incentive travel was considered to be "more durable" than other incentives', apparently because of 'the large range of destinations available and, perhaps more importantly, the way in which people's perception of these destinations may even be altered by careful theme and activity choice and presentation'.

The Ability to Travel

To be able to satisfy their desire to travel, tourists must be able to meet various conditions. In particular, they must be able to afford both the time and money to do so. Tourism requires the ability to get away from school, work, home and social and other commitments. The growth of leisure time in the past century brought about by technological and other improvements has greatly increased the amount of time available for travel, especially in many Western societies (WTO 1983b; Papadopoulos 1986). Particularly important here has been the structuring of this time in the form of designated paid holidays to which many groups have been entitled since the late 1930s. As periods of leave have lengthened and incomes have risen, tourism has become more fragmented with main holidays being supplemented by secondary vacations and short breaks. A 1985 survey of European Community residents showed that 19 per cent of respondents had been away on holiday more than once for four days or more in 1985, 37 per cent had had one such vacation trip while 44 per cent had not gone on holiday at all (Commission of the European Community 1986). The European short-break market, involving trips of three nights or less, has also been growing rapidly. Factors contributing to this include not only longer vacation entitlements but also economic uncertainties, difficulties in double-income families co-ordinating their holiday time and mitigating the stress of daily life through taking multiple leisure breaks (Poitier and Cockerell 1992). As the spatial patterns of secondary holidays and short breaks differ from those of main holidays, such trends may have important geographical implications. The coast, for example, is the dominant destination for Europeans' main holidays but urban, rural and touring holidays become relatively more important for the supplementary and short-break markets.

As standards of living have risen and incomes have increased, more people have been able to afford to travel and to take advantage of extended vacation entitlements. Spending on tourism is usually regarded as discretionary spending for although tourist travel has been expressed earlier as a response to a desire to escape, in an overall hierarchy of needs this usually comes after more basic ones such as food, housing and clothing have been satisfied. Even in some developed Western societies not all groups will yet have such discretionary power, as shown by Meunier's (1985) study of low-income families in Quebec. Increasing rates of unemployment may give rise to additional free time but this is unlikely to be translated into more tourist travel unless accompanied by increases in discretionary income, which is usually not the case. There is, however, some evidence that holidays are more and more becoming regarded as part of the household budget, not just discretionary items (Guitart 1982; Boerjan and Vanhove 1984).

Aspects of the economic ability to travel have been

studied in two main ways: through econometric analyses of the determinants of demand and through examination of profile data.

The economic determinants of tourist demand are basically income and prices. Crouch and Shaw (1992) conducted a comprehensive meta-analysis of the literature on the determinants of international tourist flows and were able to establish general income and price elasticities. They found on average that a 1 per cent change in the income of tourists results in a 1.76 per cent change in tourism demand in the same direction while a 1 per cent change in the relative price of tourism services results in a 0.39 per cent change in the opposite direction.

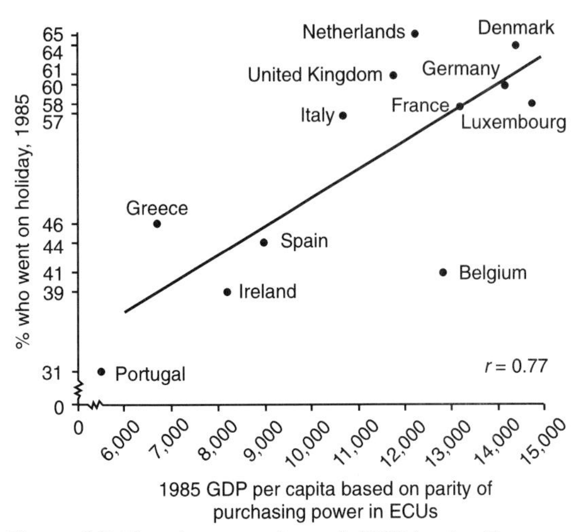

Figure 2.2 Travel propensity and GDP in the European Community, 1985.
Source: Redrawn from Commission of the European Community (1986).

The results of the European Community survey (Commission of the European Community 1986) indicate that on a national scale, rates of holidaytaking (of four nights or more) generally rise in line with the gross domestic product (Fig. 2.2, r = 0.77). As Fig. 2.2 shows, the propensity to take holidays is generally much greater in the richer members of the Community than in poorer more peripheral states such as Portugal, Greece and Ireland. Belgium exhibits a rather anomalous result; other surveys give Belgians a higher travel propensity (Boerjan 1984).

The impact of economic factors on demand is also reinforced by the socio-demographic profiles of holidaymakers and non-holidaymakers which show that the most significant discriminating factors are two interrelated economic ones, occupation and income (Table 2.2). The propensity for holidaymaking, and especially multiple departures, is much higher in the upper occupation and income categories while regular non-holidaymakers are much more numerous in the low-income groups and among farmers and fishermen. Age also has some influence, with the propensity to holiday being lowest among elderly people. The presence of children, however, does not appear to play a major role, except in families with three or more children.

Less immediately apparent is the relationship between the holiday propensity and the size of urban area and type of dwelling. Rates of holidaytaking are shown to increase with size of urban area and with multi-unit dwellings. Why this should be is not altogether clear. It may be that average incomes are greater in larger cities or that rural populations are more elderly. However, other more detailed studies from France (Baretje and Defert 1972) and Belgium (Boerjan 1984) show that the rate of holidaytaking increases with city size even when incomes are held constant. This may reflect motivational factors, with the need to escape increasing with city size, or it may simply indicate that larger cities have better transport linkages and that it is easier to get away on holiday. Whatever the reasons, the pattern of urban settlement and nature of the urban hierarchy in any country is likely to have a significant impact on the pattern of tourist travel generated there through the differential propensities to travel associated with urban size.

The European Community survey also sought to ascertain more directly why adult Europeans did not go away on holiday during 1985. Results from the survey suggest that while about one-fifth of the non-holidaymakers preferred to stay at home, others wanted to get away but were constrained in some way from doing so. Economic factors were the major constraint (44 per cent), followed by work ties (16 per cent), exceptional family circumstances, poor health, moving residence and unforeseen impediments. The importance of these factors varied from country to country. In Denmark, for example, 38 per cent said that they preferred to stay at home while in the peripheral countries the majority of non-holidaymakers could not afford to leave home (Portugal, 67 per cent; Ireland, 61 per cent; Greece, 55 per cent).

Expense (52.6 per cent) and an inability to take time off work (27.2 per cent) were also cited as the two major obstacles to travelling overseas in a 1992 Japanese survey (Japan Travel Bureau 1992). Other obstacles mentioned by the Japanese respondents relate more specifically to barriers to overseas travel, namely language barrier (26.2 per cent), anxiety about security abroad (23.4 per cent), fear of flying (14.4 per cent) and dislike of foreign food (13.7 per cent).

In addition to these socio-economic constraints, more formal immigration and fiscal barriers to international

Table 2.2 Profiles of holidaymakers and non-holidaytakers in the European Community, 1985

	Did not go away in 1985			Went away in 1985		
	Total (%)	Don't usually go away (%)	Others (%)	Once (%)	Several times (%)	Total (%)
Total EC 12	44	21	23	37	19	56
Income						
High R++	25	7	18	43	32	75
R+	40	15	25	39	21	60
R−	51	27	24	37	12	49
Low R−−	64	41	23	27	9	36
Occupation of head of household						
Executives, top management	15	2	13	42	43	85
Professionals	18	4	14	39	43	82
White collar	29	8	21	44	27	71
Business, shop owners	44	19	25	40	16	56
Manual workers	49	22	27	39	12	51
Retired	51	31	20	31	18	49
Others, non-active	56	33	23	30	14	44
Farmers, fishermen	75	51	24	20	5	25
Age						
15–24	38	11	27	41	21	62
25–39	38	14	24	41	21	62
40–55	47	24	23	37	16	53
Over 55	53	33	20	30	17	47
Number of children (<15 years) at home						
None	44	23	21	37	19	56
One	44	18	26	38	18	56
Two	40	15	25	41	19	60
Three or more	56	28	28	30	14	44
Urbanization						
Large town	34	15	19	42	24	66
Small town	41	18	23	38	21	59
Village	55	30	25	32	13	45
Type of dwelling						
Flat in block of						
over 50 flats	34	12	22	39	27	66
11 to 50 flats	35	15	20	44	21	65
up to 10 flats	40	18	22	38	22	60
Semi-detached or terrace house	47	22	25	37	16	53
Detached house	46	23	23	35	19	54
Farm or isolated country house	62	41	21	29	9	38

Notes: R++ upper quartile
R−− lower quartile
Source: Commission of the European Community (1986).

travel may exist. Governments may seek to control both inbound and outbound tourists through passport and visa requirements, foreign exchange controls and other regulations. The general tendency in recent years has been an easing of these restrictions. Part of the upsurge in Japanese travel in the 1970s can be attributed directly to the liberalization of Japanese currency exchange regulations in 1964 and to the easing of procedures to obtain passports in 1970. Similarly, the increase in Taiwanese outbound travel after 1979 reflects a gradual relaxation of travel restrictions from that date, including significant liberalization of passport regulations in 1989 (Li 1992). Increased freedom for citizens to travel abroad was one of the more immediate consequences of the political upheaval in Eastern Europe in 1989.

Political instability is also a major handicap to tourist travel for conditions of uncertainty and personal insecurity are scarcely compatible with the quest for relaxation. Sharp downturns in the tourist traffic have thus accompanied political upheaval, war and terrorist activity in different parts of the world (e.g. Greece, late 1960s; Spain, 1974 and 1976; Fiji, 1987; former Yugoslavia, from 1990). More generally, the Gulf War depressed international travel worldwide in 1991. However, tourist memories have been shown to be relatively short with visitor arrivals generally recovering reasonably quickly after events, upon the cessation of the troubles and the return to normal conditions. Certain destinations, however, may have a continuing problem, such as Northern Ireland where Buckley and Klemm (1993: 193) conclude: 'although individual outbreaks can be "got over", there is a slow deterioration of the province'.

Measuring Demand

Demand for tourism is expressed and measured in different ways. Most studies refer to what might be termed 'effective demand', that is generally to the number of people who actually participate in a tourist activity or visit a given area. Then there is deferred demand, that is 'those who could participate but do not, either through lack of knowledge, or lack of facilities or both', and potential demand which refers to 'those who cannot at present participate and require an improvement in their social and economic circumstances to do so' (Lavery 1974). This book is concerned almost exclusively with effective demand, with the emphasis being on establishing actual patterns of use and demand rather than on predicting future ones.

Characteristically, effective demand is measured in terms of the number of tourists leaving or visiting a country or region, the number of passengers using a certain mode of transport, the number of bednights spent in a particular type of accommodation, the number of people using a given recreational facility or taking part in a specific activity such as skiing and so on. Economically, effective demand might also be expressed in dollars spent on a given activity in a particular region or generated by a specific market. None of these variables lends itself very readily to measurement, and data problems continue to beset most areas of tourism analysis. These issues are examined in greater detail with respect to particular topics in later chapters.

Demand for tourism is also expressed in terms of travel propensity. Schmidhauser (1975, 1976) differentiates between two types of travel propensity, net and gross. Net travel propensity refers to the proportion of the total population or a particular group in the population who have made at least one trip away from home in the period in question (usually a year) and is calculated by the following formula:

$$\text{Net travel propensity (as \%)} = \frac{p}{P} \times 100$$

where p = the number of persons in a country or in a particular population group who have made at least one trip away from home in the given period

P = total population of the country or group.

Gross travel propensity refers to the total number of trips taken in relation to the total population studied and is expressed by the formula:

$$\text{Gross travel propensity (as \%)} = \frac{Tp}{P} \times 100$$

where Tp = total number of trips undertaken by the population in question

P = total population of the country or group.

Schmidhauser also discusses the concept of travel frequency which refers to the average number of trips taken by a person participating in tourism in a given period according to the formula:

$$\text{Travel frequency} = \frac{Tp}{p} = \frac{\text{gross travel propensity}}{\text{net travel propensity}}$$

where Tp and p have the same meaning as above.

An increase in effective demand may reflect changes in net travel propensity, gross travel propensity or a combination of the two. Where the net travel propensity is low, for example 30 per cent, growth in tourist demand may come largely through an increase in the proportion of the population who attain a standard of living enabling them to take a holiday. Where the net travel propensity is much higher, for example 60–70 per cent, an increase in demand may result mainly from existing travellers travelling more frequently, that is, an increase in gross propensity.

The Evolution of Demand

Many interrelated factors have contributed to often quite dramatic increases in the demand for tourism since the 1950s. Absolute population growth and rising standards of living which have resulted in more leisure time and greater discretionary spending have greatly boosted the numbers able to travel, both domestically and internationally. At the same time, both the public and private sectors in many parts of the world have actively fostered this demand and encouraged the expansion of the tourist industry. Tourism, particularly in the 1960s, appeared as a source of unlimited growth. In the market areas, many travel agencies and tour operators appeared, stimulating demand to increase their new businesses. The transport industry also extensively promoted tourist travel while technological improvements brought down the relative cost of travel. At the same time, the public and private sectors in many areas, attracted by the various economic benefits which tourism appeared to offer (jobs, profits, taxes, diversification, regional development, etc.) and as yet unaware of associated negative impacts, embarked on extensive destination development programmes and tourist promotion campaigns. Such activity helped convert latent and deferred demand into effective demand while further growth in travel markets encouraged more development. Table 2.3 shows aspects of the growth of the tourist industry in Australia from 1968 to 1979 and provides a good example of increased supply and demand during this period. Growth has been tempered from time to time by factors such as the energy crisis in the early 1970s, economic downturns, wars and civil strife but Lee (1987: 3) appears not to be over-optimistic in his claim that 'Global tourism has clearly proved its vitality and resistance to inflationary and currency pressures, political instability, unemployment and limitations on purchasing power'.

World Tourism Organization (WTO) estimates indicate the number of international tourist arrivals worldwide has increased steadily from 25 million in 1950, to 69 million in 1960, 160 million in 1970, 284 million in 1980 to more than 440 million in 1990 (Fig. 2.3). Throughout this period the only absolute decreases in the global arrivals were experienced in 1982 and 1983, largely as a result of recessionary conditions felt throughout the world. Worldwide growth picked up again in the latter part of the decade. A slight decline (−0.1 per cent) occurred in 1991 as a result of the Gulf crisis but recovery was soon under way. The WTO reported global increases of 5.5 per cent in 1992 and 3.8

Table 2.3 Evolution of selected aspects of the Australian tourist industry; 1968–79

Sector	Number of firms and organizations 1968	1979
Retail travel agencies	585	1557
Foreign tourism organizations	27	43
International on-line airlines	13	23
International tour operators/wholesalers	40	146
International tourists	*1968*	*1979*
Australian departures	252,000	1,132,000
International arrivals	275,800[a]	793,345

Note: [a] 1969.
Source: After Leiper (1984b).

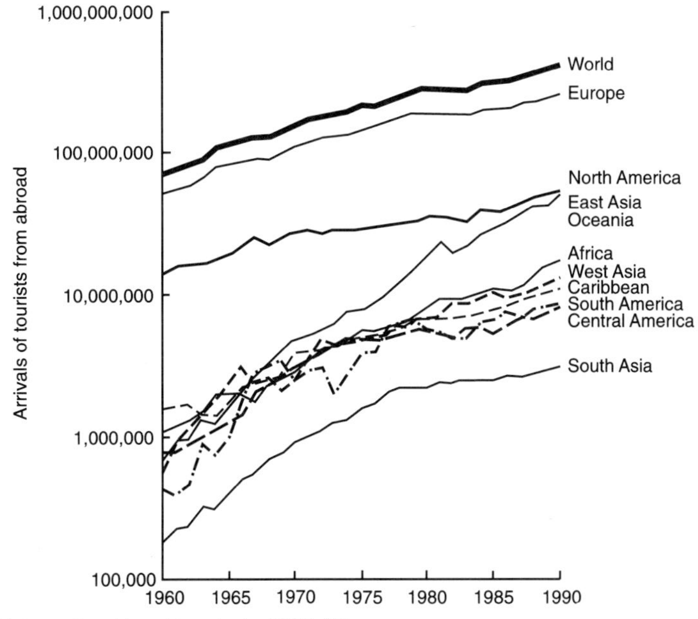

Figure 2.3 Evolution of international tourist arrivals, 1960–90.
Data source: World Tourism Organization.

per cent in 1993, with estimated international tourist arrivals worldwide that year reaching half a billion.

Throughout this period Europe has remained the dominant world region for international tourism, the number of arrivals recorded passing from 50 million in 1960 to 270 million in 1990. Despite these huge absolute increases, Europe has lost market share with the region's proportion of international arrivals dropping from 72 per cent in 1960 to 62 per cent three decades later. North America's share of the world total also dropped in the 1960s and 1970s, declining from 20 per cent in 1960 to 12 per cent in 1980, since when its share has stabilized. East Asia and the Pacific has experienced consistently high rates of growth since 1960 and has come to rival North America in world market share (1 per cent in 1960, 11.5 per cent in 1990). This expansion reflects the economic strength of Japan and the emergence of other newly industrializing countries such as Singapore, South Korea and Taiwan, together with an increase in longer-haul travel. Growth elsewhere in the world has generally been slower. The developing world still attracts only a relatively small share of the world's international tourist arrivals; Africa, the largest of the other regions, recorded only 4 per cent in 1990.

No comparative data sets similar to the WTO's international arrival figures exist for domestic tourism and it is difficult to establish as comprehensive a picture of the evolution of this sector. Time-series data from Europe and other statistics on domestic tourism in developed countries do provide some indication of recent trends. These suggest that while growth in domestic tourism has occurred, some markets are maturing or becoming saturated, as evidenced by stable or declining numbers of holidays taken internally.

Table 2.4 shows a steady increase in the number of domestic holidays taken in France since 1964. In contrast, the number of holidays of four nights or more taken in Great Britain peaked in the 1970s and progressively declined during the 1980s with the percentage of holidays taken abroad reaching 40 per cent by the end of the decade. Some reversal of this trend is evident in the early 1990s but whether this is temporary and accentuated by the Gulf crisis or more long term remains to be seen. The significance of the domestic travel component in the United Kingdom, however, does vary according to the measure used. In 1992 British residents took 118.9 million trips (of from 1 to 60 nights, all purposes); domestic travel accounted for 80 per cent of trips, 61 per cent of nights but only 43 per cent of expenditure (STB 1993). The number of main holidays spent domestically in Germany was greatest in the early 1980s after which some stabilization appears to have occurred. This decline may have been offset to some extent by growth in supplementary 'short-break' trips of less than four nights.

Sluggish growth or decline in domestic tourism is not restricted to Europe. New Zealand registered its fourth successive annual decline in domestic bednights in 1989/90 (Pearce 1993a). In Australia, domestic tourism grew by only 10 per cent in the period 1984/85 to 1989/90, most of this growth occurring in 1989/90 with small declines being registered in every other year except 1987/88 (BTR 1991). Some of this slow growth or decline can be attributed to prevailing economic conditions in Australasia. In the USA, the late 1980s were a period of overall economic expansion during which consecutive increases in travel away from home were recorded (1990 person trips up 2.3 per cent over 1989) (US Travel Data Center 1989, 1991).

The more detailed European figures in Table 2.4 indicate, however, that other factors also come into play, notably the changing relationship between domestic demand and outbound travel. In Great Britain, net travel propensities have hovered around 60 per cent since 1970 but the percentage of holidays taken abroad has risen dramatically resulting in the observed domestic decline. The proportion of Germans taking their main holiday abroad first exceeded 50 per cent in 1967 but continuing increases in travel propensity (peaking at 74.8 per cent in 1990) have partially offset this trend, resulting in stability rather than absolute decline in the number of main domestic holidays taken. In contrast, the proportion of French holidays taken abroad has remained virtually unchanged since 1973 (at around 16 per cent) with the result that higher travel propensities (and a growing population) have been translated into more domestic holidays. These different trends can partially be attributed to differences in the resource base and patterns of development. Both Great Britain and Germany are generally less endowed with attractive climates and beaches and constituted major markets for the development of package tours to Spain and other Mediterranean destinations. Not only does France have these resources but also during the 1960s and 1970s the government pursued an aggressive coastal tourism development programme and other policies designed to boost domestic tourism (Pearce 1989b).

Large size and a broad resource base appear to contribute to low levels of outbound travel in the United States. Ottersbach (1985) reports that only 14 per cent of the US adult population had taken a foreign trip in the previous three years while 61 per cent had made a domestic trip of 100 miles or more in the previous year. Regional variations in domestic trip making were not pronounced but levels of foreign travel were twice as high in the west as in the south.

Important differences may also occur in the patterns of domestic and outbound demand within countries. Table 2.5, for instance, indicates that within Great

Table 2.4 Evolution of domestic tourism in Germany, Great Britain and France, 1954–91

	Germany			Great Britain[b]			France[b]		
	% taking holiday	Holidays in Germany[a] (millions)	% holidays abroad	% taking holiday	Holidays in Britain (millions)	% holidays abroad	% taking holidays	Holidays in France (millions)	% holidays abroad
1954	24	7.9	15						
1956	26	7.9	20.2						
1958	28	8.4	27						
1960	28	8.1	31.4						
1962	32	8.2	40.2						
1964	39	9.6	42.9				43.6	31.2	12.2
1965					30	14.2			
1966	42	9.7	48.1		31	15			
1967					30	14.2			
1968	39	8.2	51.2		30	14.2			
1969					30.5	15.8	45	32.5	13.9
1970	41.6	8.5	54	59	34.5	14.2			
1971				59	34	17.5			
1972	49	9.4	56.9	62	37.5	18.4			
1973				63	40.5	16.9	49.2	31.2	16.8
1974	52.5	9.8	58.1	62	40.5	14.2	50.1	33.5	14.3
1975				60	40	16.6	52.5	35.2	15.5
1976	53	10.1	57.6	61	37.5	16.2	54	37.4	14.8
1977				59	36	17.7	53.3	36.6	16.6
1978	56.2	10.1	60.9	61	39	18.7	54.3	40.6	16.8
1979				63	38.5	21.0	55.9	43.8	14.9
1980	57.7	10.2	62.2	62	36.5	24.7	57.2	44.8	15.1
1981				61	36.5	26.6	57.2	45.7	15.3
1982	55	10.2	61.4	59	32.5	31.1	57.8	47	15.1
1983				58	33.5	30.2	58.3	48.3	14.3
1984	55.3	9.2	65.5	61	34	31.3	57.3	48.9	15.2
1985				58	33	32.3	57.5	48.8	15.4
1986	57	9.3	66.3	60	31.5	35.7	58.2	49.6	16.9
1987	64.6	9.5	69.5	58	28.5	41.2	58.5	51.5	16.6
1988				61	33.5	37.6	59.5	52	16.9
1989				59	31.5	40	60.7	52.6	17.8
1990[c]	74.8	9.9	70.4	59	32.5	39	59.1	52	17.5
1991[c]	72.1	9.9	69.6	60	34	36.8	59.8	53.7	16.0

Notes: [a] Main holidays.
[b] All holidays (4+ nights).
[c] For Germany, former W. German *Länder* only.
Sources: Dundler (1988); Gilbrich (1991); British Tourist Authority (1989, 1992); *Annuaire Statistique de la France* (1954–1991).

Britain those in the upper socio-economic groups (classes AB and C1) and residents of the South East are over-represented in terms of those taking a holiday abroad. These interrelated factors reflect the ability to travel, both in terms of wealth and readier access to the Continent.

However, as Ronkainen and Woodside (1978: 580) note: 'demographics may represent *enabling* factors only and not *explanatory* factors for travel behaviour'. In one of the few studies of its kind relating to domestic and outbound tourists, Ronkainen and Woodside report psychographic differences in their sample of Finnish and American travellers. Foreign travellers appeared more self-confident and well informed, were opinion leaders rather than seekers and were looking for more active holidays than those spent at home. In contrast, domestic tourists were found to be more traditionalist, favouring holidays centred around relaxation rather than action. The trends shown in Table 2.4, however, suggest that it is changes in the enabling factors which may be the more important in terms of the growth in international travel for it is difficult to envisage a sufficiently marked shift in the psychographics of the British and German markets over the period in question.

The recent shift from domestic to outbound travel in

Table 2.5 Profile of domestic and international holidaytakers in Great Britain, 1991

	Adult population (%)	Adults taking no holiday[a] (%)	Holidays in Britain (%)	Holidays abroad[b] (%)
Social Class				
AB (professional/ managerial)	17	8	20	32
C1 (clerical/ supervisory)	22	20	22	26
C2 (skilled manual)	29	30	29	26
DE (unskilled/ pensioners, etc.)	32	43	29	17
Region of residence				
North	6	6	5	6
Yorkshire and Humberside	9	7	10	11
North West	12	11	13	11
East Midlands	7	8	8	5
West Midlands	10	10	10	8
East Anglia	3	3	4	3
South East:				
London	13	14	10	15
Rest of South East	18	16	19	21
South West	8	10	6	6
Wales	5	5	6	5
Scotland	9	10	10	9

Notes: [a] Holidays of 4+ nights.
[b] Holidays of 1+ nights.
Source: After British Tourist Authority (1992).

developed countries appears to be matched in some developing countries by the emergence of significant domestic demand alongside the more traditional inbound market. Berriane (1993), in a comprehensive study of domestic tourism in Morocco, estimates that one out of three Moroccans took a summer holiday in 1985. While the growth in domestic tourism there can in part be attributed to the development of the Moroccan economy, Berriane cautions against too simplistic explanations, including the imitation of foreign visitors. Rather, he suggests, the growth of domestic tourism in Morocco reflects a combination of external and internal factors and exhibits its own characteristics, including a strong family dimension. In Malaysia, domestic hotel guests increased by 22 per cent in 1989 over the previous year and a further 16 per cent the following year (TDC 1991). Further studies elsewhere of these two types, in-depth analysis and the monitoring of trends, are required to establish the general nature of the evolution of domestic demand in developing countries. If such studies can be put in place the opportunity exists to analyse the evolution of demand from an early take-off phase, in contrast to many developed countries where net travel propensities were already relatively high when tourism first became the subject of serious research.

Conclusions

Clearly there is scope for more detailed research into tourist motivations and other factors influencing the demand for tourism. However, from the work to date it might be argued that the interaction between origins and destinations implicit in all tourism arises primarily out of a basic need to leave the origin. What emerges from these studies is that the fundamental motivation for tourist travel is a need, real or perceived, to break from routine and that for many this can best be achieved by a physical change of place. Thus, 'change of place' is seen to be not just one of the defining attributes of tourism, but the very essence of it. The basic need to escape may be expressed in terms of more specific motives so that particular patterns of interaction will occur depending on the extent to which different destinations can respond to or appear to cater for these different motives and the extent to which residents can realize their particular needs and desires. The next three chapters examine in detail these patterns of interaction at different scales while subsequent chapters include discussion of how the demand generated in the market areas manifests itself in the destinations. These analyses and this discussion must be seen in terms of the evolution of the demand for tourism which in general has been characterized by growth and changes in the relationships between domestic and international tourism.

CHAPTER 3

Patterns of International Tourism

International tourism can be considered at different scales. In this chapter broad global patterns are first identified then attention focuses on the nature and extent of international circuit tourism and the spatial evolution of flows from the world's four leading markets. Conclusions are then drawn in which the patterns identified are discussed in relation to the models outlined in Chapter 1. This discussion also sets the scene for the analysis of intra-national travel by international tourists in Chapter 4. At all scales, however, analysis is limited and complicated by the data available. Major data sources are therefore reviewed at the outset with more specific types of statistics being considered in subsequent sections.

Data Sources

It is important to recall that what constitutes international tourism is movement across international frontiers while domestic tourism concerns travel within national boundaries. This distinction may be especially significant at a national level, for example in terms of foreign exchange earnings, social impact and immigration control. Clearly too, the fact of being in a foreign country may add something to the appeal of the trip for the individuals concerned, with the differences in culture, customs and so on being a major attraction for many tourists, especially the wanderlust travellers. In such cases, international tourism constitutes a particular form of tourism and the distinction made by official statistics between international and domestic tourists is an especially useful one. In other instances, as with much sunlust tourism, the reason for and nature of the holiday may be essentially the same whether it involves a German travelling to the Mediterranean coast of France (international tourism) or a New Yorker holidaying in Florida (domestic tourism). With this type of tourism, the distinction between domestic and international tourist flows is perhaps more official than functional. Whatever the nature of the travel, greater distances will in fact be covered by many domestic vacationers in the United States and other large countries such as Canada and Australia than international travellers in Europe. In 1990 the average round trip distance covered by domestic travellers in the United States (on trips of 100 miles or more) was 839 miles (Waters 1992).

The most comprehensive body of international tourism statistics readily available is that published annually by the World Tourism Organization (WTO) in their series World Tourism Statistics. Other international organizations such as the Pacific Asia Travel Association (PATA) and the Organization for Economic Co-operation and Development (OECD) also collate and publish statistics concerning tourism in their member countries. In their series the WTO, an intergovernmental organization, compiles and distributes travel statistics furnished by members and some non-member states. While the WTO provides technical guidelines for the collection of these statistics in an attempt to provide internationally uniform and accurate figures (WTO 1978, 1981), it remains dependent on the respective national tourist organizations, immigration or statistics departments for the nature and quality of the data received. These bodies collect travel data in a variety of ways and for a range of purposes. Consequently the type of information and its reliability varies considerably from country to country. Researchers must therefore view the published data carefully, particularly when making comparisons among countries.

Travel statistics worldwide are most frequently expressed in terms of 'frontier arrivals', that is the number of visitors entering a country as determined by some form of frontier check and irrespective of purpose of visit and length of stay; excursionists, that is visitors staying less than 24 hours, are usually included. 'Tourist arrivals' refer to visitors staying at least this minimum length of time. Where the majority of arrivals are by air through a limited number of points of entry, as is the case for example in island states such as Japan, Australia and New Zealand, then the degree of control is usually very high and most related statistics can normally be considered reliable. Where immigration procedures require the completion of an arrival/departure (A/D) card, a range of information is usually obtained including, for example, details on nationality, age, occupation, purpose of visit (however defined) and intended length of stay. In other instances, where there is a large volume of traffic arriving overland through a number of entry points, as is the case in much of Europe, the degree of control is much less and some form of estimate may be used. Not all countries require the completion of an A/D card and, in these cases, periodic surveys may be employed to provide additional information.

A second common source of international travel statistics is that based on accommodation returns. Many countries require international visitors to complete registration cards in hotels and other forms of accommodation, which are then collated and analysed. A number of problems arise here, not the least of which is that the cards are often associated with some form of taxation so that a degree of underestimation might be expected. Conversely, visitors moving around the country would be recorded more than once, thus inflating the figures. Visitors staying in any form of non-commercial accommodation would not be recorded at all.

The WTO statistics do not usually take into account purpose of visit as all foreigners are usually recorded under 'frontier arrivals' and no distinctions are made in the accommodation figures. International comparisons are therefore limited essentially to studies of 'all travellers'. The originating national body, however, usually does provide some breakdown, for instance into 'holiday and vacation', 'business', 'visiting friends and relations (VFR)', 'education', 'sport' and so on. A more detailed study by the WTO of variations in purpose of visit by major regions in 1979 showed vacation travel accounted for almost three-quarters of all international movements. Only in the Middle East, where a large pilgrimage traffic was included in the 'other' category and where business traffic is more significant, did holidaymakers account for less than a half of all visits. The composition of visitors to particular countries within these regions may, of course, vary significantly from the regional average. Where possible, differences in the spatial behaviour of particular types of visitors should be taken into account.

Geographic analyses in particular are often complicated or handicapped by the combination of markets in published statistics. The degree of aggregation usually depends on the markets' importance at a given destination. In most European countries, other countries from within Europe are listed separately; in more distant destinations, such as much of the Pacific, they are commonly listed together under 'Europe'. Conversely, New Zealanders and Australians, who constitute major markets in the Pacific, appear separately in Pacific statistics but are frequently lumped under 'the rest of the world' as far as many European statistics are concerned. This clearly limits wide-ranging and detailed international analyses. The varying use of 'country of residence' and 'nationality' further complicates geographical and marketing analyses.

Figures on outbound travel are subject to many of the same limitations as the arrivals statistics but pose particular problems when attempts are made to disaggregate the destinations involved. Most countries routinely collect information on only one destination, usually the main one, the one in which the most time was spent or the last country visited. Where multi-destination travel is involved significant underestimates in country to country travel may result with a consequent lack of correspondence between the figures reported by the origin and destination countries. These and other technical considerations are usefully reviewed in greater length by Edwards (1991) and Dann (1993).

Table 3.1 Distribution of arrivals, person nights and expenditure in New Zealand by country of residence, 1989

	Arrivals[a] (%)	Total person nights[b] (%)	Total expenditure[b] (%)	Expenditure per person per day[b] (NZ dollars)
Australia	31.0	25.2	23.3	87.89
USA	18.5	14.3	20.3	134.79
Japan	11.0	4.2	16.8	384.03
UK	8.3	15.5	9.4	57.54
Canada	4.2	4.8	4.8	94.05
West Germany	2.4	3.8	3.4	84.53
Other	24.6	32.2	22.0	63.93
All countries	100	100	100	95.15

Notes: [a] All arrivals.
[b] Visitors aged 15 years and over.
Source: Pearce (1990a).

Consideration must also be given to what it is that is being measured. Absolute number of travellers, even if they can be classified by purpose of visit, do not tell the whole story. Length of stay and expenditure are also decisive factors in determining the impact of markets on particular destinations as Table 3.1 demonstrates. Table 3.1 shows that when demand is measured in terms of person nights and total expenditure rather than arrivals, the Australian share of international tourism to New Zealand in 1989 falls from just under one-third to around one-quarter. Conversely, the high daily expenditure of the Japanese (NZ$384 per day compared to the overall average of NZ$95) means that their contribution to total expenditure is much greater than their share of arrivals, while their percentage share is much less due to their shorter visits to New Zealand. Variations on these three measures are less pronounced for the American market, while the British are notable for their increased share of total person nights resulting from a much longer stay (Pearce 1990a).

Care must therefore be exercised in the analysis and interpretation of international visitor statistics, particularly when different sources or data from different years are being used. In many cases the resultant patterns will be indicative rather than definitive but these too can provide important insights into global patterns of international tourism.

Global Patterns of International Tourism

Tables 3.2 and 3.3, which portray the major generating and receiving countries, provide a useful starting-point for trying to order international tourist flows throughout the world.

Table 3.2 Major markets: world's top fifteen international tourism spenders, 1992

Rank 1992 (1985)		Country	Tourism expenditure[a] (US$ million)	Share of expenditure worldwide (%)	Cumulative (%)
1	(1)	USA	39,872	14.68	14.48
2	(2)	Germany	37,309	13.55	28.03
3	(4)	Japan	26,837	9.75	37.78
4	(3)	UK	19,831	7.20	44.98
5	(10)	Italy	16,617	6.04	51.02
6	(5)	France	13,910	5.05	56.07
7	(6)	Canada	11,265	4.09	60.16
8	(7)	Netherlands	9,330	3.39	63.55
9	(14)	Taiwan	7,098	2.58	66.13
10	(8)	Austria	6,895	2.50	68.63
11	(13)	Sweden	6,794	2.47	71.10
12	(12)	Belgium	6,603	2.40	73.50
13	(11)	Mexico	6,108	2.22	75.72
14	(9)	Switzerland	6,068	2.20	77.92
15	(15)	Spain	5,542	2.01	79.93

Note: [a] International transport excluded.
Data source: WTO.

Given the limitations in recording outbound flows noted above, the World Tourist Organization now depicts major markets in terms of expenditure on international tourism rather than departures. Table 3.2 indicates that over half of all international tourism expenditure in 1992 was generated by just five countries, in order: the USA, Germany, Japan, the United Kingdom and Italy. Japan's third ranking is in large part a function of heavy expenditure per tourist rather than the absolute number of departures (11 million in 1990). The leading fifteen world spenders accounted for 80 per cent of expenditure worldwide; all but two of these countries (Japan and Taiwan) are located in Europe or North America.

Table 3.3 World's top fifteen destinations, 1992

International tourist arrivals

Rank 1992 (1985)		Country	Tourist arrivals	Share of arrivals worldwide (%)	Cumulative (%)
1	(1)	France	55,590,000	12.37	12.37
2	(3)	USA	44,647,000	9.27	21.64
3	(2)	Spain	39,638,000	8.23	29.87
4	(4)	Italy	26,113,000	5.42	35.29
5	(11)	Hungary	20,188,000	4.19	39.48
6	(5)	Austria	19,098,000	3.96	44.44
7	(6)	UK	18,535,000	3.85	47.29
8	(9)	Mexico	17,271,000	3.59	50.88
9	(12)	China	16,512,000	3.43	54.31
10	(8)	Germany	15,147,000	3.14	57.45
11	(7)	Canada	14,741,000	3.06	60.51
12	(10)	Switzerland	12,800,000	2.66	63.17
13	(13)	Greece	9,331,000	1.94	65.11
14	(14)	Portugal	8,921,000	1.85	66.96
15	(15)	Czechoslovakia	8,000,000	1.66	68.62

International tourism receipts

Rank 1992 (1985)		Country	Tourism receipts[a] (US$ million)	Share of receipts worldwide (%)	Cumulative (%)
1	(1)	USA	53,861	13.45	13.45
2	(4)	France	25,000	12.59	26.04
3	(3)	Spain	22,181	9.06	35.10
4	(2)	Italy	21,577	6.69	41.79
5	(5)	UK	13,683	5.61	47.40
6	(6)	Austria	13,250	4.69	52.09
7	(7)	Germany	10,982	3.80	55.89
8	(8)	Switzerland	7,650	3.15	59.04
9	(11)	Hong Kong	6,037	2.39	61.43
10	(10)	Mexico	5,977	2.33	63.76
11	(9)	Canada	5,679	2.29	66.05
12	(14)	Singapore	5,204	2.23	68.28
13	(13)	Netherlands	5,004	2.06	70.34
14	(15)	Thailand	4,829	2.05	72.39
15	(12)	Belgium	4,053	1.87	74.26

Note: [a] International transport excluded.
Data source: WTO.

In terms of destinations, international tourism, whether measured by arrivals or receipts, exhibits a slightly more dispersed pattern, but one which remains nevertheless highly concentrated. Table 3.3 shows that the fifteen leading destinations accounted for 69 per cent of arrivals and 74.3 per cent of expenditure worldwide. The top eight countries received over 50 per cent of arrivals; the first six accounted for over half of the receipts. The relative rankings of the leaders may change depending on the measure used: France and the USA alternate for first and second places and lower down the table, destinations come and go. Some, such as Hungary and former Czechoslovakia, receive large numbers of visitors who appear to spend relatively small amounts, others, such as Hong Kong and Singapore, attract smaller volumes of large spenders. Overall, the table reinforces the broader regional patterns identified in Fig. 2.3, namely the prominence of Europe and North America and the emergence of certain Asian destinations, while also highlighting the concentration in particular countries within these regions.

While the leading fifteen destinations' share of all tourist arrivals remained unchanged over the period 1985–92 (around 89 per cent), increased concentration occurred in terms of the leaders' share of worldwide receipts (relates to the income received by the receiving countries) (66.2 per cent in 1985; 74.3 per cent in 1992) and expenditure (relates to the spending by the generating countries) (75.4 per cent to 79.9 per cent).

International tourist flows

Tables 3.2 and 3.3 have identified the major international tourist markets and destinations throughout the world. A next step is to establish global patterns of flows between generating and receiving countries. This can be attempted using several different approaches.

Table 3.4 World's top ten international tourist flows, 1990

Canada → USA	17,262,000
USA → Canada	12,267,000
Germany → France	12,097,000
Germany → Italy	10,676,000
Switzerland → Italy	10,331,000
Portugal → Spain	10,106,000
France → Italy	9,219,000
Mexico → USA	7,450,000
UK → France	7,346,000
Belgium → France	7,210,000

Data source: WTO.

Table 3.4 depicts the ten largest individual flows throughout the world using comparable, available WTO figures. Table 3.4 shows that the single largest exchanges in the world occur across the Canadian–United States border, the flow from north to south being by far the greatest. With the exception of the Mexican flow into the United States, the other largest flows are between neighbouring countries in Europe, with Italy and France being the major recipients. The difficulties of recording accurately large volumes of traffic in these should be noted, but Table 3.4 appears nevertheless to provide a good indication of where the major flows do occur.

A more comprehensive global picture might be established by identifying dominant flows, that is first- and second-order flows, towards each destination throughout the world. Figs 3.1 and 3.2 portray these flows for each of the 134 destinations in 1990 for which appropriate data are available from the WTO (some destinations had to be excluded as their major markets were not disaggregated to a country level).

On this basis, United States visitors are clearly the most dominant group, constituting the major market and generating first-order flows for 38 destinations, or one-quarter of those portrayed, while ranking second at a further 20 destinations. Three-quarters of the destinations where the United States constitutes the leading market are found in the Americas, the inclusion of a number of small Caribbean states clearly affecting the overall importance of the country on this increase. At the same time, United States visitors generate first-order flows to a range of other destinations including the United Kingdom, Germany and Israel. In addition to generating the single largest flow – to the USA – Canada takes second place to its larger neighbour in half a dozen Caribbean destinations and is the dominant market for Cuba, from which Americans are excluded by their foreign policy.

Germany (12 first-order flows, 12 second-order flows), France (14, 9) and the United Kingdom (5, 12) constitute an important set of major European markets. Many of these flows are intra-European but longer-haul traffic is also evident, particularly to former colonies and overseas territories in the cases of France and the United Kingdom. Japan (7, 6), Australia (5, 4) and South Africa (4, 1) and Argentina (4, 0), are the only other countries to constitute the dominant market for four or more destinations. Each of these constitutes a secondary regional source in Asia, the South Pacific and southern Africa and South America respectively, regions which are more remote from and thus less influenced by the major markets of the United States and Europe. Virtually all the other first-order flows are between pairs of neighbouring countries, for example Spain and Portugal, India and Bangladesh.

The overall impression to emerge from this brief examination of first- and second-order flows is thus one of comparatively short movement between countries in the same region of the globe. Distance clearly plays a major role in shaping international tourist flows. On top of this general pattern are superimposed some more

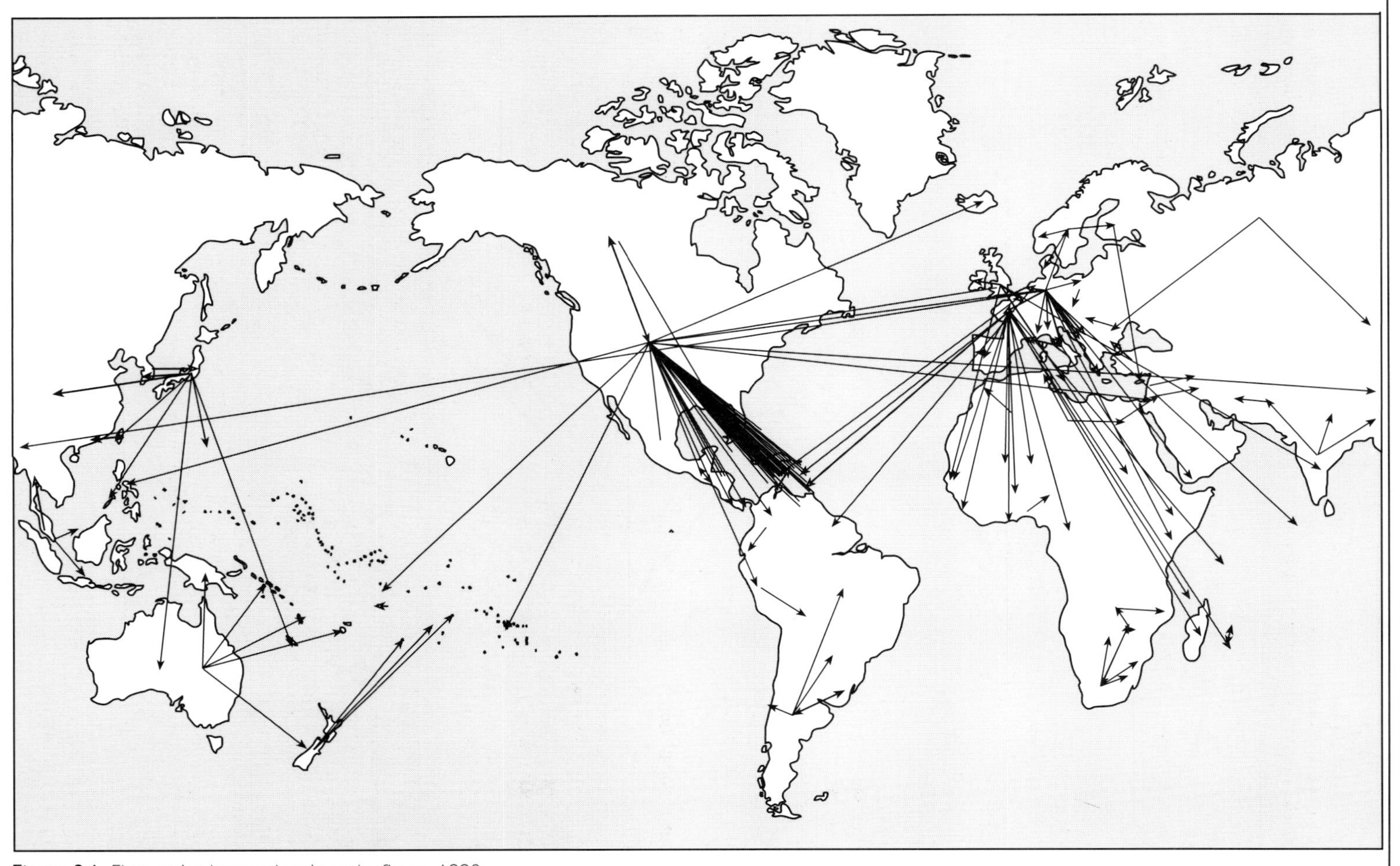

Figure 3.1 First-order international tourist flows, 1990.
Data source: World Tourism Organization.

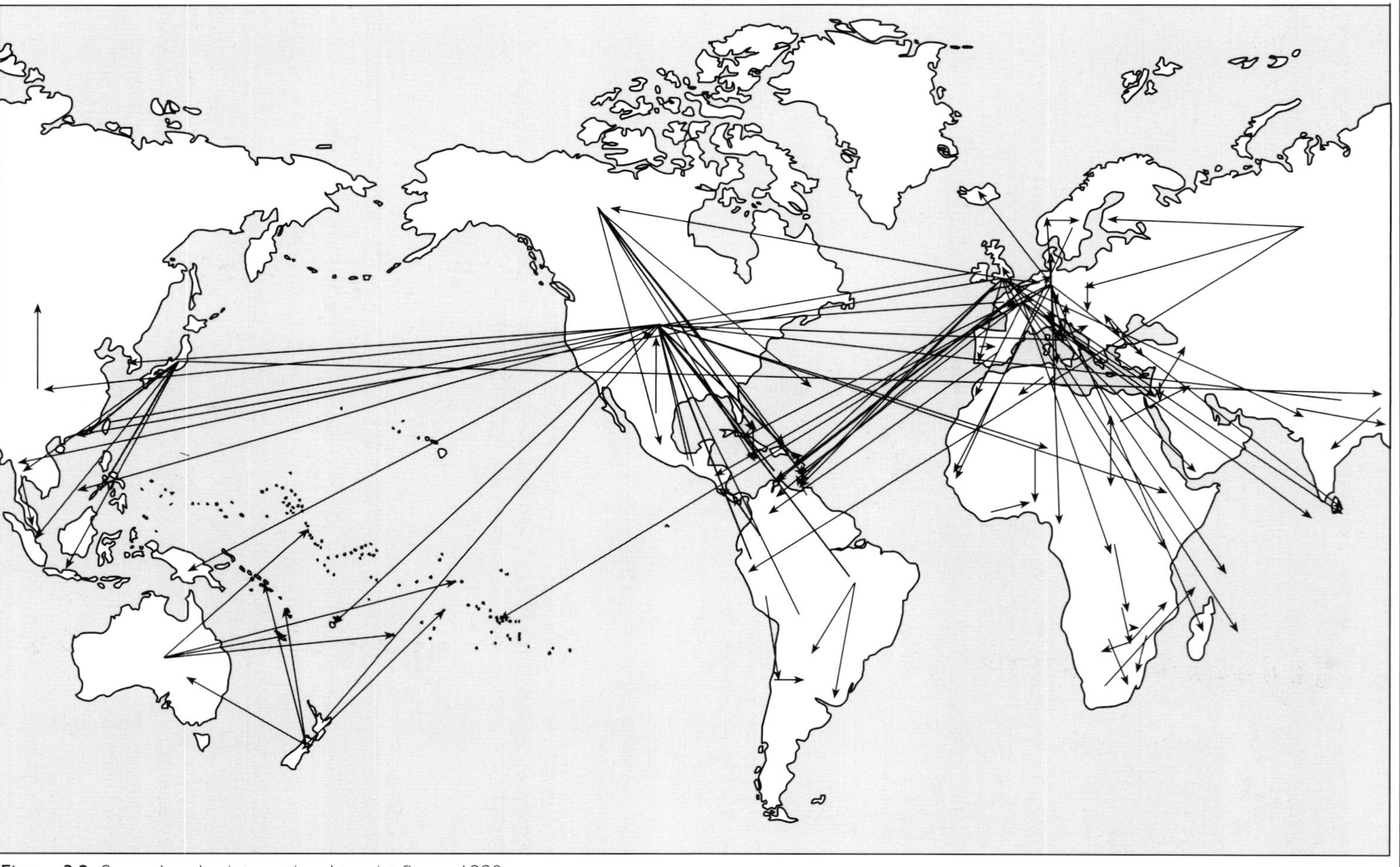

Figure 3.2 Second-order international tourist flows, 1990.
Data source: World Tourism Organization.

selective longer-haul flows, particularly from the United States with its large and relatively affluent population. Africa stands out as a continent less frequented by American visitors but one where flows are influenced by former colonial ties and selective charter tourism from Western Europe.

Concentration ratios

Clearly an analysis of first- and second-order flows does not bring out all the complexities of global patterns of tourist travel. It is important to note, however, that market-wise international tourist flows are more often highly concentrated than dispersed. A straightforward measure of concentration applied in the manufacturing industry, the concentration ratio, might usefully be employed here. In manufacturing, the concentration ratio simply expresses the share of any sector controlled by the largest few enterprises in that sector, for example by the top three (Ellis 1976), and is commonly measured in terms of gross output or employment. In the case of tourism, the concentration ratio might be used to express the percentage of a region's market, as measured by the number of visitors coming from say the three largest markets, or the percentage of the total number of visitors from one country who go to the three most favoured destinations. Calculation of such ratios is limited in some countries where the published data aggregate individual national markets into, for example, 'Europe' or 'Africa'.

Table 3.5 represents the concentration ratios of 129 WTO destinations, the majority of which are based on tourist arrivals at frontiers for 1990, together with comparable data for 1979 where available. The general pattern is one of dependence on a small number of markets. In 1990 two-thirds of all the destinations derived half or more of their traffic from only three markets, while 45 per cent received 60 per cent or more of their visitors from just three countries. Some deconcentration occurred over the period shown; for the 102 destinations for which data were available in 1979 as well as 1990, three-quarters had a concentration ratio of 50 or more and 47 per cent of 60 or greater. However, little systematic variation could be found in these changes and only a weak relationship with absolute growth over the period existed.

In 1990, 11 destinations had a concentration ratio in excess of 90, most of which were Caribbean destinations heavily dependent on the United States markets. Other high-ranking destinations such as San Marino and Lesotho are small enclaves surrounded by their major markets. Top-ranked Mongolia is scarcely a major tourist destination (124,000 arrivals) but one which is located between two of the world's largest countries, China and Russia. At the other end of the scale, the 20 destinations having a concentration ratio of 40 or less, with the notable exception of Germany, are all developing countries, many of which have a small total volume of visitors. Overall, however, there is no correlation between the concentration ratio and the size of the tourist traffic.

Circuit Tourism

International tourism is not simply limited to travel between pairs of origins and markets. As shown in Thurot's model (Fig. 1.5), travel to one destination may be combined with visits to other destinations in what is variously known as circuit tourism or multi-destination travel. The essence of the concept was clearly outlined by Cullinan *et al.* (1977) in their Central American study, namely 'A circuit tour is a pleasure trip which includes two or more countries by a resident of a third country'. This basic concept has been expressed and measured in a variety of ways throughout the world, but in most of these instances the analysis involves all travellers not just those on a pleasure trip (Pearce 1994). Many countries, however, continue to record arrivals in isolation, taking account only of the part of any trip spent within their frontiers.

In French Polynesia a distinction is made between destination and circuit travel. Classification of visitors into these two categories is based on cross-tabulating ports of embarkation and disembarkation from arrival/departure (A/D) cards. Where the two are the same, then French Polynesia is considered to be a visitor's sole destination; where they differ, the territory is deemed to be visited as part of a larger circuit. This logic is not infallible but for most travellers the matching of ports of embarkation and disembarkation provides a reasonably reliable measure of circuit travel.

Tourist organizations in other countries, for example Hong Kong (Hong Kong Tourist Association 1993a) and Vanuatu (Tourism Council of the South Pacific 1989), use the term 'multi-destination travel'. In both cases the data are drawn from sample surveys of departing visitors. In the case of the annual Hong Kong survey, reference is to 'overnight destinations during current trip away from home'. In one of the few examples of this information being collected for outbound travellers the large (n = 40,000) inflight survey of United States travellers to overseas countries (Canada and Mexico excluded) seeks information on the number of countries visited (USTTA 1991).

Elsewhere the concept of stopovers is employed. In its international visitor survey the Australian Bureau of Tourism Research (1990) defines a stopover 'as a stay of one night or more on the way to or from Australia in any country including stopovers made in the visitor's country of residence'. By extension, when a stopover is

Table 3.5 Concentration ratios of 127 destinations, 1979 and 1990

Destination	1990	1979	Destination	1990	1979	Destination	1990	1979
Mongolia	97.3		Ecuador	65.3	56.1	Norway[nh]	50.1	66.5
US Virgin Is.[ah]	97.3		W. Samoa	64.8	65.9	France	50.1	56
Bermuda	96.1	97.4	Papua New Guinea	64.5	70.4	Grenada	49.7	37.2
Lesotho[vf]	95.9		Pakistan	64.3	32.2	Indonesia	49.3	33.5
Guam	93.6	85.9	Denmark[nh]	64	61.7	Finland	49.1	52.6
Bahamas	93.3	96.3	Iraq[vf]	63.7	51.8	Zambia	49	41.1
N. Marianas[vf]	92.9		Jordan[vf]	63.7	57.4	Philippines	48.9	61.1
San Marino[vf]	91.9	93.4	Antigua & Barbuda	63.5	75.6	Netherlands	48.8	50.8
Cayman Is.	91.4	92.9	St Kitts & Nevis	63.1	59.5	Myanmar	48.8	
Jamaica	90.6	89.1	Paraguay	62.4	80	Niger	48.3	54.6
Uruguay	90.4	93.5	Tonga	61.7	64.3	Burkina Faso[ah]	47.8	
Réunion	88.5		Malawi★	61.2	63.7	Greece	47.2	33.1
Botswana	87.4	90.1	Monaco[ah]	61		China	46.1	
Canada	86.8	91.4	Fiji	60.9	76.8	S. Africa	46.1	
Guadeloupe[ah]	84.9	81.1	New Zealand[vf]	60.4	72.3	Cuba★[vf]	45.5	55.3
Swaziland[aa]	84.7		Solomon Islands	60.2		Israel	44.8	48.8
Niue	83.2		Costa Rica	59.2	65.6	Kenya	44.7	40.7
Martinique	82.8	71.9	Cook Islands	58.9	78.8	Belgium	44.3	59.5
Czechoslovakia[vf]	82.4	64.9	Japan	58.8	50.5	Brazil	44.3	53.7
Poland[vf]	80.8	80.5	Cyprus	58.8	56	Algeria[vf]	43.7	62.7
Ireland	80.6	82	Austria[aa]	58.8	74.6	Iceland	43.7	51.4
El Salvador	79.3	68.6	St Lucia	58.4		Venezuela	40.7	60.7
Vanuatu	79	72.2	Fr. Polynesia	58.3	67.1	Sweden[nh]	40.6	45.4
S. Eustatiu	76.5		Leichtenstein[ah]	58.3		United Kingdom	40.1	37.1
Aruba	74.9	89.6	Madagascar	58.2	72.2	Curacao	39.9	52.5
Malaysia	74.8	59.2	Seychelles	58.2	45.6	Bangladesh	39.3	64.2
Malta	73.9	78.7	Mauritius	57.7	64.4	Mali[ah]	39.2	
Zimbabwe	73.4	85.7	Guatemala	57.7	60.1	Sri Lanka	38.5	43.8
Senegal	70.8	55.6	Hungary	55.5	49.8	Kiribati[vf]	38.3	
Bhutan	70.7		Hong Kong[vf]	55.4	46	Togo[ah]	37.8	33.2
USA	70.3	70.1	Spain[vf]	55	62	Germany[aa]	37.6	44.4
Portugal	69.9	60.9	Iran	54.2	30.4	Peru	37.5	35.9
Bulgaria	69.3	58.8	Tunisia	54.1	57.1	Yemen[ah]	36.9	26.9
Comoros	69.2	69.1	Macau[vf]	54	69	Nepal	36.6	30.1
New Caledonia	68.9	67.5	Maldives	53.4		Thailand	36.3	36.6
Barbados	68.5	67.2	Australia[vf]	53.1	61.5	India	35.9	32.1
Korea	67.5	75.6	Morocco	52.6	56.8	Bolivia[ah]	33.5	39.3
Syria	67.2	70.1	Dominica	51.9	21.6	Turkey[vf]	31.6	31
Chile	66.8	67.1	Panama[vf]	50.7	55	Egypt[vf]	28.3	35
Nigeria★	66.6		Yugoslavia[ah]	50.5	51	Nicaragua	28	
Argentina	66.3		Switzerland[aa]	50.2	55.6	Honduras[vf]	24.9	
Taiwan	66.2	77.3	Luxembourg[ah]	50.2	63.2	Sudan	18.4	
Romania[vf]	65.4	47.3	Italy	50.1	53.9	Ethiopia	15	16.8

Notes: ★ 1989.
[aa] Arrivals of tourists from abroad in all accommodation establishments.
[ah] Arrivals of tourists from abroad in hotels and similar establishments.
[nh] Nights of tourists from abroad in hotels and similar establishments.
[vf] Arrivals of visitors from abroad at frontiers.
Data source: WTO.

made, circuit tourism or multi-destination travel would normally occur, though the inclusion of the possibility of stopping over in one's home country blurs the definition which Cullinan *et al.* presented. The term stopover is also used in Singapore but in absolute rather than relative terms as there 'A maximum length of stay of four days has been used as the cut-off point, to determine "stopover" traffic' (Singapore Tourist Promotion Board 1992). As such a definition fails to distinguish adequately between those visitors staying less than four days but on a single destination trip or five days or more when Singapore is still part of a longer itinerary, Singaporean stopovers cannot accurately reflect the extent of circuit tourism experienced by the state.

Leiper (1989) adopts a different approach, deriving a Main Destination Ratio. This he defines as 'the percentage of arrivals by tourists in a given place for whom that place is the main or sole destination in the current trip, to the total arrivals in that place'. The approach combines the use of several sets of official statistics, not only those on arrivals in the selected destinations but also departures from the markets. The latter are usually expressed in terms of the outgoing residents' 'main destination'. These figures are then expressed as a percentage of all arrivals from these markets recorded at the destination(s) analysed to give the Main Destination Ratio. This approach has the advantage of drawing on more commonly available data, thereby enhancing the scope for comparative studies – Leiper uses the examples of outbound travel from Australia, New Zealand and Japan. However, the clear distinction between mono- and multi-destination travel is lost as 'main destinations' include both sole destinations and those which constitute the main but not exclusive object of a trip.

Table 3.6 shows that circuit tourism is not particularly pronounced in the world's largest outbound market. In 1990 three-quarters of United States travellers visited only one country on each trip overseas, only a small percentage of travellers visited three or more countries and the overall mean was just 1.4 countries per trip. Little variation occurred by type of traveller but the propensity to make multi-destination trips did differ by destination. Virtually all United States travellers to the Caribbean confined their visit to a single island. In contrast, the opening up of Central Europe appears to have occasioned more itinerant visits there, with a mean of 3.2 countries visited. Africa and the Middle East also generate higher than average multi-destination trips. The mean reported for Western Europe was 1.6 countries in 1990, compared to 1.9 in 1977 and 3.9 in 1967 (O'Hagan 1979).

Table 3.6 Number of countries visited by United States travellers overseas, 1990

	Number of countries visited				
	1 (%)	2 (%)	3 (%)	4+ (%)	Mean
Type of traveller					
All US travellers	77.7	12	5.4	4.9	1.4
First visit	75.9	12.4	5.0	6.6	1.5
Repeat visit	74.5	13.7	6.1	5.7	1.5
Vacation	73.2	13.8	6.6	6.4	1.5
Business	73.1	14.0	7.1	5.9	1.5
VFR	81.9	11.2	3.8	3.1	1.3
Destination					
Western Europe	66.9	17.1	8.1	7.9	1.6
Eastern Europe	20.1	22.1	17.7	40	3.2
W. Indies/Caribbean	96.4	2.0	0.6	0.9	1.1
Central America	76.9	15.0	3.8	4.3	1.4
South America	88.9	5.0	3.0	3.1	1.2
Africa	36.8	33.4	18.7	11.1	2.1
Middle East	52.0	24.6	9.6	13.8	2.0
Far East	73.8	12.1	6.9	7.2	1.5
Oceania	69.0	19.0	5.5	6.4	1.6

Source: After US Travel and Tourism Administration.

Similar low values are recorded in Hong Kong, one of the few destinations which appears to collect such information. In 1992 visitors to Hong Kong stayed overnight in only two different international destinations, that is, on average they were visiting Hong Kong plus one other country. Visitors from the long-haul markets of North America, Europe and Australasia tended to combine a visit to Hong Kong with about two other countries, while Asians, particularly Japanese, averaged only one and a half countries per trip, including Hong Kong.

Table 3.7 provides a basic measure of the extent of circuit tourism in four Asia Pacific destinations for which relatively comparable data were obtained (Pearce 1994). Table 3.7 shows that the overall proportion of circuit visitors ranged from about one-third in Vanuatu to just under two-thirds in Hong Kong. Part of the explanation for this variation appears to lie in the function of these four destinations. Hong Kong, as both a well-connected regional hub and a physically small destination with a limited range of attractions, relies primarily on circuit visitors. In contrast, the small South Pacific destinations of French Polynesia and Vanuatu are more isolated and have a 'sun-sand-sea' image that is perhaps more conducive to stay-put holidays. Between these lies Australia, a large destination with a diverse range of attractions which encourages single destination travel, but also a well-developed air network which facilitates circuit tourism. Distance and accessibility (the presence or absence of direct routes) also appear to play a role in influencing the proportion of circuit tourism at each destination generated from the different markets, though the number of destinations in Table 3.7 limits the generalizations which can be made. Table 3.7 indicates a tendency for higher levels of circuit tourism to be associated with long-haul travel (e.g. the North American and European ratios in the region) and destination tourism with short-haul direct travel (e.g. New Zealanders to Australia and French Polynesia, Australians to Vanuatu). Several interrelated factors may affect this pattern. Greater

Table 3.7 Circuit travellers in Hong Kong, Australia, French Polynesia and Vanuatu by major markets

	Market							
	Asia (%)	Japan (%)	USA (%)	Canada (%)	W. Europe (%)	Australia (%)	NZ (%)	All visitors (%)
Hong Kong (1991)	46[a]	34	91		93	78		62
Australia (1990)	34	27	58	63	64[b]	—	16	40
French Polynesia (1991)	—	10	40	67	61[c]	39	22	38
Vanuatu (1990)	94[d]	—	86	—	78	11	48	32

Notes: [a] Excluding Taiwan.
[b] Excluding UK and Ireland.
[c] Excluding France.
[d] Including Japan.
Sources: Hong Kong Visitors Association; Bureau of Tourism Research; Service de Tourisme de la Polynésie Française; Tourism Council of the South Pacific.

Table 3.8 Percentage of total trip being spent in Hong Kong, the USA and Australia

	Market							
	Asia (%)	Japan (%)	USA (%)	Canada (%)	W. Europe (%)	Australia (%)	NZ (%)	All visitors (%)
Hong Kong (1992)	36.6[a]	43.9	11.1		13	16.7		21
USA (1990)	79	76	—	—	64.9	55.5	58	68.6
Australia (1990)	84	76	61	63	71[b]	—	61	70

Note: [a] Excluding Taiwan.
[b] Excluding UK and Ireland.
Sources: Hong Kong Tourist Association; US Travel and Tourism Administration: Bureau of Tourism Research.

lengths of travel will increase not only the need for technical stopovers but also a desire to break one's journey and perhaps derive a better return on transport costs by combining visits to two or more destinations on the one trip. Japanese circuit tourism depicted in Table 3.7 exhibits the opposite trend to that just outlined, a function perhaps of the generally shorter holidays taken by this market. That is, the more time spent travelling, the less desire there is to visit more than one destination. Given these variations, the market mix of any one destination will be a major determinant of the overall level of circuit travel experienced there – the large share of short-haul Australians, for example, depresses the total proportion of circuit tourism in Vanuatu.

By way of comparison, in 1992 42 per cent of visitors to Great Britain from outside Western Europe reported their intent to combine a visit to Britain with at least one other European country (BTA/ETB 1992). The propensity to make a combined visit was highest among those from English-speaking countries outside North America, making relatively short stays (less than a week) on their first trip to Britain.

An indication of the relative importance of any destination on a circuit can be obtained by deriving the proportion of any trip spent there. This requires details not only on the length of stay in each destination but also the length of the total trip. While the former figures are readily available, not only from surveys but also A/D cards, the latter are recorded less often. Table 3.8 shows the sort of results that can be obtained and the variations which may occur from segment to segment. Visitors to Hong Kong, for instance, were on average spending only one-fifth of their trip in that destination compared with around 70 per cent of those visiting the USA and Australia. Unfortunately the published tables do not enable separate values for circuit tourists alone to be established but such values do contribute to the identification of different countries' tourism functions, in particular the hub and staging post role of Hong Kong, and raise questions about the role of size and resources.

Other data from the Hong Kong survey enable aggregate patterns of the circuits incorporating the territory to be established (Fig. 3.3). For the total vacation traffic passing through Hong Kong, China, Thailand and Singapore stand out as the three most important other nodes on the visitor circuit. However, significant variations occur from market to market, particularly between the long-haul markets. Hong Kong is clearly part of a broader Asian circuit for North American and

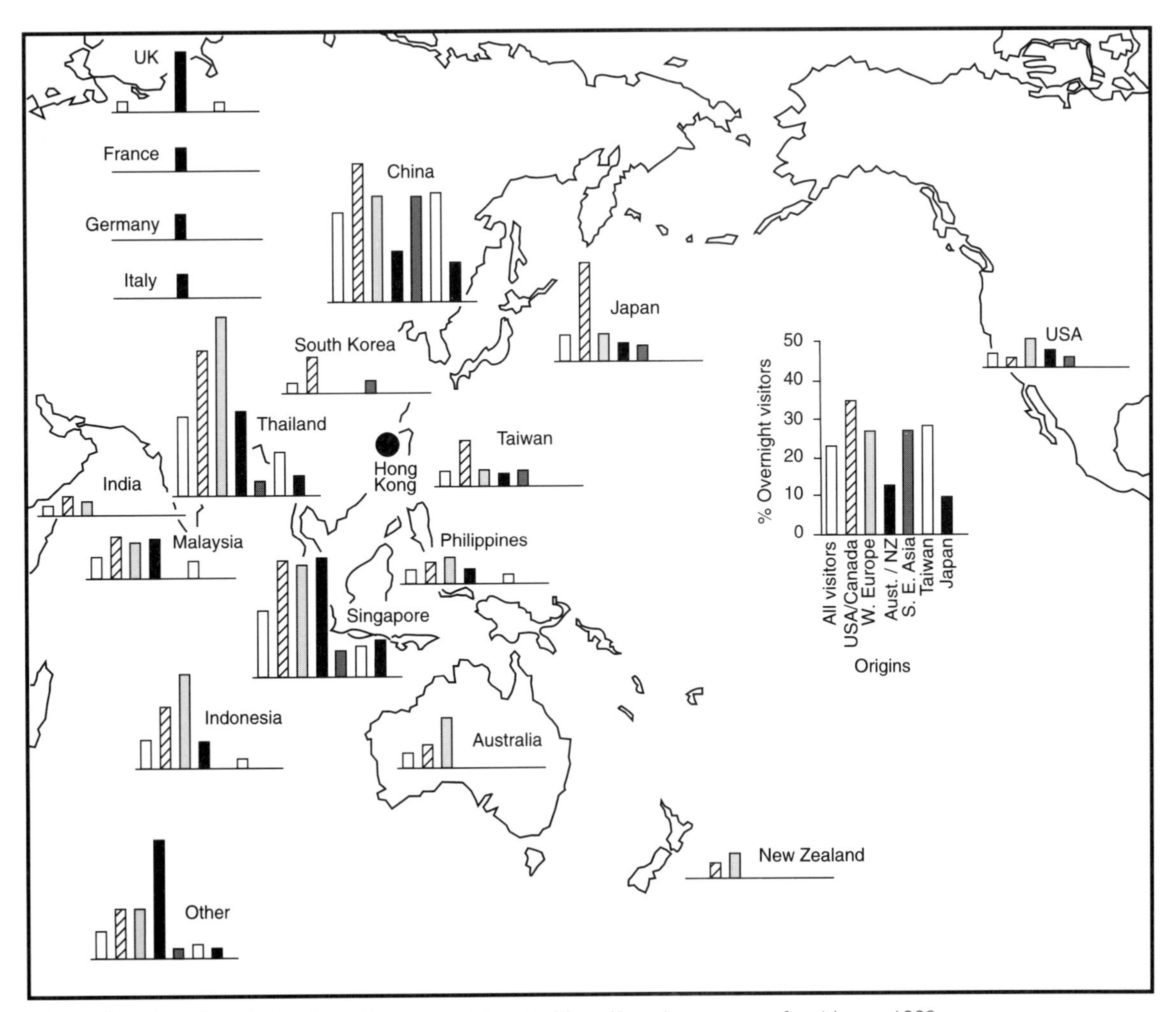

Figure 3.3 Overnight destinations of vacation visitors to Hong Kong by country of residence, 1992.
Data source: Hong Kong Tourist Association (1993a).

European travellers, but Japan and Taiwan are linked more closely into the circuits of Americans than Europeans. Regional links, particularly with Singapore, are also part of Australasian travel through Hong Kong but the prominence of European destinations also points to its role as a midway stopover for through traffic between Australasia and Europe. This also occurs to a lesser extent with the European market, for whom some travel to or from Australia and New Zealand is recorded. For most Asian travellers, Hong Kong is part of a more restricted intra-regional circuit, incorporating in particular trips to China.

Route plan data from Fiji also highlight differences between short-haul intra-regional circuits and midpoint stopovers on longer end-to-end routes. These data, it should be noted, relate only to sectors directly linked to Fiji, that is the immediate flights into and out of Fiji, and thus do not provide coverage of the full circuits involved. The majority of New Zealand visits to Fiji consist of destination travel or circuits involving other Pacific destinations (Fig. 3.4). Australia also follows this pattern. In contrast, less than one-fifth of American visits to Fiji involve mono-destination trips with the most popular circuits incorporating travel to and from the United States and either New Zealand or, to a lesser extent, Australia (Fig. 3.5). European circuits show a broadly similar pattern to that of the Americans but visiting Fiji en route to or from Australia and New Zealand is also complemented by a greater amount of travel along circuits involving linkages elsewhere in the South Pacific.

The examples here tend to dispel the notion of

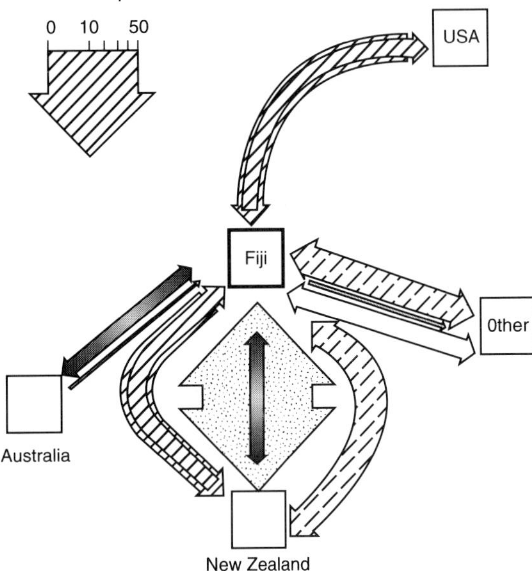

Figure 3.4 New Zealand visitors' route plans to and from Fiji, 1990.
Data source: Fiji Visitors Bureau.

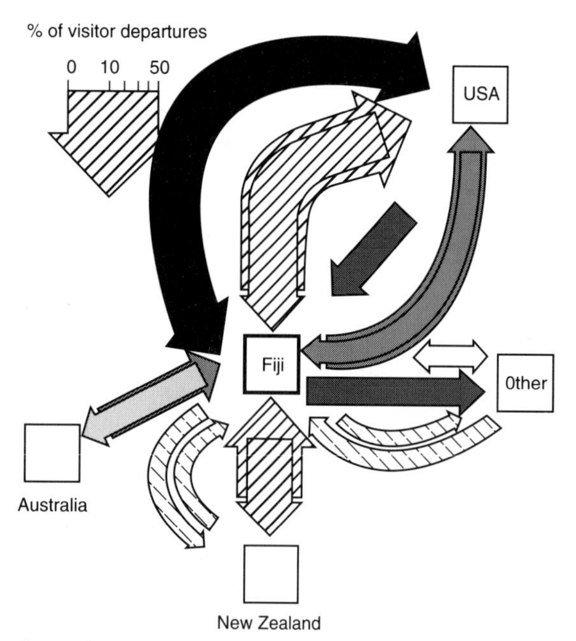

Figure 3.5 American visitors' route plans to and from Fiji, 1990.
Data source: Fiji Visitors Bureau.

globetrotting tourists 'collecting' countries as they flit from one destination to the next. This may be true of a small segment of the market but the overall pattern to emerge from this limited sample is one in which international tourist travel is characterized by a large proportion of single destination trips or of circuits limited to two or three countries. Little remains known about the motivations and preferences of circuit tourists and what distinguishes them from those who visit only one country. More work is required on these aspects along with extending the analyses discussed above to a wider range of markets and destinations.

The Evolution of International Tourist Flows

The evolutionary models discussed in Chapter 1 suggest that the volume of traffic to more distant destinations will increase over time and that the character of the tourists visiting given destinations will also change. Few attempts have been made, however, to verify these trends through empirical analysis. The studies that do exist generally focus on one country, with the evolution in outbound travel being but one component of the markets reviewed (e.g. Schnell 1988; Cook 1989). A more systematic approach is attempted here in which the spatial evolution of the world's four largest outbound markets (Table 3.2) is analysed with the aim of deriving some general observations. Data on the four markets – the United States, Germany, the United Kingdom and Japan – are not strictly comparable as they are drawn from different sources, they are based on different definitions and they are presented in varying degrees of geographical detail. Nevertheless, using a common approach – spatial variations in market share – it is possible to draw out some common trends.

Good longitudinal data on the leading European markets – Germany and the United Kingdom – are provided by two longstanding surveys, respectively the Reiseanalyse of German households undertaken by the Studienkreis für Tourismus (Dundler 1988; Gilbrich 1991) and the British International Passenger Survey (BTA 1989). The results of the German surveys refer to main holidays abroad (five nights or more) while the British findings are more comprehensive, being based on a survey of all overseas trips. The spatial evolution in these two markets for the periods 1970–90 and 1965–90 are shown in Figs 3.6 and 3.7.

Considerable growth occurred in German outbound travel over this period as the propensity for holiday-taking in general and international travel in particular grew (Table 2.4). Important changes in the relative composition of German travellers also occurred. Neighbouring Austria saw its share of the German market for main holidays decrease from 28 per cent to 10 per cent in two decades. Italy also experienced a marked drop in

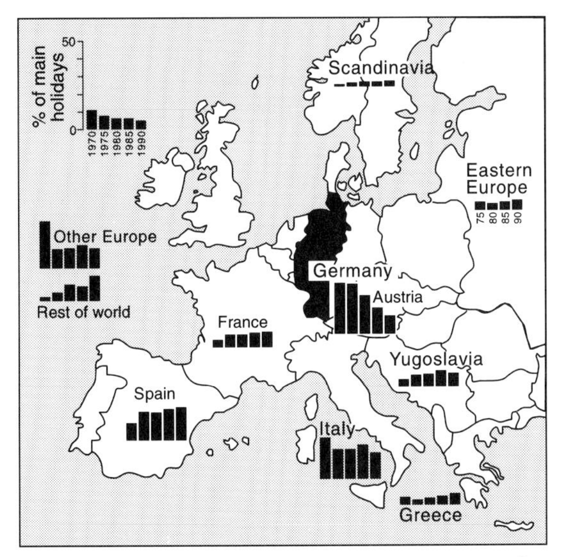

Figure 3.6 Spatial evolution of main holidays from Germany, 1970–90.
Data source: Studienkreis für Tourismus.

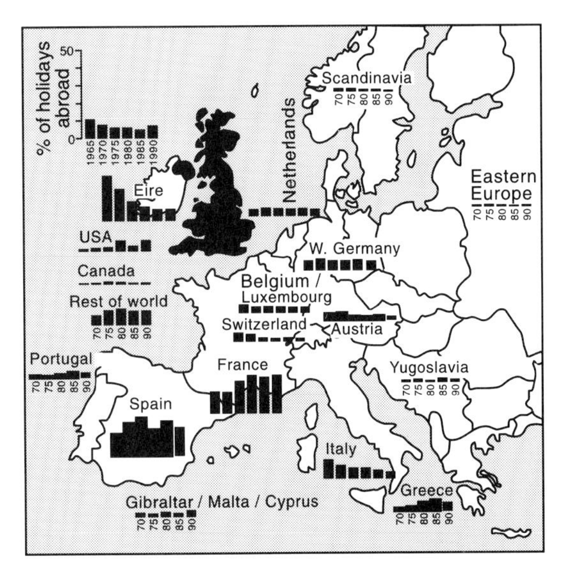

Figure 3.7 Spatial evolution of holidays abroad from the United Kingdom, 1965–90.
Data source: British Tourist Authority (1981a, 1992).

market share from from 22.6 per cent to 14.5 per cent. In contrast, Spain's share doubled to over 18 per cent, with similar rates of growth being experienced by France and Yugoslavia but from a smaller base. While most of the German traffic is still intra-regional, steady increases were recorded in the traffic beyond Europe, with the rest of the world's share growing from just on 2 per cent in 1970 to almost 14 per cent in 1990. Turkey and the USA/Canada were the most significant of these destinations.

From 1965 to 1990 the number of British visits abroad increased from 6.5 million to 31.2 million. In relative terms, a major redistribution of the British traffic occurred as Ireland's share dropped from one-quarter to under 7 per cent. Italy's share also decreased steadily from 10.9 per cent to 3.8 per cent and smaller declines were registered by the Alpine destinations of Austria and Switzerland. The major beneficiaries of these changing patterns were France (12.8 per cent in 1965 to 22 per cent in 1990) and Spain although the latter experienced fluctuating fortunes in the latter part of the period. Steady increases were also being recorded by Greece, Portugal and former Yugoslavia. Other important European destinations such as Germany and the Netherlands experienced stable market shares throughout this period. The 'Beyond Europe' share doubled in the period from 1970 (7.5 per cent) to 1980 (16.2 per cent; of which North America, 7.9 per cent) and retained this level in 1990 after falling away in the mid-1980s due to some decline in the transatlantic traffic.

Figure 3.8 depicts the evolution of the Japanese outbound traffic for the years 1965–90, a period which begins with the market at a much more immature phase than either the British or German markets. Japanese Ministry of Justice figures record only 163,000 departures in 1965 compared to almost 11 million in 1990. More than half of all this growth occurred between 1985 and 1990, when outbound travel was boosted by a strong yen and the Japanese Ministry of Transport's Ten Million Programme which aimed at doubling the number of Japanese going abroad over five years as a means of reducing Japan's trade surplus. The Ministry of Justice figures record departures by main destination only but given the relatively low propensity for multiple destination visits noted in the preceding section and the tendency for these to involve nearby countries, these figures should still be reasonably representative of the overall geographical pattern of Japanese travel, especially at the regional scale.

Throughout the period 1965–90, Asia has accounted for about half of the Japanese departures, peaking at 59 per cent in 1978, but over time some redistribution of flows within the region has occurred. Hong Kong's share has steadily declined and been replaced by Korea as the single largest Asian destination (Plate 3). In relative terms Taiwan rose rapidly in the late 1960s before declining, the decrease in the late 1980s being particularly pronounced. Similarly, but at a lower level, the Philippines' share increased gradually then dropped back. In contrast, smaller and more distant Singapore has experienced consistent increases from a smaller base.

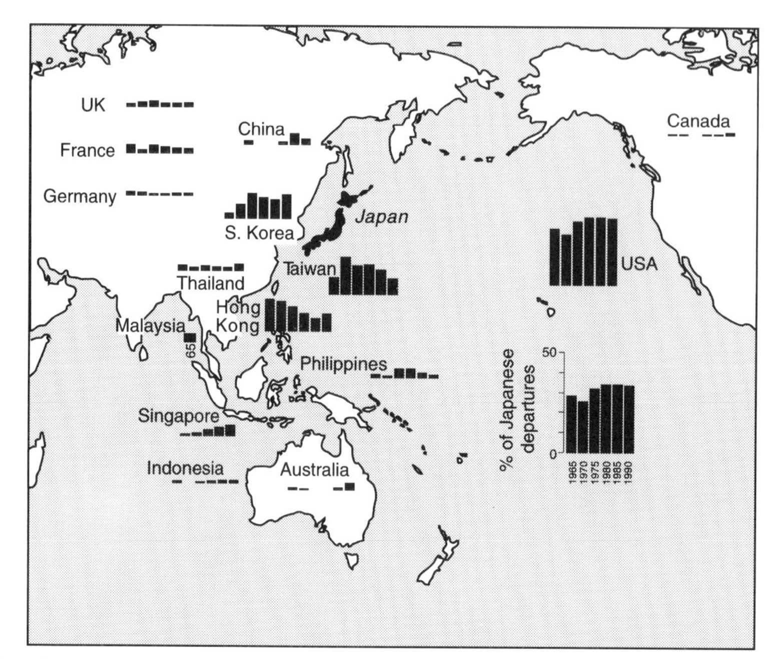

Figure 3.8 Spatial evolution of Japanese departures, 1965–90.
Data source: Japanese Ministry of Justice.

Plate 3 Escorted by their flag-carrying guide, a group of Japanese tourists visits the Korean Folk Village near Seoul.

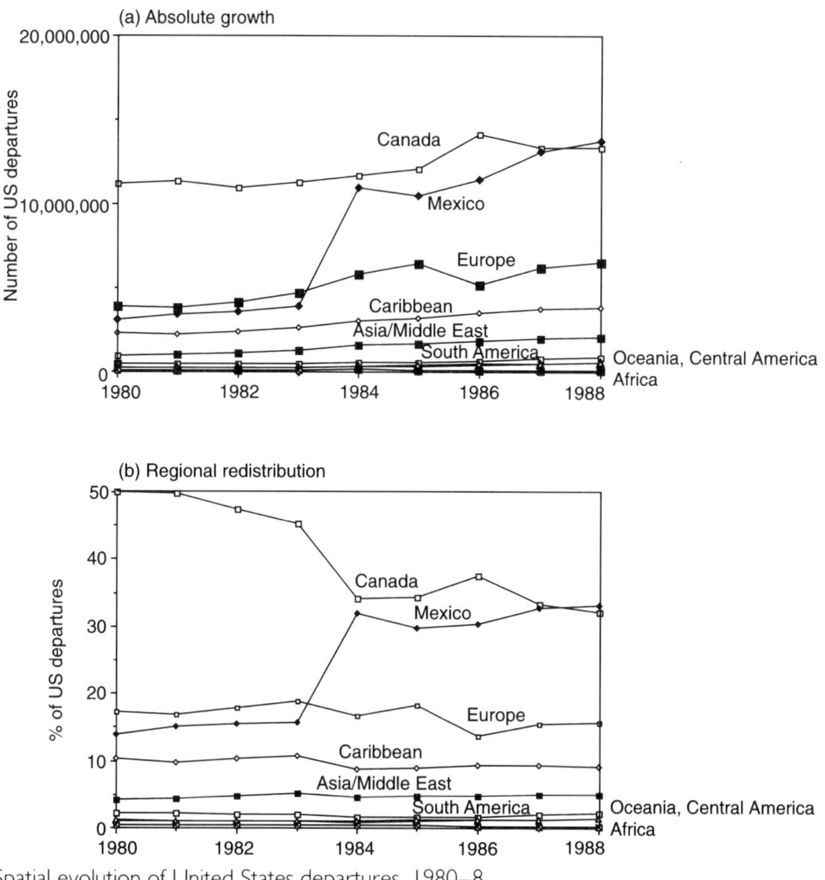

Figure 3.9 Spatial evolution of United States departures, 1980–8.
Data source: Cook (1989).

Figure 3.8 shows the largest destination for Japanese travellers to be the United States which by the late 1970s accounted for about one-third of the total traffic. More detailed arrival figures show that in 1980 Hawaii recorded about half of Japanese travellers to the United States and Guam, about 20 per cent (Bailey 1986). Europe has registered about 10 per cent of Japanese outbound travel since 1980, with France (2.7 per cent in 1990), the United Kingdom (2.6 per cent) and Germany (1.6 per cent) being the three leading destinations there. In the latter part of the period Australia started to show some growth, accounting for almost 4 per cent of Japanese departures in 1990.

Figure 3.9 depicts the trends from a less comprehensive set of data for United States citizens' departures for the period 1980–8. The data set is based on the National Travel Survey and arrivals figures for Canada and Mexico, the latter being complicated by a change of definition from 1984 (Cook 1989). At the end of the period Canada, Mexico and overseas destinations combined each held about one-third of the US market. The major redistribution involves the number and share of the visitors going to the United States' northern and southern neighbours, the analysis of market share clearly being skewed by the definitional change noted. However, in absolute terms Cook reports that United States travel to Canada weakened significantly during the 1970s, and although it recovered somewhat in the mid-1980s the 1988 figure was still below the 1973 peak of 14.3 million departures. Conversely, United States travel to Mexico grew steadily in the 1970s, dropped in 1980 but picked up again throughout the rest of the period. In absolute terms United States travel to most overseas regions grew reasonably steadily throughout the period shown, growth in travel to Europe, the largest destination, being temporarily checked in 1986 by terrorist attacks against Americans. Other world regions attract only a minor and relatively stable share of US travellers.

Comparison of the patterns and trends among the evolution of the four largest markets revealed in Figs 3.6 to 3.9, draws attention to several major points.

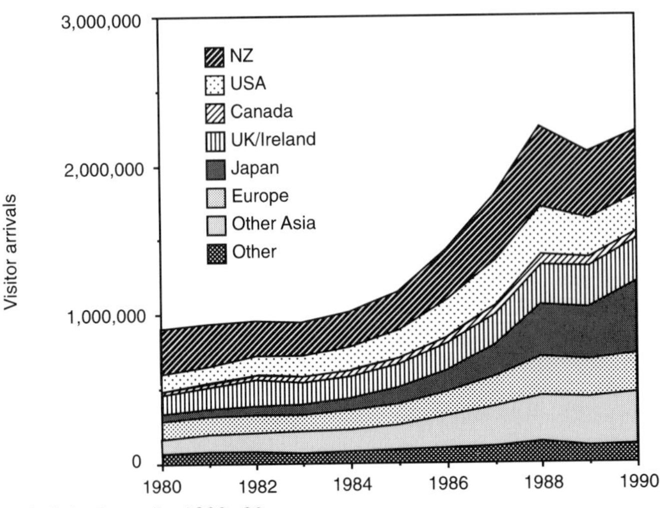

Figure 3.10 Evolution of arrivals in Australia, 1980–90.
Data source: Bureau of Tourism Research.

First, Figs 3.6 to 3.9 further highlight the importance of the intra-regional flows noted earlier, particularly for the two British and German markets for whom Europe holds a much more dominant place than does Asia for Japan or North America (Canada and Mexico) for the United States. The proportion of Germans travelling 'beyond Europe' is increasing rapidly but from a much smaller base than extra-regional travel from either Japan or the United States. The British 'beyond Europe' share in 1990 was similar to what it had been a decade earlier (16 per cent). The Japanese extra-regional traffic was reasonably constant throughout the period as was the American (definitional problems noted) in the 1980s. Some notable exceptions such as the increase in Japanese departures to Australia aside, what this means in most cases is that the growth in the longer-haul traffic from each of these four markets is primarily a function in the overall growth in demand rather than a relative redistribution to more distant destinations outside each region.

The changes in market share that do occur are essentially recorded within each respective region. The most fundamental is a decline in departures to traditional neighbouring 'non-sun' countries and a corresponding growth in the traffic to 'sun-sand-sea' destinations, most of which are characterized by lower living costs. Thus the German traffic to Austria declines as holidaymakers head towards cheaper Mediterranean destinations, as do British vacationers forsaking Ireland. Mexico gains in popularity for United States travellers as Canada's appeal is weakened. The Japanese pattern is less straightforward, there being few readily accessible sunlust destinations. Hong Kong declines in importance, much of the 'sun-sand-sea' traffic is oriented to Guam, Hawaii and later to Queensland in Australia, but part of Korea's growth can be attributed to developments on the semi-tropical Cheju Island (Bailey 1986). In the cases of the United Kingdom and Germany this near-neighbour decline extends the drop in domestic demand noted earlier (Table 2.4). Italy also loses market share in both these markets, a function perhaps of lack of price competivity in comparison with other Mediterranean destinations.

Overall it appears that the growth in demand in outbound tourism from the world's four major markets in the 1970s and 1980s was largely associated with a boom in departures to sunlust destinations. This may reflect a fundamental underlying change in holiday preferences, the basic tastes in the new groups acceding to international travel and generating mass tourism, and changing price structures reflecting the relative strength of the markets' currencies, technological developments and the emergence of lower cost destinations. Particularly important in this latter respect have been the development of inclusive charter tours and aggressive tourist development programmes in Spain, Mexico and to a lesser extent other Mediterranean destinations (Pearce 1983b, 1987a, 1987b). Similarly, the upsurge in Japanese travel to Hawaii followed the introduction of contract inclusive tour fares in 1970 (Bailey 1986).

Destinations will be effected by these and other market trends in different ways and to varying degrees depending on their location and individual market mix for (as Fig. 1.4 outlined), each is likely to be subject to overlapping spheres of influence. Figure 3.10, for example, shows the effect of the expanding Japanese market on the growth and composition of visitor arrivals in Australia. By 1990 the number of Japanese arrivals had eclipsed those from neighbouring New Zealand, traditionally Australia's largest market but in absolute terms a small one and one which was experiencing irregular increases in its outbound traffic during the 1980s. Strong growth is also shown from other parts of Asia and, to a lesser extent, Europe. An earlier study of demand in the South Pacific showed how the larger destinations of Australia and New Zealand drew proportionately more of their visitors from markets outside the region while the small destinations (in terms of visitor arrivals) depended heavily on intra-regional travellers. Given the low generating potential of the South Pacific, growth was seen to lie in the destinations' ability to attract longer-haul visitors (Pearce 1983a). By 1990 the Japanese had also become the leading market for New Caledonia, Japanese visits remaining relatively stable during the mid-1980s while Australian numbers never fully recovered from their 1984 peak after the civil unrest of 1985 and subsequent years.

The evolutionary models discussed in Chapter 1 also suggested that over time the character of the tourists will change (Figs 1.9 to 1.11). However, beyond the evolution in the geographic attributes noted here and one-off studies of profile characteristics (Cockerell 1989; Cook 1989; Oum and Lemire 1991) it is generally difficult to assess more systematically long-term changes in demand due to a basic lack of appropriate time-series data, for example of psychographics. Figures 3.11 and 3.12 provide two examples of how the Japanese market has evolved which might serve as examples for studies in other locations where similar information is available.

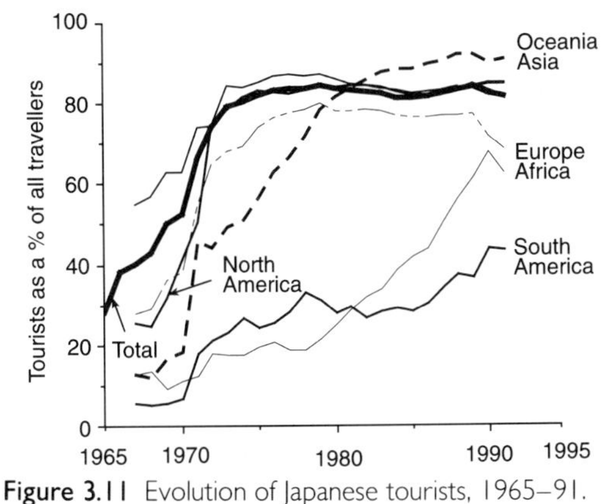

Figure 3.11 Evolution of Japanese tourists, 1965–91.
Data source: Japanese Ministry of Justice.

Figure 3.11 depicts the regional breakdown of purpose of visit for the period 1965–91 and provides some evidence of the evolution of tourist travel in the total travel picture. In this case, tourist travellers have been defined as those giving 'sightseeing' as their prime reason for travelling. Other travel purposes include business, both private and official, study and research, immigration, accompanying spouse or parents. Over the period examined, tourists increased from 19 per cent of all travellers in 1964 to 83.9 per cent in 1989 before dropping back slightly. The overall increases in the Japanese outbound traffic noted earlier thus seem to be largely a function of bigger tourist flows. It also appears, however, that an increase in the tourist traffic is related initially to existing exchanges between Japan and any given region as evidenced by the volume of other traffic. In the cases of South America and Africa where the total volume of traffic is small, the proportion of tourists has increased much more slowly and still remains the lowest overall. As well as distance and cost involved, lack of information appears to have played a role here (Tokuhisa 1980). On the other hand, in Asia, North America and Europe where the total traffic was much greater at the beginning of the period, the proportion of tourists has evolved more rapidly. In Asia and America the proportion of tourists has stabilized at over 80 per cent whereas in Europe the figure falls below 70 per cent in 1991 – whether this reflects the Gulf crisis or some other trend is not yet clear. A steady growth in travellers to Oceania throughout this period has been accompanied by a constant increase in the proportion of Japanese tourists visiting the region, with the highest levels of all regions being recorded (in excess of 90 per cent since 1987).

Changes in the demographic structure of Japanese arrivals in one Oceania destination, New Zealand, are shown in Fig. 3.12. In 1974/75 Japanese tourists in New Zealand were overwhelmingly male (66 per cent) and predominantly middle-aged to elderly. By 1986/87 the genders were evenly balanced and the 20–34 age group constituted half of all Japanese tourists (34 per cent in 1974/75). Much of this change can be attributed to the growth in the honeymoon market, together with the emergence of other segments, notably the 'office ladies'. While package tours remain dominant, there has also been a significant increase in independent travel. The changes evident in New Zealand appear to reflect more general changes in the Japanese market which has matured to include a wider range of segments than in its early phases.

These demographic segments tend to favour different destinations, with the Japan Travel Bureau (1992) reporting that a disproportionate number of middle-aged and elderly people go to Asian destinations, single working women to beach resorts in places such as

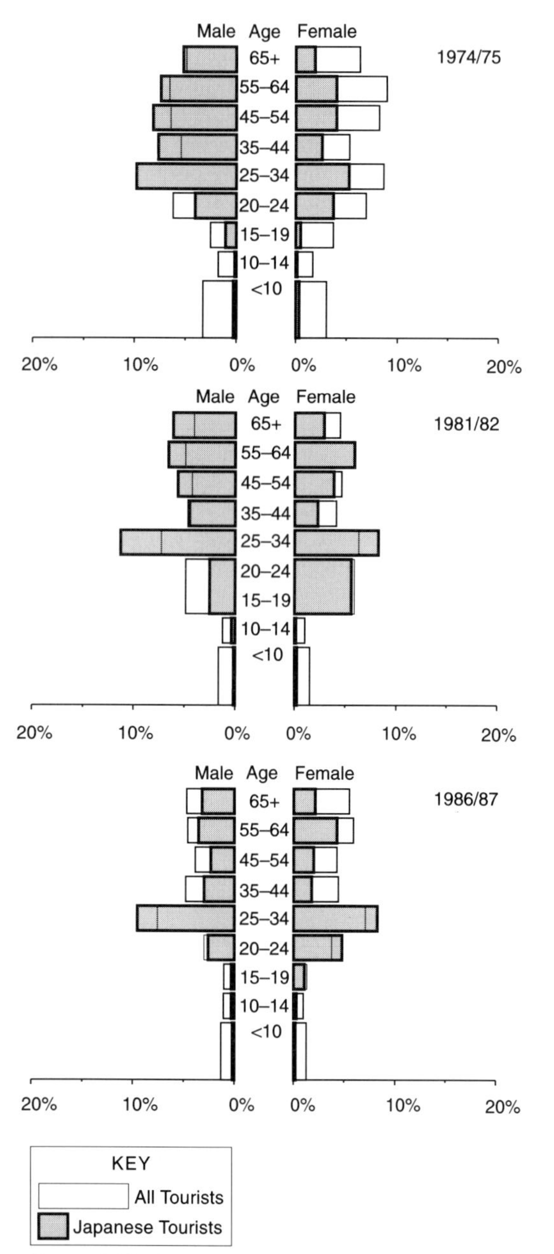

Figure 3.12 Changing demographic structure of Japanese tourists in New Zealand, 1974/75–1986/87.
Data sources: Statistics of Migration and New Zealand Visitor Statistics.

Hawaii and Guam, a higher proportion of newly weds to Australia, and of mature married couples and middle-aged people to Europe. The same report also notes that while 'nature' destinations (mountains, lakes, etc.) remain the most popular, from 1987 to 1991 preferences for urban cultural elements or easy-to-reach destinations have strengthened relative to nature.

Conclusions

Many of the general patterns identified in this chapter conform with the ideas proposed in the models discussed in Chapter 1, particularly that of Miossec (Fig. 1.4). Throughout, a broad distance decay effect in international tourist flows can be observed, with intra-regional traffic predominating in many parts of the globe and reciprocal flows developing between countries, particularly neighbouring ones. The size and affluence of the major West European countries along with the United States also give rise to long-haul international flows which are superimposed on more general regional patterns. While some dispersion in outbound travel from these markets and that of Japan has been apparent since the 1970s, these markets remain dominated by intra-regional travel with most of the growth in longer-haul flows being a function of an overall increase in the volume of travellers rather than a marked redistribution of the destinations favoured by them.

Globally, international tourist travel remains very concentrated, both in terms of the shares of the major markets and destinations and the selectivity evidenced by the concentration ratios. The analysis of circuit travel also highlighted the limited amount of interaction which occurs between origins and multiple destinations at a country-to-country level (Fig. 1.5) – mono- rather than multi-destination travel was the norm in the countries examined.

Data drawn primarily from A/D cards have enabled global patterns and some more specific trends to be identified. What is required now are more detailed special-purpose surveys and studies to establish more fully the characteristics of tourists forming particular flows to develop the links between motivations and actual travel patterns.

CHAPTER 4

Intra-national Travel Patterns

Research on international travel patterns has largely been limited to analyses of flows between sets of generating and receiving countries. There continues to be comparatively little recognition in the literature that these international movements are complemented by two sets of intra-national movements, one in the generating country as tourists leave home and make their way to a port of departure (and subsequently return home via the same or a different port of entry), the other in the receiving country (or countries) as the tourists move to a destination or visit a variety of destinations. In a strict sense these intra-national movements may not be international in the same way that the actual linking trips between countries are, but logically they form integral parts of an international travel system and have been so considered by geographers dealing with other less temporary movements, particularly those associated with international migration (P. White and Woods 1980).

A certain form of this intra-national travel has been conceptualized by Hills and Lundgren (1977) and S. G. Britton (1980a) in their work on the impact of international tourism in the Caribbean and in the Pacific (Fig. 1.8). In each case the market is concentrated upwards through the local-regional-national hierarchy by integrated metropolitan-based enterprises and little dispersion occurs in the destination country where development is essentially concentrated in resort enclaves. To date, these models have been little tested by empirical research. Nor has travel between developed countries been considered systematically from this perspective. At a more local level, Fig. 1.7 depicts the various functions and flows which might be associated with international tourism in major metropolitan areas. Such areas may act as points of entry into and exit from a country, constitute major destinations in themselves or act as a staging post sending visitors out into the surrounding region or along a tour circuit.

Concentration on flows between rather than within countries is to a certain extent a function of the data which are readily available. Whereas immigration statistics or visitor arrival figures are compiled systematically at a national scale enabling the types of analyses presented in Chapter 3, more detailed survey data are usually required for intra-national studies. Such data may be both difficult to compile and to analyse.

Lack of attention to intra-national flows also appears to be due to the rather narrow outlook held by many national tourist organizations and other institutions which see their role largely in terms of increasing total arrivals and boosting gross foreign exchange earnings. An understanding of intra-national movements becomes particularly important when other goals are pursued, such as regional development or spreading the impacts, both positive and negative, of tourism (Pearce and Johnston 1986; Pearce 1992a). After all, a significant share of the regional economic impact of tourist development results from the direct transfer of money which accompanies the movement of tourists from one region to another. Information on internal travel may also contribute to overall expansion. Surveys in French Polynesia, for example, show that the level of satisfaction is generally much higher among tourists visiting more than one island than among those restricting their stay to Tahiti (Pearce 1984). From a marketing perspective, a growing number of tourist organizations and businesses are also beginning to identify the subnational origins of many of their visitors and, faced with the costs and challenges of tapping huge markets such as the United States, are increasingly developing more targeted campaigns aimed at the specific regions within those countries which have the greatest propensity to visit their destination. As a result, more information has become available recently dealing with regional origins but analysis of where international tourists go and what they do between reaching their port of entry and leaving their port of departure continues to be limited by available data.

This chapter focuses first on the outbound traffic and seeks to establish systematic variations in the regional origins of major markets. Little is known about the actual routes taken by international tourists within their own country to reach the port of embarkation. The second part of the chapter concentrates on intra-national travel within destination countries, reviewing the different approaches which have been employed to date. Most studies have focused selectively on specific aspects of intra-national travel. Some, for instance, have concentrated on flows into the gateways (particularly in multi-gateway countries), or related ports of arrival to ports of departure. Others still have considered the places visited between the two and a few have attempted to map routes and itineraries. A limited amount of attention has also been focused on patterns of movement within particular cities or resorts. This provides a particularly challenging area of research and one which has generated a variety of innovative methods.

Various demand and supply factors might be expected to influence both the outbound and inbound travel patterns. Spatial variations in the general factors effecting the propensity for international travel discussed in Chapter 2, for example Table 2.5, are likely to generate regional biases in travel to particular destinations. In larger countries, questions of propinquity and accessibility are also likely to come into play, the latter being in part determined by the distribution of points of departure. For the inbound traffic, demand factors will include the volume, origin and types of visitors while decisive supply variables may include the number and location of international gateways, the nature and distribution of the attractions and the development of the transportation network and accommodation. These in turn may reflect government policies and private sector initiatives. Such factors will influence the extent to which travel is concentrated or dispersed and whether itineraries are varied and complex or confined to a limited and well-defined circuit. The travel patterns of sightseeing visitors in a well-developed continental country accessible by a variety of gateways are likely to be more varied and diffuse than those of 'sun-sand-sea' tourists visiting a developing tropical island served only by a single international airport.

Intra-national Variations in the Demand for International Travel

Surveys on variations in international travel from regions within generating countries have been undertaken both in countries of origin and at destinations, although neither is particularly common. Data from the USTTA inflight survey on outbound American travellers to major world regions shown in Fig. 4.1 are complemented in Table 4.1 by the more specific figures for as comprehensive a range of destinations for which appropriate data could be obtained from the respective national tourist organizations. Such detail tends to be more readily available for the United States than other markets, not only because of its overall importance as a source of visitors but also because the sheer geographic size of the country suggests that a regional breakdown is an effective means of market segmentation. Once general patterns have been established for the United States, brief comparisons are made with other markets.

First, Fig. 4.1 shows that major differences occur in the percentage of all travellers generated by each region and their population base. The general pattern is one in which proportionately more overseas travellers (travel to Canada and Mexico is excluded) are generated in the Pacific and Atlantic (New England, Mid-Atlantic and South Atlantic) seaboard regions and relatively fewer in the remaining regions of the interior. While other factors such as regional wealth may play a role, regional variations in the propensity to travel would appear to reflect ease of accessibility (proximity plus a greater range of gateway airports) and perhaps some tendency to be more outward looking on the seaboards.

When the regional destinations are considered, these geographic biases become more pronounced. A disproportionate share of departures for Europe and the Caribbean originate on the Atlantic seaboard and to the Far East and Oceania from the Pacific Coast states. Other stronger than expected linkages also appear to reflect ethnic ties, notably from the Middle Atlantic to the Middle East, to Central and South America from the South Atlantic and to Central America from the Pacific.

These broad patterns often become more acute when the regional origins of United States visitors to specific destinations are examined (Table 4.1). Within the general Atlantic seaboard–Caribbean and Pacific Coast–Oceania split certain flows are more pronounced. Bermuda, for example, derives a particularly large share of its visitors from New England and the Middle Atlantic while Jamaica extends its market to draw heavily on the East North Central region. Within Oceania, Australia and New Zealand uniformly draw just over one-third of their visitors from the Pacific Coast states while Fiji and French Polynesia derive over half of their traffic from this same region. Distance and direct linkages are clearly factors here but the character of the destinations also appears to play a role. Although strong, the Pacific Coast bias for the two larger and more general sightseeing destinations (Australia and New Zealand) is less prominent than for the two small tropical island groups (Fiji and French Polynesia) for whom the Caribbean destinations might be considered more direct competitors for visitors from other United States regions. The Asian destinations also have a disproportionately high share of Pacific Coast visitors though it is not clear why Singapore and Hong Kong should vary so much. Few European countries report the regional origins of their United States visitors. The United Kingdom pattern in Table 4.1 approximates that of all departing United States travellers, a reflection perhaps of the general appeal of this destination to

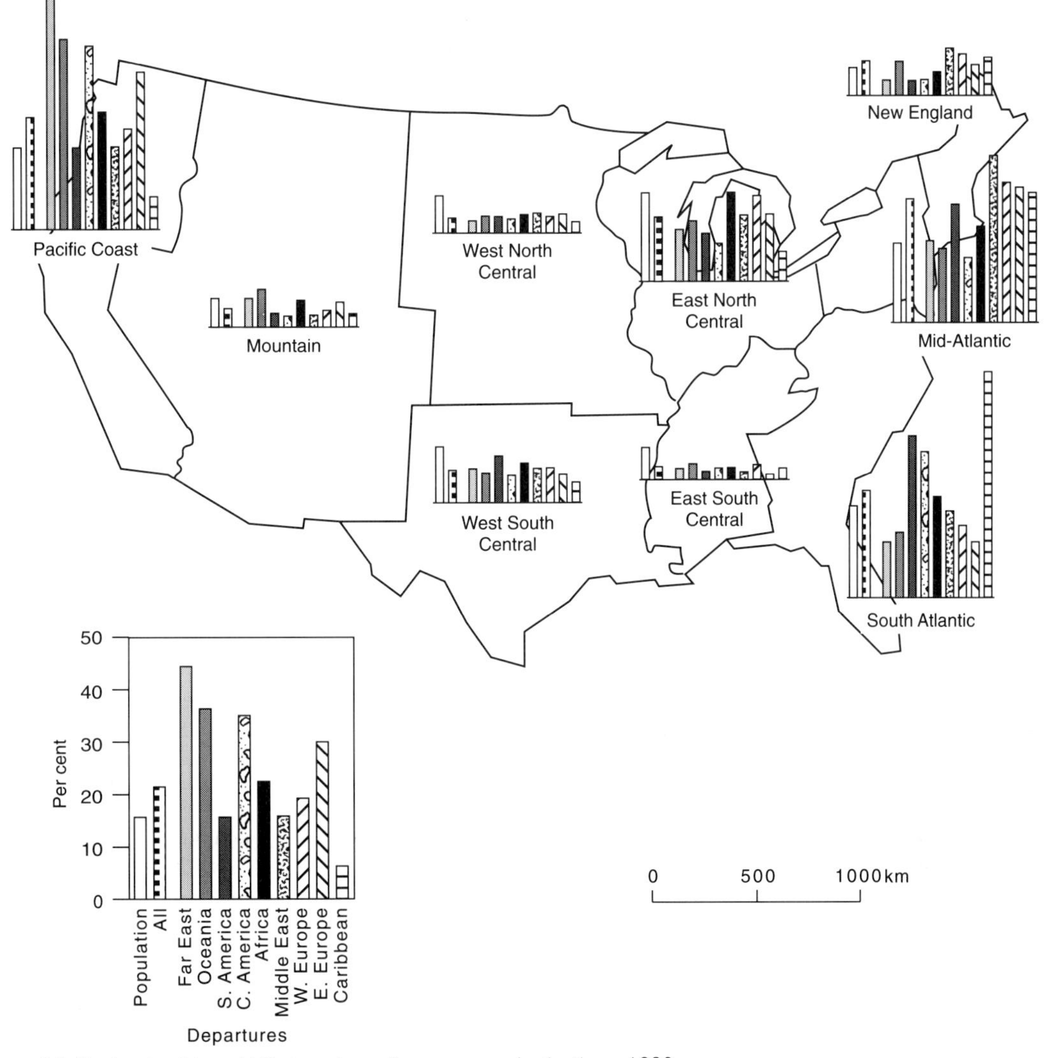

Figure 4.1 Regional origins of US departures for overseas destinations, 1990.
Data source: USTTA (1991).

the American market. In contrast, the strong concentration of the Jewish community in New York and New Jersey is mirrored in the pronounced Middle Atlantic flow to Israel.

Further work is required to establish more precisely the factors underlying these flows but it is clear that for many destinations the United States cannot be seen as a single uniform market. Rather, its constituent regions are of greater or lesser importance to individual destinations whose marketing campaigns must identify and target those exhibiting the greatest propensities for visitation or the most potential. Faced with severe competition and budgetary constraints, many small destinations will be limited to building up existing flows, so the greater the knowledge of these, the more effective the resultant campaigns might be.

The Pacific/Atlantic split identified in Fig. 4.1 and Table 4.1 is accentuated by the continental scale of the United States market. Similar patterns are also found in other large countries for which fewer data are available, notably Canada and Australia. For instance, 55 per cent of Australian visitors to Fiji and French Polynesia in 1990 originated from New South Wales and only 1.3 and 9.8 per cent respectively from Western Australia,

Table 4.1 Regional origins of United States travellers

		New England (%)	Mid-Atlantic (%)	East North Central (%)	West North Central (%)	South Atlantic (%)	East South Central (%)	West South Central (%)	Mountain (%)	Pacific Coast (%)	Others (%)
US population	1990	5.3	15.1	16.9	7.1	17.5	6.1	10.7	5.5	15.7	
All overseas departures	1990	6.6	23.5	12.3	2.9	20.4	2.5	6.2	3.6	21.4	
Hong Kong	1992	4.7	14.3	10.7	5.4	11	1.8	6.9	10	34.2	1
Thailand	1986	4.4	15.2	8.8	3.5	13.7	1.9	9.1	5.8	37.6	
Singapore	1991	4.5	11.9	5.8	9.2	6.5	0.2	1.9	4.4	45.2	8.6
Australia	1990	5.6	11.8	11.1	7.2	10.5	2	6.5	8.7	35.6	
New Zealand	1992	4.8	10.1	11.5	5	13	2.2	6	9.3	35.5	2
Fiji	1990	2.5	10.3	8.8	4.1	8.8	1.4	4.2	7.6	52.3	
Fr. Polynesia	1991	3.7	10.7	7.9	2.5	9	1.6	4.2	7.7	52.6	
Jamaica	1992	7	29.4	16.7	4.4	22.9	2.7	6.3	1.2	8.7	
US Virgin Is.	1993	10.4	27.8	13	3.6	24	3.5	6.6	3.3	7.7	
Puerto Rico	1991–2	7.3	38	9.4	3.3	23.9	2.3	4.4	2.4	7.4	1.5
Bermuda	1992	24.3	38	6.6	1.6	20	2	2.7	1.2	3.6	
UK	1991	10	22.7	11.5	4.8	17.2	2.1	7.4	4.3	20	
Israel	1990	6.5	42.2	9.8	3.1	14.1	1.6	4.3	3.3	15.1	

Data sources: National tourist organizations; USTTA.

whereas the latter state generated 23.9 per cent of Australian visitors to Singapore in 1991. But what of smaller geographical markets having no such continental divide? No large data sets comparable to Table 4.1 have been able to be compiled but the figures available suggest other factors might be at work.

Figure 4.2 depicts the regional origins of Japanese visitors to Australia, New Zealand, Singapore and Thailand, together with the distribution of the population and departures from Japan. Kanto (the greater Tokyo region), with one-third of the country's population, generates almost half of all departures. However, for the four destinations shown, the Kanto region's share ranges from 61 per cent for Singapore to 38 per cent for Australia. In the latter case, the Chubu region generates a disproportionately high share of departures. More detailed purpose of visit figures for New Zealand reveal more business (52.8 per cent) and VFR (49.7 per cent) arrivals originate in the Kanto region than do holidaymakers (40.7 per cent), the latter group constituting 87.5 per cent of all Japanese arrivals.

Data from Australia, Hong Kong, French Polynesia and Martinique for 1991 and 1992 indicate that over 40 per cent of their French visitors resided in the Ile de France (Paris) region, approximately twice the percentage that might be expected on a purely population basis. In the latter two cases, some of this concentration might reflect a significant proportion of visitors on official business but this should not greatly influence the traffic to Australia and Hong Kong.

The urban hierarchy of Germany is more balanced than that of France and the pattern which emerges for the destinations shown in Table 4.2 is less consistent. Arrivals from North Rhine-Westphalia, the largest of the German *Länder*, are over-represented in the United Kingdom but under-represented in the three more distant destinations of Australia, Singapore and Hong Kong. Bavaria, with the second largest population, generates proportionately more visitors not only to neighbouring Austria but also to Australia and Singapore; relatively fewer Bavarians, however, visit the United Kingdom. Baden-Wurtenburg is under-represented throughout. The city states of Hamburg and West Berlin are over-represented in all four destinations but Bremen only marginally so. Time-series data of German visitors to the United Kingdom show the regional distribution has remained largely unchanged over the period 1974–91.

While the destinations analysed above cannot be considered representative – in particular, data on the largest flows to neighbouring countries were unavailable – some general points might still be made. Significant regional variations in the origins of visitors do exist. Country-to-country analysis can be considered only as a first step, particularly for marketing purposes. Larger countries such as the United States, Canada and Australia exhibit a 'continental divide' in which distance, accessibility and awareness generate East–West biases in travel patterns, and, in the case of the overseas travel from the United States, an interior basin of lower propensities. In smaller countries large urban areas may generate a disproportionate share of international departures. This may be a function of greater accessibility, the concentration of tour operators and market-

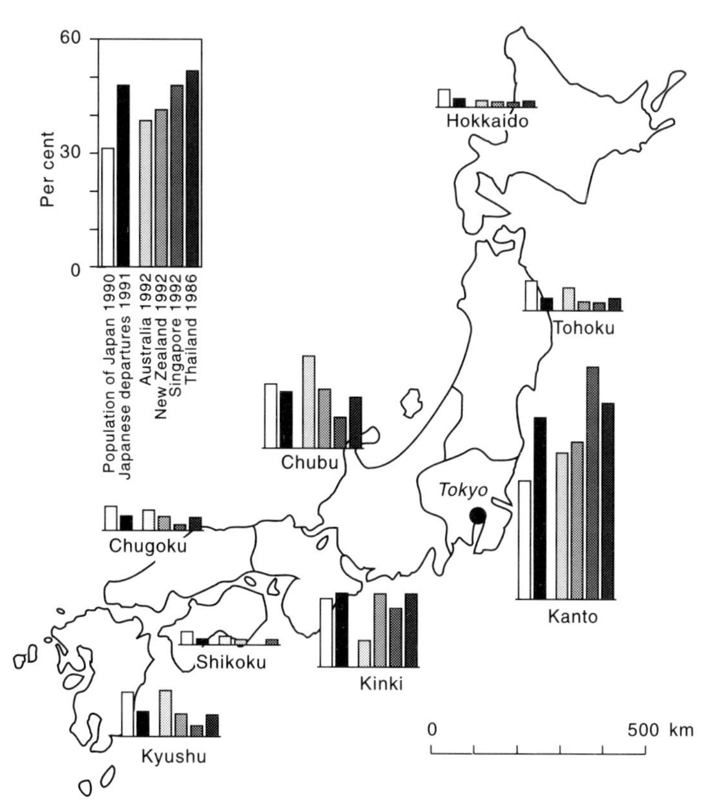

Figure 4.2 Regional origins of Japanese departures.
Data sources: Japan Travel Bureau (1992) and national tourist organizations.

Table 4.2 Regional origins of German arrivals

Land	Population 1990 (%)	Austria 1991 (%)	UK 1991 (%)	Australia 1991 (%)	Singapore 1986 (%)	Hong Kong 1992 (%)
Baden-Württemberg	15.0	15	12.9	11.6	10.1	12.9
Bavaria	18.0	23	11.8	23.4	29	18.3
Bremen	1.1		1.0	1.6	1.3	1.6
Hamburg	2.6	3.2	4.3	10.6	6.5	7.7
Hesse	8.8	7.4	11.8	10.0	10.9	10.9
Lower Saxony	11.8	8.5[a]	10.7	7.5	6.3	6
N. Rhine-Westfalia	27.5	27.4	36.5	19.5	21.2	19.6
Rhineland-Palatinate	5.9	9.9[b]	4.3	3.7	3.1	3
Saarland	1.6		1.0	1.0	0.6	0.8
Schleswig-Holstein	4.2	2.4	2.1	1.7	2.5	2.2
West Berlin	3.1	2.9	4.3	5.1	8.0	6.2
Other or unstated	0.0			4.0		10.9

Notes: [a] Includes Bremen.
[b] Includes Saarland.
Data source: National tourist organizations.

ing activities by foreign organizations and perhaps a reflection of the general tendency for travel propensities to increase with urban size noted in Chapter 2. Other factors have also been identified, such as special links induced by ethnic ties. Exceptions to some of these patterns can also be observed – Hesse in Germany is not especially prominent in Table 4.2 despite the location of Germany's busiest airport in Frankfurt. Further work is now required to compile and analyse a more comprehensive data set and to elucidate further the factors underlying regional flows.

Intra-national Trade Patterns in Receiving Countries

Port of entry/port of departure

The gateway through which international visitors enter the intra-national system is often a useful point of entry into the analysis of intra-national travel patterns in receiving countries. Where visitors enter the system may reflect the main objectives of their visit, particularly if a choice of gateways is available. Moreover, immigration statistics are often broken down by port of entry (and departure) so that the basic dimensions of the traffic through each gateway are frequently available. Studies have been undertaken of single and multiple gateways and of the links between points of entry and departure.

Janin (1982) found the Aosta Valley in Italy was largely a transit zone for many of the motorists using the Mt Blanc and Grand-Saint-Bernard tunnels which link Italy to France and Switzerland respectively. By relating the number of vehicles passing through the two tunnels to the number of different nationalities recorded in the various accommodation establishments in the valley, he observed that the proportion of motorists staying there increased with distance from the country of origin. Likewise in France, Soumagne (1974) has shown that the majority of British motorists arriving by car ferry at Cherbourg were in transit, primarily making for Spain, Portugal and Brittany, and not heading for Normandy itself (2.5 per cent).

Multiple gateway studies include that by Levantis (1981) who used immigration data to examine the arrival patterns at Greece's 25 main gateways. His analysis showed marked differences in the traffic at each gateway: one-third of the arrivals at Athens airport were Americans, Yugoslavs accounted for a similar proportion at Evzoni and Scandinavians constituted over half of all arrivals at Rhodes airport. These differences appear to reflect propinquity and purpose of visit (the American 'cultural' tourist and the Scandinavian sunseeker?). The sunlust traffic itself also appears to be spatially differentiated for Pearce (1987b), in a more specific study of inclusive charter tours to the Mediterranean, found significant regional biases occurred in the regional distribution throughout Greece of British, German and Swedish arrivals. In 1985 one-third of the traffic from the UK was concentrated on Corfu, Crete was the most favoured destination for Germans and Rhodes attracted the largest number of Swedes. Time-series data for the Swedish market indicated dispersion of the traffic away from Athens and then Rhodes to some of the other island airports and a general deconcentration of charters. Similar regional biases were also reported for non-scheduled arrivals in Spain and some redistribution of the traffic also occurred over time, with the Canaries' share generally growing at the expense of the mainland coastal airports.

As package holidays associated with charter tours are normally localized in nearby resorts, the distribution by airport of the non-scheduled traffic usually gives a very reliable picture of the dispersion of such visitors. In other cases where different forms of tourism occur the resultant patterns may be less clear-cut. Figure 4.2 depicts the distribution of direct arrivals at Indonesia's four major gateways for 1991. Jakarta, the capital, had the most balanced composition, attracting almost 40 per cent of all direct arrivals, particularly from a range of origins in Asia and Europe. The three other gateways are characterized by the prominence of particular market segments. Australians constituted over one-quarter of the arrivals in Bali (Ngurah Rah airport), with Japanese, British and other Europeans also being important; Batam is dominated by Singaporeans while Medan (Polonia) airport, the gateway to North Sumatra, is characterized by its mix of Malaysian, Singaporean and Dutch arrivals. Propinquity is clearly a major factor in the cases of Batam and Polonia, with each attracting short-haul visitors for economic, short-break, stay-put holidays. Similarly, Bali attracts many Australians and other visitors for longer but essentially stay-put beach holidays. For many of the Dutch visitors, however, Medan is the port of entry for a sightseeing tour of Indonesia which may begin with North Sumatra and subsequently include Jakarta, Jogyakarta and other destinations, often ending in Bali. Jakarta may play this gateway and redistributing role for other markets. Bali thus attracts a number of stay-put sunlust tourists for whom the direct arrivals figure is a reasonable indication of intra-national travel as well as other, mainly sightseeing, tourists who will have entered Indonesia elsewhere and are not included in this statistic for Bali. While field observations of the structure of the industry and the nature of the demand enable general observations along these lines to be made, more detailed surveys and other research is needed to establish patterns more precisely.

Elsewhere writers have related port of arrival to port of departure in an attempt to get some measure of penetration and a broad picture of routes taken. Pollock, Tunner and Crawford (1975) cross-tabulated the entry/exit points used by US and Canadian visitors to British Columbia and identified four basic types of routes:

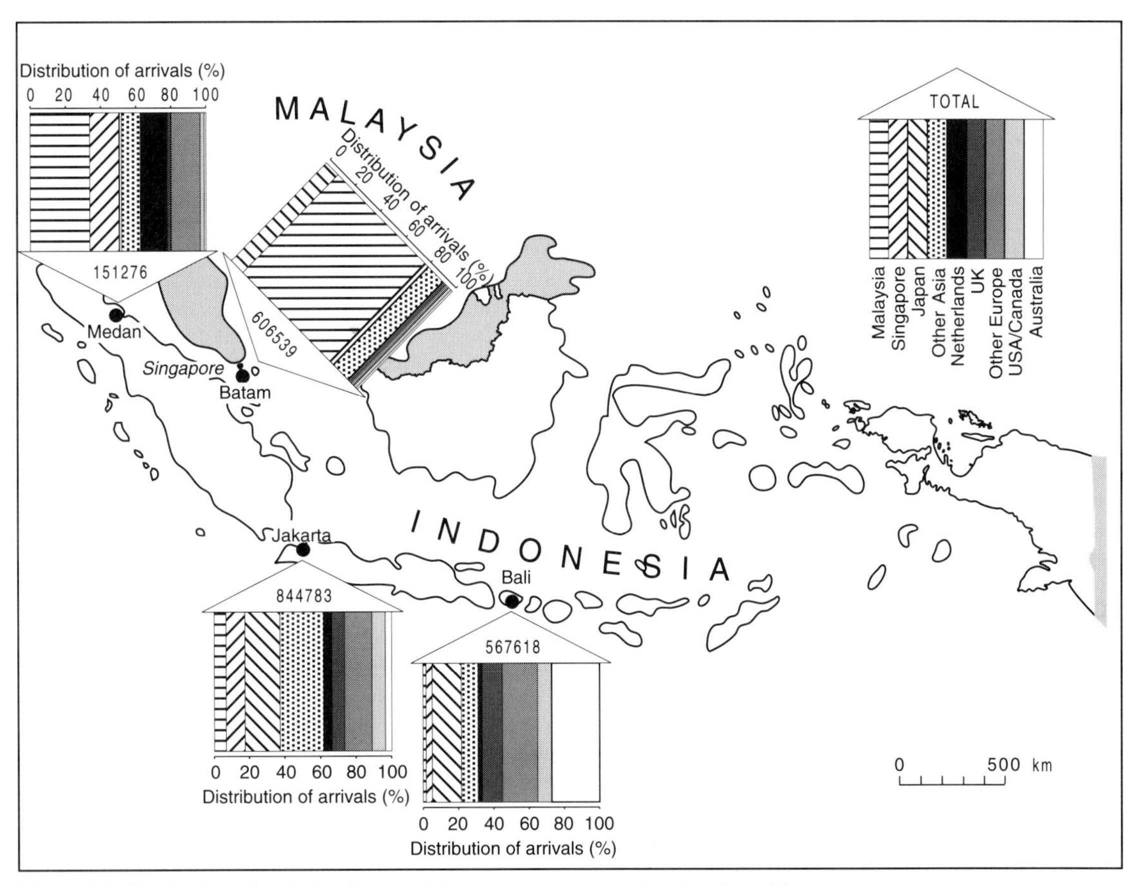

Figure 4.3 Distribution of arrivals in Indonesia by major gateway and nationality, 1991.
Data source: Statistical Report on Visitor Arrivals to Indonesia (1991).

1. the direct route, whereby visitors entered and left through the same zone
2. the long traverse, whereby visitors entered at one end of the province and departed at the opposite end
3. the short traverse, where visitors made diagonal-type routes across the edge of the province
4. the loop, where visitors returned to their point of origin via a different zone to the one they had entered.

The direct route, comparable to Mariot's access and return routes (Fig. 1.1), was found to be the most common, especially with those visitors (mainly Americans) leaving via the Lower Mainland, of whom 66 per cent had also entered British Columbia by the same zone.

The British Columbia data were subsequently analysed further by Murphy and Brett (1982: 159) using discriminant analysis. They found

> The significance of the entry point for most individual regions and its respective signs suggest some peripheral regions act as interceptors to the rest of the province. ... Once the visitors entered British Columbia there was limited evidence of extensive touring ... these results suggest that the southern regions were not only attracting most visitors but they also absorbed them, allowing relatively few to penetrate further into the province.

The attractive power of the southern regions was seen to lie in the presence of the major metropolitan areas close to the United States border and thus readily accessible to the important West Coast traffic. The authors also showed that these centres attracted a large number of people visiting friends and relations whereas independent vacationers were found in the more isolated northern regions of the province.

In Japan too, many visitors arrive and leave by the same airport. A 1979 survey of over 1000 foreign visitors found that 61 per cent of all foreign arrivals followed the pattern of 'to and from Tokyo' (Shirasaka 1982). Then came Tokyo–Osaka (17.5 per cent),

Osaka–Osaka (7.5 per cent), Osaka–Tokyo (7.3 per cent). Tokyo's dominance is not surprising given the concentration of international air carriers on that city. This dominance is also reflected in the degree of foreign visitor penetration, for other information in the Japanese survey showed that many visitors did not travel beyond the capital: over 40 per cent of those in the Tokyo–Tokyo category, or one-quarter of all arrivals, visited no other place.

A different pattern is evident in New Zealand where there is a greater tendency for visitors to depart from a different airport to that at which they arrived. In 1981–2, just over half of the arrivals at Auckland airport, which accounted for two-thirds of all arrivals, left via Auckland, with about one-third exiting via Christchurch and one-tenth leaving by Wellington. Arrivals at Christchurch exhibited a comparable pattern (59 per cent of departures via Christchurch, 31 per cent via Auckland and 10 per cent via Wellington). A larger proportion of those arriving in Wellington left via the same airport (73 per cent) with the remaining traffic being shared almost equally by Auckland and Christchurch. This latter pattern is in part a function of the larger share of business and official travel to the capital and its central location. The Auckland and Christchurch patterns reflect the nature of international tourism in New Zealand, much of which is characterized by visits to a number of scenic attractions in both the North and South Islands. Many vacationers thus arrive at one gateway, tour the country, and leave via a second gateway. About one-quarter of the arrivals did not even stay overnight in the gateway at which they arrived.

Places visited and routes followed

Other information on the intra-national travel patterns of overseas visitors comes from surveys which ask respondents to list regions or places visited. The type of detail collected and the degree of analysis undertaken varies from country to country and survey to survey, depending on the particular context and ease with which the data can be compiled and analysed. A first measure of the extent of intra-national mobility based on such data consists of the average number of regions or places visited. The information that is available from a number of countries suggests that the degree of intra-national travel may be rather limited, though strict comparisons are constrained by the use of different areal units.

In 1990 overseas visitors to the United States visited on average just two states each (USTTA 1991). About one-half restricted their stay to a single state, one-quarter visited two, and the remainder three or more. Little variation in this pattern occurred by purpose of visit but some differences are found by regional origin. Visitors from Central America (1.4 states), the Caribbean (1.4) and South America (1.7) had the most focused visits, the smaller more distant markets of Oceania (2.3), the Middle East (2.3) and Africa (2.2) the most dispersed, while Europeans, the largest market, on average visited two states each.

A similar pattern occurs in Australia (BTR 1990), with overseas visitors there in 1990 visiting just under two states each on average (though from a smaller total, eight including the ACT (Australian Capital Territory) and Northern Territory). Again the longer-haul visitors tended to be slightly more mobile (Europeans visited 2.5 states each, British 2.3, Canadian 2.4, American 2.3) while closer markets were more focused (New Zealanders averaged 1.3 states each, Asians 1.7). Some variation by purpose of visit also occurred, ranging from a mean of 2.2 states for holidaymakers to 1.6 for those visiting friends and relatives.

The limited amount of multi-state travel in these two cases may reflect the sheer size of these countries and of individual states. However, when regional patterns of travel within the United Kingdom are examined an even more constrained pattern appears, for in 1991 overseas visitors there stayed on average in only 1.4 regions (Table 4.3). The near-neighbour Belgian/Luxembourg and French segments fell below this average, with Australians and Canadians being the most mobile, but the overall range is relatively limited (1.1 to 2.0 regions). This restricted mobility is also borne out by other results from Britain's International Passenger Survey (BTA 1989). Only one-quarter of all visitors in 1986 were reported as having a touring holiday (incorporating overnight stays in more than two towns) though this proportion rose to 36 per cent for independent holidaymakers. Almost one-third of all visitors restricted their stay to London.

Table 4.3 shows only a very marginal decrease in the average number of regions visited over the period 1978–91, with the Australian mean declining the most (from 3 regions to 2). However, a significant redistribution in the specific regions visited is recorded, notably some deconcentration from London. Two-thirds of all visitors went there in 1978 while just over half did in 1991. Wales, the South East, Cumbria and North also experienced significant decreases while many of the central regions were beneficiaries of this changing pattern. The deconcentration shown may reflect promotional strategies designed to redistribute growth away from London and testify to the success of marketing and development efforts of individual regions.

In the case of island micro-states, which often consist of groups of small but scattered islands, the emphasis in terms of measurement and marketing typically involves the nature and extent of inter-island movement (Pearce

Table 4.3 Regional pattern of staying visits in the United Kingdom by major market, 1978 and 1991

Region	All visitors		USA		Canada		France		Belgium/ Luxembourg		Australia	
	Country of residence											
	1978 (%)	1991 (%)	1978 (%)	1991 (%)	1978 (%)	1991 (%)	1978 (%)	1991 (%)	1978 (%)	1991 (%)	1978 (%)	1991 (%)
London	66.1	53.6	83.9	69.7	65.3	51.8	41.2	33.8	45.7	32.8	78.0	61.6
South East	12.7	11.5	8.5	11.2	19.0	13.8	17.9	10.4	12.4	12.8	12.1	15.7
West Country	7.1	8.0	8.7	9.0	13.2	13.4	6.6	7.5	1.2	4.1	13.9	19.0
Thames and Chilterns	6.0	7.9	7.0	9.4	7.5	9.4	7.0	9.6	5.0	5.8	12.0	13.1
Heart of England	5.6	7.0	8.8	9.2	10.9	10.0	6.0	6.6	1.3	4.5	14.3	11.0
North West	5.3	5.8	5.9	6.5	9.9	11.9	1.1	3.1	1.3	3.1	13.0	11.5
East Anglia	5.0	5.5	5.0	5.9	6.6	6.8	5.2	5.2	3.0	5.4	12.2	8.1
Southern	4.7	4.3	4.1	4.0	6.8	5.9	7.0	4.7	2.3	3.0	8.2	7.6
Yorkshire and Humberside	4.4	5.1	4.7	6.2	7.9	8.4	2.9	3.1	2.5	2.7	9.8	11.1
East Midlands	2.5	3.4	2.5	3.1	3.4	4.9	3.4	3.3	0.3	3.5	5.9	6.5
Northumbria	2.4	1.9	1.5	1.8	4.6	3.6	0.2	1.1	0.7	1.2	3.8	3.4
Cumbria	1.8	1.3	2.8	2.2	4.1	2.5	0.4	0.7	0.2	0.9	4.1	4.6
Channel Islands	0.1	0.2	0.1	0.1	0.5	0.4	—	*	—	*	0.1	0.4
Isle of Man	0.1	0.1	—	0.1	0.1	0.1	—	*	—	*	0.3	0.2
Unspecified	2.5	0.2	2.8	0.3	3.1	0.2	1.7	*	0.5	*	5.7	0.2
Scotland	9.7	9.5	14.6	14.0	22.6	19.7	4.3	5.5	2.7	6.5	25.6	18.9
Wales	5.2	3.7	5.4	5.3	7.9	6.1	3.7	2.4	0.5	3.0	21.0	10.3
Northern Ireland	0.8	0.6	0.6	0.9	2.4	1.9	0.6	0.2	0.4	0.2	1.5	1.1
Nil nights in UK	4.9	6.8	0.4	0.9	0.3	0.6	14.5	18.9	27.5	26	0.3	0.3
Average number of regions visited	1.5	1.4	1.6	1.6	1.9	1.7	1.3	1.2	1.1	1.2	3.0	2.0

Note: * Staying visits (overnight) only.
Sources: Based upon British Tourist Authority (1981c, 1992).

1984, 1990b, Pearce and Johnston 1986). International tourism within French Polynesia, for example, is centred on three main islands – Tahiti, Moorea and Bora Bora – with less than 5 per cent of all international visitor movements in 1984 including travel to other islands within the group. Tahiti, with Faa international airport and over half of all hotel accommodation, is the prime focus (Fig. 4.4). Nine out of ten visitors in 1984 intended spending at least one night there, with 59 per cent of all arrivals confining their visit to Tahiti. Very few limited themselves to just Moorea or Bora Bora. About one-third of all visitors made multiple island visits, with the near neighbour combination of Tahiti and Moorea being the most popular, followed by visits to all three islands.

Figure 4.4 also shows a significant reduction in multiple island visits over the period 1980–4. Some of the decrease in Moorea-bound traffic in 1984 is the direct result of the temporary closure of Club Med that year. The more general decline in travel beyond Tahiti to other parts of French Polynesia is largely associated with the changing patterns of destination/circuit travel. Those having French Polynesia as their sole destination show greater mobility within the territory than those visiting the islands as part of a larger circuit. In 1980, twice as many circuit travellers stayed only on Tahiti compared to destination visitors (51 and 24 per cent respectively), with the former group also being more likely to stay on more than one island. As the proportion of destination travellers in the total traffic decreased after 1980, so the overall level of inter-island mobility declined. Differences also exist in the internal travel patterns of different nationalities. The Americans and Europeans are more mobile and are more likely to include a visit to Bora Bora while most Australians and New Zealanders limit their stay to Tahiti and Moorea. The 'Club Med effect' on the Australian pattern is most marked, with only 2.6 per cent making a single island visit to Moorea in 1984 compared to 54 per cent in 1980 and 38 per cent in 1981. Few significant differences appear to exist between the internal travel patterns of group and individual travellers nor first time and return visitors.

Travel within Tonga is more spatially constrained (Pearce and Johnston 1986). Some 80 per cent of a sample of visitors at Fua'amotu airport in 1981 restricted their stay to the main island of Tongatapu, with virtually all of the remainder combining a stay there with a visit to one or more of the other major islands or groups, notably Vava'u, 'Eua and Ha'apai. The two

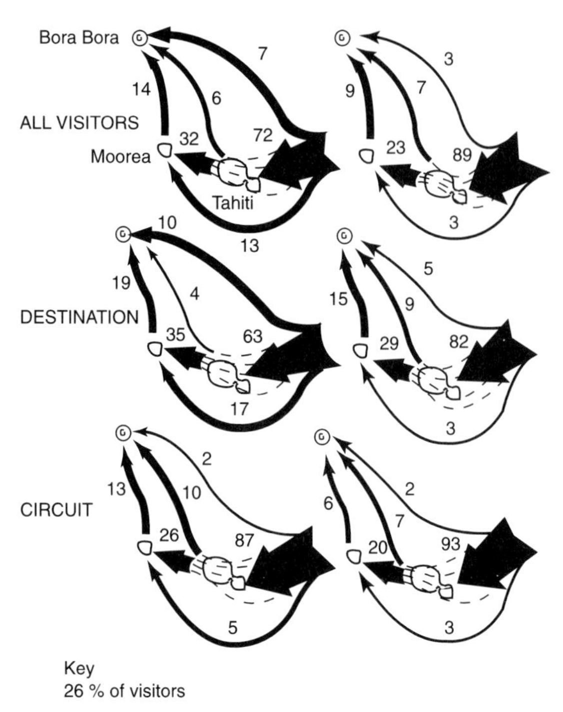

Figure 4.4 Patterns of travel within French Polynesia, 1980 and 1984.
Source: Pearce (1990b).

groups of visitors have distinct profiles. Some of the differences between the two groups can be attributed to the varying purposes of visit but it seems that those going beyond Tongatapu constitute a group of more adventuresome, independent and informal travellers whereas those limiting their stay to the main island are much more conventional tourists who rely more heavily on the tourist industry infrastructure. The latter group also generated a disproportionately high share of bed-nights and total expenditure, distributing both more widely within the kingdom.

The location of the international airports and the distribution of accommodation suggests the pattern of internal travel within other South Pacific states would follow the more limited Tongan model, with the majority of tourist visits being concentrated on the main island in each group. Most tourist activity in Fiji, for example, is concentrated on Viti Levu; in the Cook Islands there is relatively little travel beyond Rarotonga and in Vanuatu the dominance of stays on Efate is scarcely challenged by side trips to Tanna and Santo. Several interrelated factors account for this pattern, including more general contrasts in the levels of development between the main and outer islands, difficulties and costs of internal transport, government policies, the distribution of attractions and the total volume of tourist traffic.

Elsewhere, intra-national mobility has been measured in terms of individual places visited rather than regions or islands. In 1982, international visitors to New Zealand on average stayed overnight in five places (NZTPD 1983). Holidaymakers averaged almost seven places each, twice the number of the VFR market, while the mean for business visitors was just two. The greater mobility recorded here for holidaymakers is to some extent a function of the more localized measure used, that is places rather than regions, but it also reflects a long-standing pattern of sightseeing tourism. In contrast, Oppermann (1992b) reports visitors to Malaysia average only one and a half overnight destinations, the majority of visitors staying put in just one location, particularly Kuala Lumpur or Penang.

Where multiple destination visits are important, analysis may focus on establishing the combinations of places visited and the routes followed. Forer and Pearce (1984) used a variety of techniques to analyse the tour itineraries published in the brochures of New Zealand coach tour operators. The major and minor nodes were mapped then amalgamated at a regional level to facilitate the analysis of interregional flows. This revealed a pattern in which the North Island was characterized by a series of linear routes between the major cities of Auckland and Wellington while the South Island pattern was much more complex and based on a series of loops. Further analysis of net flows identified northbound or southbound biases in the flows between particular regions. A marked relationship was also found between the importance of each node (expressed in the number of coach nights recorded) and its relative connectivity in the total system (determined by the number of links to other nodes). The larger nodes were linked to a greater number of other nodes than the smaller ones which had fewer linkages. Finally, a typology of nodes was derived using a *k*-mean clustering algorithm. The five variables employed were tour nights, connectivity, average length of tour visiting the node, stopover function (proportion of second-night stays) and relative impact (ratio of tour nights to local population). Six major groups of nodes were identified on the basis of the nodes' location relative to other major nodes and their own limited attractions (Table 4.4).

This more comprehensive approach, involving the classification of different types of nodes as well as an analysis of the flows between them, begins to offer insights into the functioning of the tourist circuit which might be incorporated into tourism plans. Nodes depending on different lengths of tour, for example, will see their fortunes vary with changes in the composition of the market as well as with changes in total demand. In terms of supply, bottlenecks in the gate-

Table 4.4 A typology of New Zealand coach tour nodes

Group 1	Gateways and major generators *Wellington, Dunedin, Christchurch, Auckland, Rotorua, Queenstown*
Group 2	Isolated major attractions *Milford Sound, Mount Cook*
Group 3	Secondary centres with strong impacts and long tour involvement *Te Kuiti, Masterton, Taupo, New Plymouth, Picton, Nelson, Greymouth, Fox Glacier, Haast, Invercargill, Oamaru, Tekapo, Blenheim, Timaru, Geraldine, Palmerston North*
Group 4	Secondary centres with short tours and above average stopover proportions *Kaitaia, Waitangi, Paihia, Russell, Lake Ohau, Franz Josef Glacier, Te Anau, Otematata*
Group 5	Secondary centres with weaker impacts and shorter tours *Whakatane, Napier, Gisborne, Chateau Tongariro, Wanganui, Westport, Hokitika, Hari Hari, Wanaka, Gore, Alexandra, Omarama, Hicks Bay, Tauranga, Kaikohe, Waitomo, Whangarei*
Group 6	Small centres with short tours and high relative impact *Te Aroha, Whitianga*
Nodes with an overflow function	*Omarama, Te Anau, Manapouri, Tekapo, Otematata, Te Kuiti, Gore, Lake Ohau*

Source: Forer and Pearce (1984).

ways and major generators will have greater repercussions on flows throughout the circuit than shortages in the secondary centres.

As more detailed itinerary data have become available from New Zealand's international visitor survey so the analysis of intra-national travel patterns has been able to be extended beyond coach tourists to the overall visitor traffic or specific segments of it. Oppermann (1993b) presents a comprehensive analysis of tourist flows and the spatial pattern of demand based on 1991 data. Figure 4.5 depicts the intra-national flow patterns for five leading markets (only flows accounting for at least 2 per cent of each country's visitors are shown, 3 per cent in the case of Germany). In each case flows from and between the gateways of Auckland, Wellington and Christchurch are prominent, the degree of dispersal beyond these centres varying considerably from country to country. Japanese travel is the most confined, being focused on an Auckland, Christchurch, Queenstown (the major South Island resort) triangle. The small but emerging German market is characterized by the most complex circuits involving smaller nodes, particularly those in Nelson and on the West Coast. This pattern is repeated to a lesser extent by British travellers. Australians and Americans, who make up the two largest markets, exhibit an intermediate pattern of travel behaviour, with strong flows between Auckland and Rotorua and a more developed Southern Lakes circuit out of Christchurch. Such differences reflect in large part variations in length of stay: the Japanese typically stay only a few days in New Zealand while Germans visit for several weeks. Pearce (1990: 38) argued short staying visitors 'opt for the most accessible sites and what are perceived to be the main tourist attractions. Those on longer visits take in other secondary and small centres and attractions'. Gray and Kearsley and Gray (1994) have adopted a different approach to analysing an earlier New Zealand data set, undertaking a complex night-by-night breakdown of itineraries to show the temporal diffusion of visitors away from the gateways.

Similar studies elsewhere have been less common. Early surveys of motorists in Britain (BTA 1976) and Scotland (Carter 1971) revealed regional and directional biases but these appear not to have been repeated. Using his own data from exit surveys undertaken at Kuala Lumpur and Penang airports in Malaysia, Oppermann (1992b) shows that overseas visitors to that country exhibit a dominantly linear pattern of travel along the west coast of Peninsular Malaysia. Weightman (1987) analysed a selection of package tours to India. Examining not only itineraries but also tour content, Weightman stresses the encapsulated and directed nature of such tours which, she suggests, 'render the mass tourist an outsider'. Weightman notes that tourists spend much of their tour either encapsulated in planes, buses or cars or confined within deluxe hotels. She further observes (p. 232):

> The tour is selective with spatial and temporal biases. Tourists are directed towards the North, the city and the past. Current realities provide a dim or even unnoticed context for directed experience.

Where resources do not allow a detailed analysis of tour circuits, specific places might be examined in terms of whether or not they constitute a main destination in themselves or function as a stopover on a larger itinerary. The place of a given destination in a total trip or vacation might be established by use of the Trip Index (Pearce and Elliott 1983). This relates the nights spent at the destination (Dn) to the total number of nights spent on the trip (Tn) according to the formula

$$\mathrm{Ti} = \frac{Dn}{Tn} \times 100$$

Thus where the Trip Index is 100, the entire trip is being spent at the one destination. On the other hand, were the index 10, the destination would account for only 10 per cent of the total trip, indicating it was but

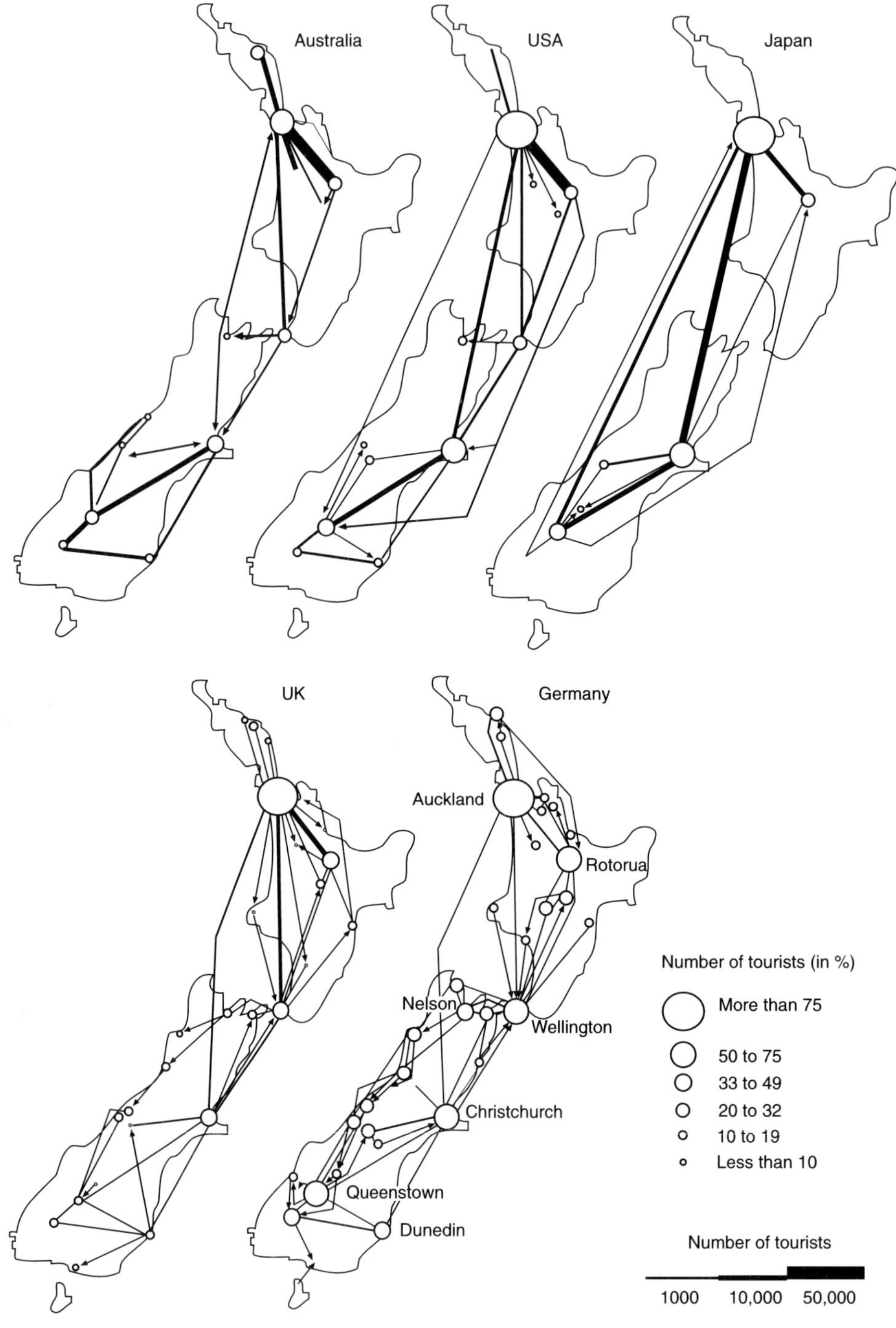

Figure 4.5 Intra-national flows in New Zealand by country of residence, 1991.
Source: Redrawn from Oppermann (1993b).

one stop among several or played only a small role in the overall holiday. An index of 0 would show, of course, that no overnight stay at all was made in the study area and that it was being visited either in transit or on a day trip.

Table 4.5 Trip Index: Westland National Park (1979–80) and Christchurch (1981)

Trip Index	Westland National Park (*n* = *1028*) (%)	Christchurch (*n* = *756*) (%)
0	20.8	2
1–10	55.5	23
11–20	16.3	25
21–50	4.5	21
51–99	0.7	9
100	2.2	17
Not stated	—	3

Source: Pearce and Elliott (1983).

Table 4.5 gives Trip Index values for two different places in New Zealand, one a rather isolated national park, the other a major urban centre (Pearce and Elliott 1983). It highlights the role of Westland National Park as a stopover on the South Island tourist circuit. Over half of the respondents recorded a Trip Index of from 1 to 10 while one-fifth did not stay overnight, the majority of this group visiting the park en route between two different overnight stops. Only 2.2 per cent of all respondents were spending their entire holiday in the park. All of these were South Islanders (as opposed to North Islanders or overseas visitors) and, with one exception, all spent seven nights or less in the park. Christchurch is also visited by many people as part of a tour but proportionately more time is being spent by many of these visitors in the city than by visitors to the park. Many stopovers in the city are associated with Christchurch's role as a gateway to the South Island (Fig. 4.5). This is particularly so with overseas visitors of whom only 4 per cent make Christchurch their sole destination, compared with half of the South Island visitors. The Trip Index also revealed different visitation patterns according to purpose of visit.

Other studies have also demonstrated the potential of the Trip Index in functionally differentiating tourist areas and in segmenting markets for promotional and planning purposes. Oppermann (1992a, 1993b) has extended the application of the Trip Index to thirty destinations in New Zealand and also applied it to destinations throughout Malaysia. His results reinforce the differences in intra-national mobility between the two countries noted earlier. Most New Zealand destinations recorded values below 10, indicative of their stopover functions. Results from Malaysia were much higher, attaining values of up to 100 in the case of the Club Mediterranée resort and over 60 for Penang and the hill resort of the Genting Highlands.

Murphy (1992) incorporated seasonal variations in his application of the Trip Index to Vancouver Island, British Columbia, and was able to develop a functional classification of areas (e.g. year round destination, mixed market, transit area) and identify a hierarchical spatial pattern influenced by the location of the island's two principal gateways. Uysal and McDonald (1989) found the Trip Index useful in distinguishing the functions of destinations within South Carolina; higher values were recorded in coastal cities with other areas being more important as stopovers and/or short trip destinations. They also explored its value in market segmentation, establishing profiles for those with low (1–30), medium (31–60) and high (61–100) Trip Index values.

Intra-destination Travel

Comparatively little is known about what international visitors do and where they go once they arrive at a particular resort, city or node on a tourist circuit. Figure 1.7 suggests a pattern of within destination travel might be complemented by day or half-day trips away from the resort or city. These movements might also be seen in terms of the visitors' activity space which, following Aldskogius (1977), might be defined as the aggregate spatial pattern of places and areas visited by individuals or groups of individuals during their visit to a given locality.

If an attempt is made to examine this issue, it usually takes the form of asking visitors what attractions or sites they have visited or intend to visit. This approach can shed some light on the most frequented attractions but rarely provides details of the combination or sequence of visits, or the amount of time spent at each site. A typical example is Table 4.6 which lists the tourist attractions and 'areas of London' most frequently visited by overseas visitors during the summer of 1987. It is interesting to note that places such as Piccadilly Circus and Oxford Street are visited more frequently than many of the designated 'attractions'. The most frequented areas are found in inner London which, of course, reflects the structure of the city and the nature of its attractions with many of the historic sites, major shopping streets and cultural centres being found there.

Deriving a representative sample framework can be difficult in surveys of tourism in urban areas. In an attempt to reduce any bias in her survey of visitors to Christchurch, Elliott (1981) limited her analysis of attractions to respondents interviewed in different

Table 4.6 Places visited by overseas tourists in London, summer 1987

Tourist attractions	%	Areas of London	%
Tower of London	72	Piccadilly Circus	92
British Museum	51	Oxford Street	89
National Gallery	49	Regent Street	76
Madame Tussauds	47	Covent Garden	65
Tate Gallery	35	Leicester Square	65
Victoria and Albert Museum	30	Knightsbridge	63
Trocadero	28	Soho	47
Science Museum	22	Bond Street	44
Natural History Museum	22	High Street Kensington	42
London Zoo	18	Kings Road	37
London Dungeon	17	Carnaby Street	35
Cabinet War Rooms	11	Portobello Road	30
Imperial War Museum	11	Greenwich	28
Museum of London	10	Barbican	20
HMS Belfast	7	Petticoat Lane	17
London Transport Museum	6	St Katherine's Dock	14
Theatre Museum	5	Richmond	11
None of these	5	South Bank	7

Source: After London Tourist Board.

types of accommodation and at various transport termini. Elliott found that one-third of her respondents had visited three to six of twenty designated sites and attractions, with a further one-third visiting seven or more of these. Inner-city attractions were visited more frequently than outer-city ones. This was especially the case with overseas visitors for whom the accessibility of the clustered inner-city attractions appears to be an important factor given their short visits to Christchurch. However, 13 per cent of all respondents had not visited any of the designated sites and attractions, with this figure increasing among those visiting friends and relations (21 per cent), attending conventions (28 per cent) or on business (34 per cent). Some 9 per cent of the overseas respondents had visited none of the attractions listed compared with 16 per cent of domestic visitors, many of whom were making repeat visits to the city. Sixty per cent of the respondents staying in Christchurch did not make any day trips out of the city. Of those who did, most made a single trip. The frequency of day trips increased with length of stay and with repeat visits to the city. Day tripping was also more popular among sightseers and those visiting friends and relations, but few differences occurred between domestic and overseas visitors.

Pearce (1982b) estimated the number of visitors to four major places of interest in Westland National Park, two glaciers and two visitor centres, and was able to determine the popularity of different combinations of places visited. A sample of visitors was interviewed at each site and asked to indicate which of the four places they had visited or intended to visit. From the percentages derived for each combination of visits and the total recorded number of visitors at that site, it was possible to calculate the total number of visitors for each combination of sites visited (for a more detailed explanation of this procedure, see also Pearce 1981b). Just under one-quarter of all park visitors visited all four sites, with the next popular combinations involving visits to one glacier and the corresponding visitor centre, visits to both glaciers, and visits to both glaciers and one visitor centre. Altogether, 95 per cent of the park visitors visited a glacier and approximately three-quarters of all visitors went to a visitor centre. In Yosemite National Park, van Wagtendonk (1980) cross-tabulated entries and exits to build up a picture of visitor use patterns within the park. The traffic was fairly evenly distributed among the four major entry/exit points surveyed, with approximately 40 per cent of the visitors leaving via their point of entry.

A completely different approach was used in a study of short-term visitors to Victoria, BC (Murphy 1980). A survey of pedestrian traffic at various sites in the central area of Victoria showed a strong correlation between the arrival of the ferries and an increase in pedestrian volumes within the Old Town. A decline in afternoon pedestrian volumes with increasing distances from the Inner Harbour was also recorded. The pedestrian survey was complemented by the 'unobtrusive' following of visitors as they left the ferry terminal, which revealed them walking very slowly, making an extensive tour of the district difficult. The slow progress of the visitors was attributed (p. 67) to the 'confusion, lack of sense of direction, crowds, pamphlet pushers and some problems with vehicles'.

Chadefaud (1981) provided details on the activity space of visitors to Lourdes by mapping the intensity of pedestrian use of the town's streets. His maps (Fig. 4.6) show that pilgrims travelling in organized groups confine their movements, or have them confined, to the lower part of the town, to the zone between the major hotels and the religious sites. The activities of pilgrims travelling independently and of other visitors are also concentrated in the lower part of Lourdes, although their movements are more dispersed, particularly around the commercial centre. Chadefaud attributes these differences to a variety of factors. In addition to a full programme of pious activities, many of the group pilgrims were aged and infirm and put off by the climb up to the *ville haute*. The independents, on the other hand, incorporate a greater element of sightseeing and non-pilgrimage activities and may need to support themselves by shopping for food and items other than the religious souvenirs found in abundance in the lower zone.

Aerial photography has been used in coastal areas,

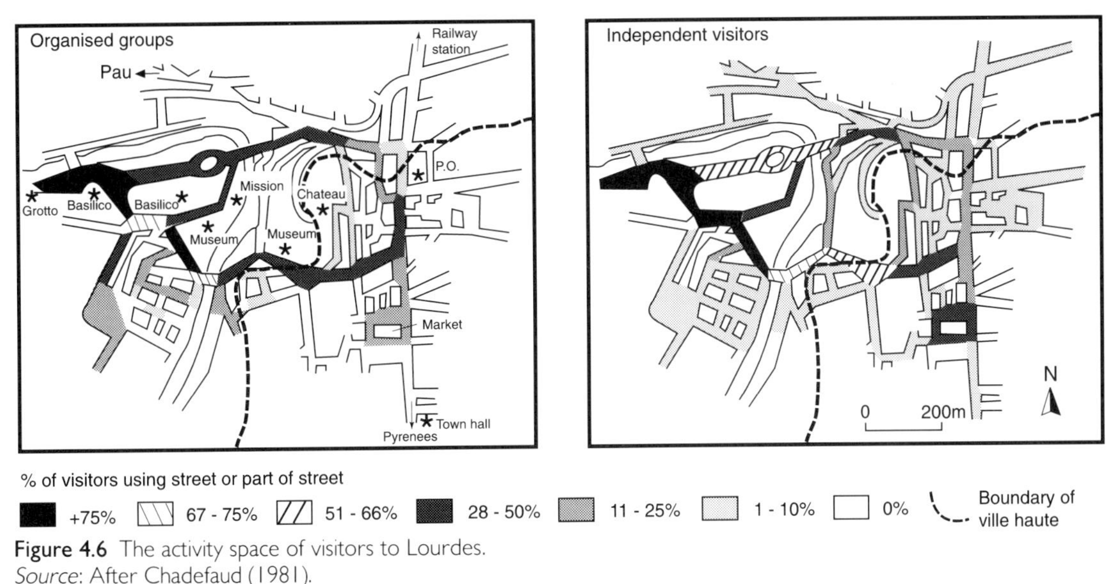

Figure 4.6 The activity space of visitors to Lourdes.
Source: After Chadefaud (1981).

for example Brittas Bay in Ireland (An Foras Forbatha 1973), to record directly the distribution of visitors at particular times. This technique, particularly if combined with a visitor survey, has potential for further use in other coastal areas, as well as in ski resorts and the countryside, but obviously has limitations in urban areas where many tourist activities occur indoors and where the presence of residents confuses the issue. In general, the literature says nothing about the pattern of movements of visitors within seaside and ski resorts. Within the latter, it should at least be possible to determine gross patterns through the use of ski-lifts.

What is lacking in most of these studies, particularly the 'places visited' surveys, is a comprehensive picture of tourist activities, especially less formal ones such as relaxing. There is also little understanding of the interrelationships of different places and activities together with the relative importance of each. Time-budget studies have the potential to provide a more complete view of tourist activity patterns (Pearce 1988a) and, in particular, to record 'behaviour patterns which are not directly observable because of their spatial and temporal extent' (Anderson 1971: 359). As Anderson notes (p. 353):

> A time-budget is a systematic record of a person's use of time over a given period. It describes the sequence, timing, and duration of the person's activities, typically for a short period ranging from a single day to a week. As a logical extension of this type of record, a space-time budget includes the spatial coordinates of activity locations.

Information for time-budget studies is usually recorded by having respondents complete detailed diaries. Such an approach generates a number of basic and interrelated methodological considerations: what is to be recorded; how and from whom is the record to be obtained, over what period; and how is the record to be analysed and presented. Resolving these issues presents many methodological challenges (Pearce 1988a) but the time-budget studies undertaken so far underline the potential which the approach offers for examining tourist behaviour at the local and regional level.

Murphy and Rosenblood (1974) incorporated the use of time diaries with interviews of first-time visitors to Vancouver Island in an attempt to analyse their spatial search behaviour. In a much larger study, Gaviría *et al.* (1975) surveyed over 3000 beach users in 16 Spanish resorts, requesting them to record their activity patterns and to provide additional attitudinal information. M. Cooper (1980) attempted to chart the spatial behaviour of tourists on Jersey over time, while P. L. Pearce (1981) examined changes in tourists' moods and activities during the course of their holiday on two small Australian tropical islands. D. G. Pearce (1988b) undertook a time-budget study in Vanuatu as part of a broader investigation into patterns of tourist circulation at different scales in the South Pacific. Hottola (1992) carried out a survey of Finnish holidaymakers in the Israeli beach resort of Eilat.

Although few in number and somewhat tentative in their conclusions, these studies have started to provide new insights into various aspects of tourist behaviour. The studies in Spain, Vanuatu and Israel, for example, indicate that the beach or water-based activities account for only a certain amount of the tourists' time, with

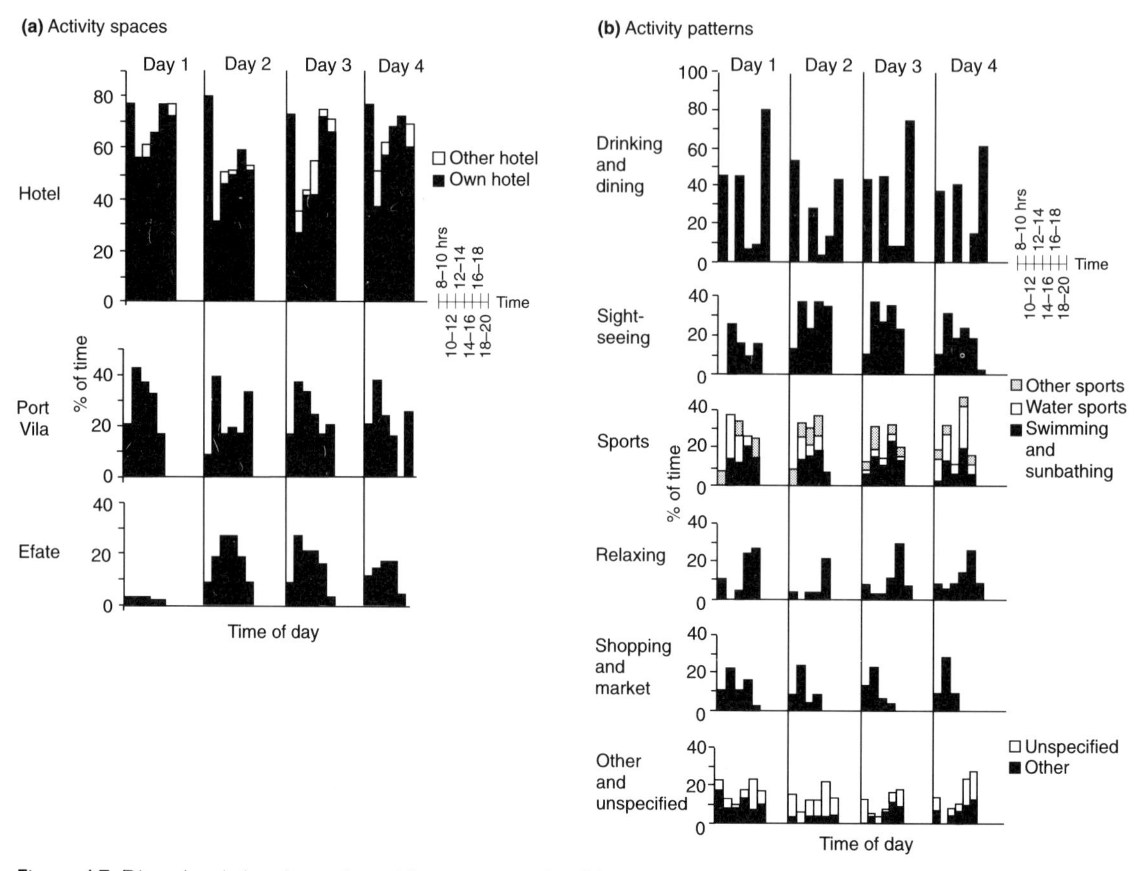

Figure 4.7 Diurnal variations in tourist activity patterns and activity spaces, Vanuatu, August 1985.
Source: Redrawn from Pearce (1988a).

much of their visit being concentrated on their accommodation. Gaviría *et al.* (1975) found that their respondents allocated only 26 per cent of their non-sleeping time to the beach, with 30 per cent being spent in and around their accommodation, 22 per cent on the streets (including sidewalk cafes), and 14 per cent in places of entertainment. Figure 4.7 shows that sightseeing (20 per cent) accounts for as much time among the Vanuatu respondents as sports (notably water sports, swimming, and sunbathing) with drinking and dining appearing as the single largest category (29 per cent of the time). Relaxing (9 per cent) and shopping and going to the market (8 per cent) are also important activities. Almost 60 per cent of the recorded time periods were spent in and around the tourists' own hotel, with a further 4 per cent being spent at other hotels. Port Vila, the islands' main centre, accounted for a further 22 per cent of the time, with 12 per cent being spent elsewhere on Efate. Similar patterns were revealed in Hottola's (1992) study of Finnish tourists in Eilat; some 44 per cent of their non-sleeping time occurred at their hotels, 20 per cent at the beach, 27 per cent in the city and 8 per cent beyond the resort.

Some day-to-day variations in locations and activities do occur in Fig. 4.7 (tourists are out and about sightseeing on Efate to a greater extent on Days 2 and 3), but regular diurnal rhythms also appear. Shopping in Vila and visiting the market are morning activities, with the late afternoon (4–6 p.m.) being given over to relaxation at one's hotel following a burst of sporting activity or sightseeing. The evening slot (6–8 p.m.) is devoted to dining, a significant share of which occurs away from the tourist's own hotel.

P. L. Pearce (1981: 278) reports an increase in self-initiated activities on Days 4 and 5 among his respondents on Australian resort islands where a programme of structured entertainment and outings is provided. The timing of this change, which involved 'more activities such as walks around the island, hunting for coral, flirting with new acquaintances, watching sunsets and reading' corresponded with the tourists' recovery from a significant dip in mood on Day 3.

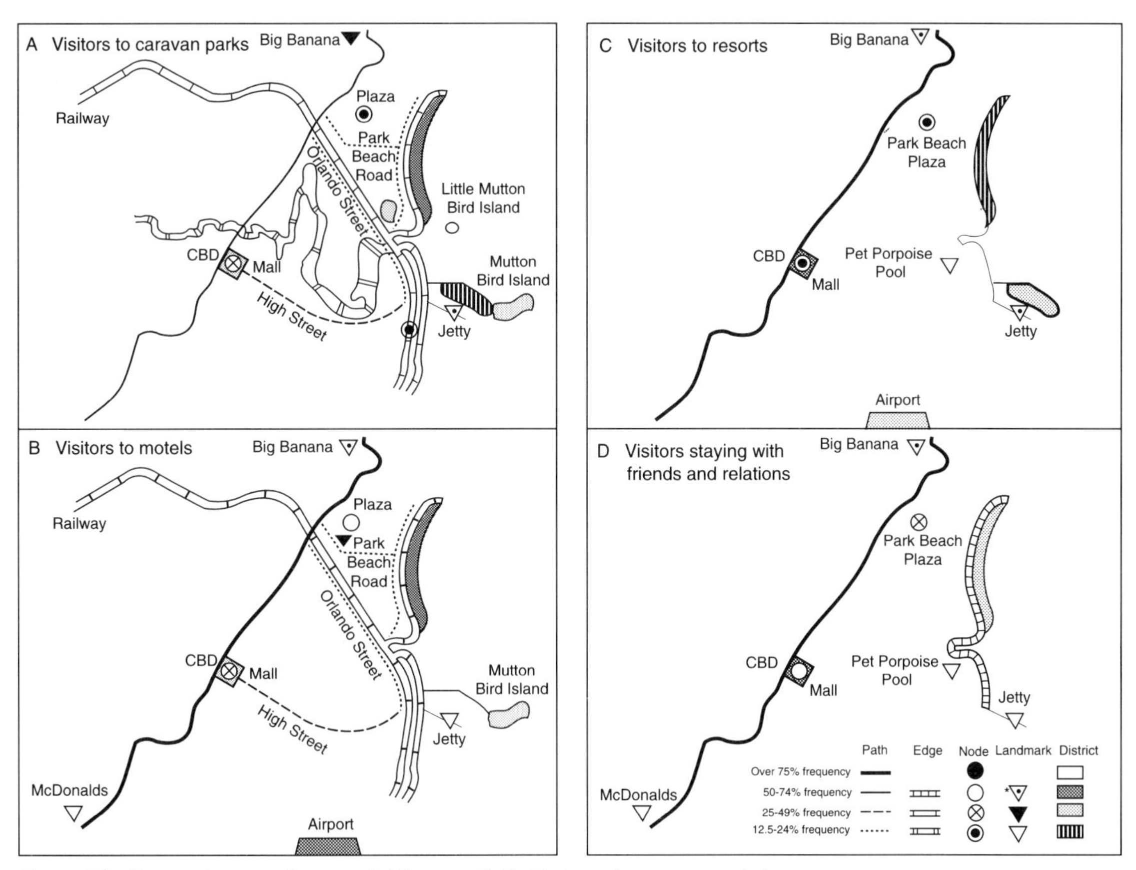

Figure 4.8 Composite mental maps of visitors to Coffs Harbour by accommodation type.
Source: Jenkins and Walmsley (1993).

M. Cooper (1980) emphasizes changes in tourists' spatial patterns over the first five days of their visit to Jersey. It is interesting to note that touring peaks on Day 2 (cf. sightseeing on Efate in Fig. 4.7). After determining a hierarchy of sites on the basis of facility provision and analysing the pattern of visits using a time-budget approach, Cooper found (1980: 365):

> The sequence of visits over the 5 days represents a progressive filtering down the hierarchy of sites. . . . The tourist's search strategy appears to use the hierarchy of site size as a surrogate for information. Tourists progressively reduce uncertainty at the expense of effort.

He likened this pattern to 'a wave of visits which spreads down the hierarchy, decreasing among increasing numbers of sites'. The decision to visit the largest (that is, the most important) sites first, also suggests that although tourist time is discretionary, it is nonetheless a valuable commodity and one that is indeed budgeted carefully.

Debbage (1991) also highlights the importance of temporal constraints in his large-scale time-budget study of the spatial behaviour of visitors to Paradise Island in the Bahamas. After devising a complex measure of overall trip mobility based on the number and range of trips to attractions visited on and outside Paradise Island, Debbage found length of stay to be a major factor in distinguishing between more mobile and more sedentary groups of visitors, noting (1991: 264):

> regardless of the personal characteristics of the individual, the temporal constraint plays a fundamental role in dictating what a person will (and will not) do while visiting the Bahamas.

This finding parallels some of the patterns of circuit tourism in New Zealand noted earlier, but as with them the nature of the relationship remains unclear. Are longer staying tourists more mobile because they have more time at their disposal or do they visit destinations for lengthier periods because in the first instance they have in mind to do and see more. Debbage also notes some association between distance travelled to the resort and length of stay, and between intra-destination mobility and type of accommodation used (c.f. Pearce and Johnston's (1986) results for travel within Tonga). However, socio-economic characteristics and degree of repeat visitation were not significant indicators of local spatial behaviour.

Cognitive studies exploring the spatial awareness of tourists in the same way that the activity spaces of day-tripping recreationists have been analysed would appear to be another fruitful avenue of future research, although obtaining a representative sample of visitors poses more problems than interviewing day-tripping householders (Elson 1976; Aldskogius 1977). In an innovative application of mental maps to tourism, Jenkins and Walmsley (1993) analysed freehand sketch maps of 145 visitors to Coffs Harbour, Australia (Fig. 4.8). Overall, they found that their respondents had a rather limited spatial knowledge of Coffs Harbour, being aware of the Park Beach area, and to a lesser extent the jetty and harbour, while other prominent features and attractions often went unnoticed. Knowledge of the area varied from group to group: visitors staying in caravan parks were the most aware, while the most rudimentary maps were drawn by those staying in resort hotels or visiting friends and relatives.

Conclusions

Research on intra-national travel patterns is relatively recent but is becoming increasingly recognized as an important complement to the more traditional country-to-country analyses. In particular there is a growing awareness of its potential for further segmenting markets and for clarifying the roles and functions which particular places have. As a result, more and more data sets and studies are becoming available in terms of intra-national variations in demand within both origins and destinations.

Diversity is a major characteristic of the data analysed and techniques used. In part this diversity stems from the lack of a common source of statistics such as the A/D card which provides the cornerstone for much of the work on international tourist flows at a more global level. The different scales at which intra-national travel occurs and can be analysed also generate a range of different data needs and types of techniques as does the rich array of problems which might be addressed. However, it is now possible to start to draw together the results of similar surveys, for example Table 4.1, and the application of particular techniques, such as time-budget studies and the use of the Trip Index, so as to begin to identify more general patterns, whether in terms of regional markets or how tourists spend their time in coastal resorts. More research along the lines reviewed here is now required in what is one of the more challenging but also rewarding fields of travel pattern analysis.

CHAPTER 5

Domestic Tourist Flows

The volume of domestic tourism world-wide has been estimated by the WTO to be about ten times greater than international tourism (WTO 1983a). Although some decline in domestic tourism has been noted in recent years (Chapter 2), this sector continues to be the most important, at least in volume terms, in many countries. Despite the magnitude of this activity, comparatively little systematic research has been undertaken into domestic flows. In part, this might be attributed to the less visible nature of much domestic tourism, which is often more informal and less structured than international tourism, and a consequent tendency by many government agencies, researchers and others to regard it as less significant. At the same time, research on domestic tourism has been plagued by data problems similar to those relating to intranational travel by foreign visitors. Quite simply, there are few regular records kept of domestic travel and none comparable to the international A/D card, inadequate as that may be for many purposes. Lack of a common, readily available data base means not only that domestic tourism is frequently ignored but also when studies of this sector are undertaken the results are often not directly comparable, limiting the identification of general patterns and trends. The emphasis in this chapter then is on the methodological issues which underlie the development of a more analytical approach to domestic tourist flows.

As with the analysis of international tourism, domestic flows can be examined at a range of scales from the national to the local. Given that both demand and supply are unlikely to be evenly distributed throughout a country, it may be useful to begin a study of domestic tourism with a nationwide perspective. A national study might identify the basic dimensions of domestic tourism and outline its broad geographic structure through an analysis of inter-regional flows. This information can then provide the background against which more detailed regional and local studies might be carried out (Owen and Duffield 1971; Rogers 1977). Techniques for analysing national patterns are illustrated first with particular reference to New Zealand. Regional analysis is exemplified by a variety of state-level studies from the USA. In each case the emphasis is as much on the procedures as the specific findings. Local-level studies are more common as data collection at this scale is frequently more manageable; consequently a wider range of examples are drawn on from the literature.

National Procedures and Patterns

Despite an early interest in national domestic travel studies (Piatier 1956), there are still few examples of comprehensive inter-regional studies where the analysis is based on a complete matrix of both origin and destination regions, however broadly defined they might be. The reason for this is not difficult to find. Unlike migration or commodity flow data (R. H. T. Smith 1970; P. White and Woods 1980), few appropriate and reliable sets of tourism statistics exist which might be used to construct such a matrix. Smale and Butler (1985) note that in Canada data were traditionally collected as either origin or destination data, with different regions being used for each, thus making comparisons of flows into or out of regions difficult and limiting analyses involving other variables such as population. Considerable investment in time, money and effort is required to undertake nationwide surveys which are sufficiently reliable that sample results might be weighted by some population factor to arrive at an estimate of total flows between regions. The scheduled 1982 National Travel Survey (NTS) in the USA, for example, was cancelled as a result of federal budget cuts. Methodological difficulties with analysing the results of the 1977 NTS are outlined by Perdue and Gustke (1985).

One example of a nationwide survey generating a useful origin-destination matrix is the New Zealand Domestic Travel Survey (DTS) which was established in its present form in 1983 (NZTD 1991). Carried out as part of a national omnibus survey, respondents are drawn from a stratified national sample over 47 weeks of the year, with information being sought on trips taken within New Zealand during the four weeks preceding the time of the survey. A trip is defined as 'a journey outside a person's home locality ... which involved a minimum of one night away from home'.

For example, in the year from April 1989, 12,354 people aged 15 years and over were interviewed, of whom 3443 people (28 per cent) had completed a trip in the four weeks prior to the interview. The raw data are then weighted to represent the population of New Zealand aged 15 years and over. Data from the DTS are presented at a regional scale (i.e. of the 22 United Council boundaries) but caution is urged in their use when they are based on small cell sizes. To reduce the effects of annual fluctuations in regional figures, which appear to be due to sample considerations rather than changes in visitor behaviour, an origin–destination matrix comprised of three-yearly averages for the period 1987/88, 1988/89, 1989/90 was derived and used by Pearce (1993), following a procedure adopted elsewhere (Secrétariat d'Etat au Tourisme 1977). Even doing this, the scope of the analysis may be limited with only indicative results being produced.

As in the analysis of other large sets of tourism data, it is useful to begin the analysis of a national matrix by identifying broad patterns before proceeding to a more detailed examination of flows. The identification of major generating areas and destinations is a first step here. Origin/destination data can then be brought together by calculating net flows into or out of each area. While all regions will both generate and receive visitors, those regions with a net positive flow, that is where the number of nights spent within the region exceeds those spent by that region's residents elsewhere, might be thought of primarily as destinations. Conversely, those regions experiencing a net negative flow might be considered essentially as sources or generating areas.

Figure 5.1 presents a composite picture of the demand for domestic travel in New Zealand, showing the total number of domestic bednights generated and spent in each region, together with a measure of the net balance, either positive or negative. Figure 5.1 shows that demand for domestic tourism in New Zealand is highly concentrated. The top five regions – Auckland, Wellington, Canterbury, Bay of Plenty and Waikato – accounted for 53 per cent of all bednights generated in the country. Auckland alone generated 8.5 million bednights or approximately one-fifth of the national total. Conversely, the bottom five regions combined generated only 6.5 per cent of all bednights, Wairarapa, the smallest, producing just under 1 per cent of the total.

What factors produce such patterns? Absolute population size is clearly a major determinant of demand. Ranking the 22 regions in terms of population size and total bednights gives a Spearman rank correlation coefficient of 0.9447. Rankings, however, do not tell the full story. Auckland, for example, had 27.3 per cent of the country's population in 1986 but is recorded as generating only 19.8 per cent of all bednights.

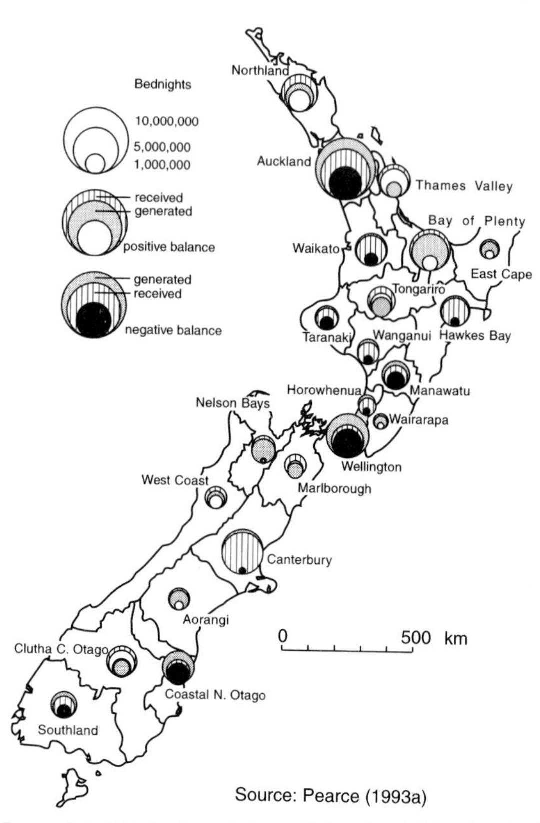

Figure 5.1 Distribution of domestic tourism in New Zealand (three year average 1987/88–1989/90).
Source: Pearce (1993a).

Although sampling procedures meant that data on the regional propensity to travel were not available from the DTS, the average amount of time that residents spent away from home could be calculated. The national average for the period considered was 18 nights, with considerable variation occurring from region to region (12 nights in Thames Valley to 30 for Tongariro), in some cases compounding population-based patterns, in others moderating them.

The pattern of bednights spent in each region is slightly more dispersed than for bednights generated. The top five regions accounted for 51.1 per cent of the total; the bottom five, 10 per cent. Despite its large net deficit, Auckland is still the leading region in terms of the absolute numbers of bednights spent there, but it is less dominant on this measure, receiving only 14 per cent of all bednights. Significant changes occur in the rankings, however, with Northland rising from ninth to third place and Thames Valley from second lowest to sixth.

When net flows are considered, eleven regions are shown to have positive balances, or be primarily desti-

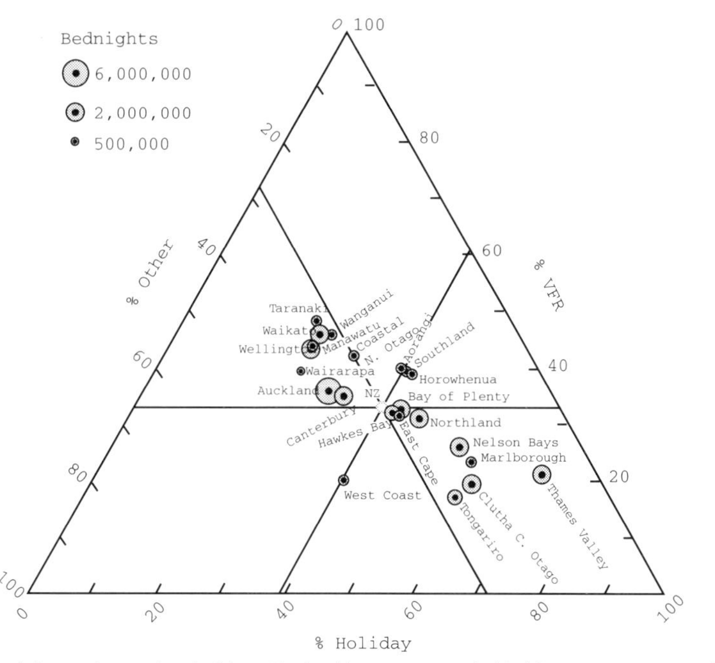

Figure 5.2 Distribution of domestic tourism in New Zealand by purpose of visit (three year average 1987/88–1989/90). *Source*: Pearce (1993a).

nations, and eleven to have negative balances or be predominantly origins or generating regions (Fig. 5.1). In some cases the balance is minimal, as with Nelson Bays (slightly positive) and Canterbury (slightly negative). Population size is again a major factor in the case of the origins, Auckland having the largest deficit. The small geographical extent of the Auckland and Wellington regions, compared to Canterbury, also appears to be significant. Resources play a role in those regions with a positive balance. A clear north to north-easterly coastal pattern is evident in the North Island (Northland to East Cape) with the central Tongariro region (having both Lake Taupo and the ski-fields of the national park) also being a major destination. Thames Valley has the largest positive flow, a consequence of its small population base, proximity to Auckland and its natural attractions. Clutha Central Otago, with Queenstown and the lakes district, has the largest positive balance within the South Island.

The characteristics of different regions and the factors influencing visits to them become more apparent when the composition of travel to each of the regions is analysed more closely. Holidays generated 39 per cent of all bednights, visiting friends and relatives (VFR) 33 per cent, and 'other' (e.g. business travel, sports trips, etc.), 27 per cent.

Population size affects demand for each of these purposes quite differently. When regional bednights received in each of these categories are ranked against population, the following Spearman rank correlation coefficients are obtained:

VFR	0.9277
Other	0.6736
Holiday	0.3055

The VFR traffic is essentially determined by where most friends and relatives are, that is in the more heavily populated regions. The holiday traffic is influenced by other factors, however, notably attractive features accessible to the major centres of population: Thames Valley, Bay of Plenty, Northland, Clutha Central Otago. 'Other' travel constitutes an intermediate case. Auckland remains the leading region in each of these categories, though its market share declines from 18 per cent for 'other', to 15.1 per cent for VFR and 10.5 per cent for holiday (only just edging out Thames Valley, 10 per cent).

Distinct clusters of regions can be identified when the percentages for each purpose of visit are plotted on a ternary graph (Fig. 5.2). Five major holiday regions stand out in terms of their overwhelming dependence on holiday visitors: Thames Valley, Clutha Central Otago, Tongariro, Nelson Bays and Marlborough. A second set of 'holiday' regions with slightly above average holiday values and lower than average percentages for the other two categories is made up of the eastern

North Island regions of Hawkes Bay, East Cape and Bay of Plenty, together with Northland. Three other regions – Aorangi, Southland and Horowhenua – receive about the same proportion of VFR and holiday bednights; each of these lies adjacent to a region containing a metropolitan centre. Another cluster of western North Island regions – Taranaki, Waikato, Wanganui, Manawatu and Wellington – is characterized by a low share of holiday bednights and an above average percentage of VFR and 'other' traffic. Taranaki is the most dependent on visits to friends and other relatives and attracts the lowest proportion of holiday visits (and is second lowest overall in absolute terms). Auckland and Canterbury have the most balanced pattern of demand of all regions, drawing in almost equal proportions on each of the three categories of visitor.

Comparison of net flows and purpose of visit (Figs 5.1 and 5.2) reveals some interesting and important relationships. Net positive flows are clearly associated with an above average share of holiday visitors. All bar one (Wairarapa) of the regions receiving more bednights than they generate have a higher than average share of 'holiday' bednights (though only marginally so in the case of the West Coast). Conversely, all regions except Hawkes Bay in the two holiday clusters in Fig. 5.2 have a net positive balance. Although generating regions often attract large numbers of visitors, especially in the more heavily populated regions, proportionately more bednights here originate from visits to friends and relatives and for other purposes. In other words, while regional population size may play a major role in determining overall flows, what appears to make the difference between a net loss or gain in total bednights is a region's ability to attract holiday visitors.

With these broad patterns of demand established, more detailed analysis of the actual flows can be undertaken based on the 22-by-22 origin-destination matrix. Figure 5.3a depicts the twenty largest flows in the national system, that is those flows accounting for at least 1 per cent of the national total. Figures 5.3b and 5.3c show the results of dominant flow analysis, that is analysis of the largest and second largest flows out of and into each of the country's twenty-two regions. Auckland's importance in the national system is further highlighted by Fig. 5.3a. Seven of the twenty flows greater than 1 per cent, including the four largest, originate in Auckland; three other such flows terminate there. Population size is again a major factor in determining the patterns shown in Fig. 5.3a though the extent to which location and available resources influences the concentration of travel would also appear to play a role, particularly in such cases as Northland and Southland. Intra-regional flows are also significant in Northland, Auckland, Bay of Plenty and Canterbury. Together, the twenty major flows account for only one-third of the national demand, indicating the diversity of domestic travel.

This diversity, however, is not without some underlying structure as closer examination of flows out of and into each region reveals (Figs 5.3b and 5.3c). In the North Island, destination flows exhibit a marked northwards bias to Auckland, the Bay of Plenty and Tongariro. In the South Island, destination flows are characterized by exchanges between neighbouring regions and intra-regional flows, with two major regional groupings in the northern and southern halves of the island being apparent. No inter-island interaction occurs at this level. When the two major flows to each region are considered, the importance of the major population centres is again highlighted (Fig. 5.3c). Auckland and Wellington have their respective spheres of influence in the North Island while Canterbury is clearly the major market in the South Island. Linkages are found between these three regions but overall inter-island exchanges are weak.

While the preceding analysis has identified major patterns in the actual flows, it is also useful to measure the magnitude of these against some predicted value in order to identify flows which might be stronger or weaker than expected (R. H. T. Smith 1970; J. N. H. Britton 1971). The Relative Acceptance (RA) Index, employed by A. V. Williams and Zelinsky (1970) in their study of international flows, might therefore be applied here, with the advantage that it is being used with reference to a total domestic tourism matrix. In this instance, the relative success of a region in attracting tourists from a generating region will be a function of all domestic tourist flows not just a selection of them.

The RA index is obtained by the formula

$$RA_{ij} = \frac{A_{ij} - E_{ij}}{E_{ij}}$$

where RA_{ij} is the relative acceptance from origin i to destination j
A_{ij} is the actual flow from origin i to destination j
E_{ij} is the expected flow from i and j.

Calculation of the expected flow (E_{ij}) is based on an assumption of origin destination independence or indifference which holds that the flow from i to j reflects the total flow to j. It is obtained by the formula

$$E_{ij} = \frac{n_j n_i}{n}$$

where n_i is the observed number of visitors from region i in the country as a whole
n_j is the observed number of visitors in region j
n is the total number of visitors in the country as a whole.

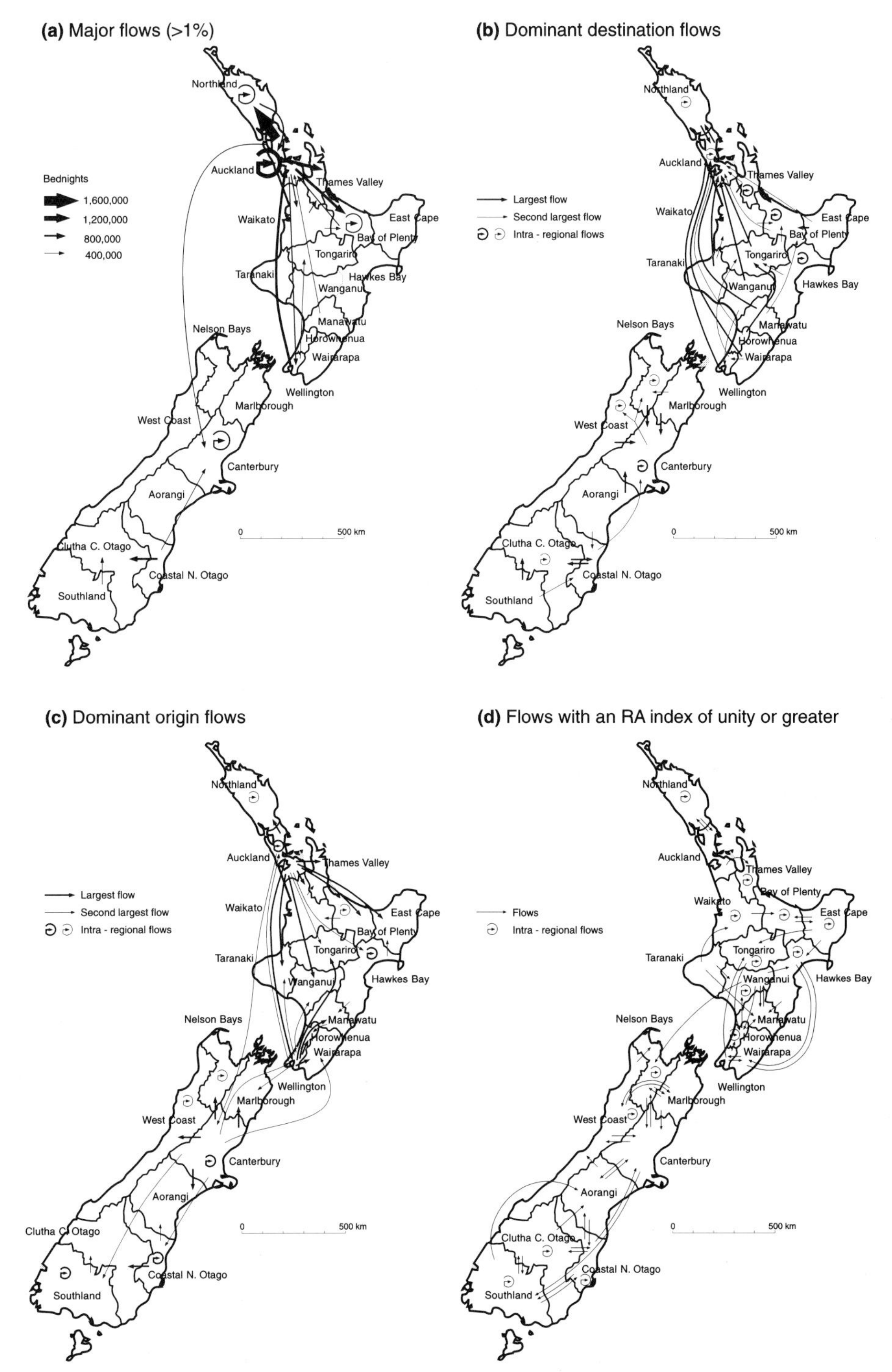

Figure 5.3 Domestic tourist travel flows in New Zealand (three year average 1987/88–1989/90).
Source: Pearce (1993a) and Domestic Travel Survey.

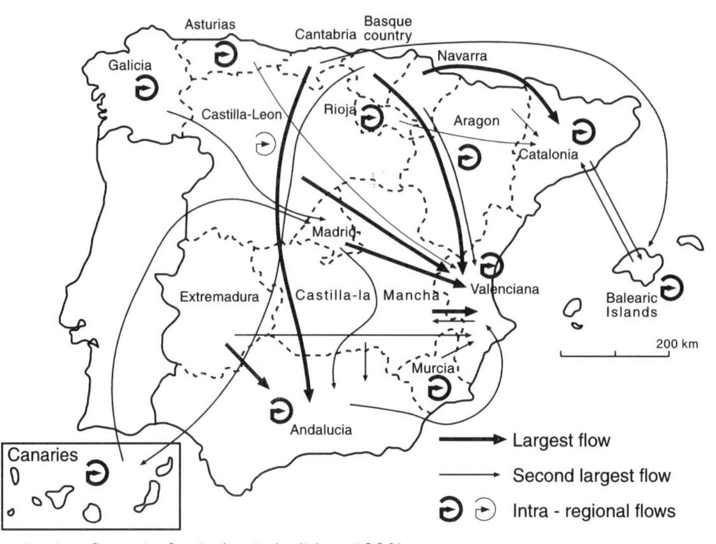

Figure 5.4 Dominant destination flows in Spain (main holiday, 1990).
Data source: Dirección General de Politica Turistíca (1991).

Thus in the case of New Zealand, as Auckland received 14 per cent of all domestic bednights, the model predicts that it should receive 14 per cent of the bednights from each of New Zealand's 22 regions.

This technique has the advantage of eliminating the effects of absolute size and enables the identification of unusually high or low flows. The RA index has a range from −1 to plus infinity, with positive values indicating a greater than expected flow and negative values the reverse. Given this range, determination of threshold values for salient positive and negative flows is, however, arbitrary.

Figure 5.3d depicts the flows of those regions in New Zealand having an RA index of unity or greater. Compared with Figs 5.3 and 5.4, the patterns shown in Fig. 5.5 are much less influenced by the major metropolitan regions. Distance, rather than population size, appears to be the significant factor in determining larger than expected flows. First, 15 of the 22 regions record strong intra-regional flows, the exceptions being Canterbury, Auckland, Wellington and some adjacent North Island regions. Second, distinct clusters of flows between neighbouring regions are evident in the upper and lower halves of the South Island and in the southern and central North Island.

Unfortunately the DTS has not been repeated in New Zealand since 1989/90. The New Zealand Tourism Board, with its increased international focus, has discontinued the survey and in its place has encouraged the regional tourist organizations to set up their own tourism monitors. While these may generate some useful regional information, the regional network is not yet complete and it is unlikely that the data collected will be sufficiently comprehensive and compatible to enable further national analyses of the type which were made possible by the DTS.

Similar procedures to those outlined here might be applied to other national data sets in order to build up a more comprehensive picture of interregional flows and ascertain any generalities in their occurrence. Figure 5.4, for example, depicts the dominant destination flows in Spain for major holidays (of four nights or more) in 1990 (Dirección General de la Politica Turística 1991). More than 60 per cent of Spanish holidays that year were spent on the coast, reflected in the popularity of the autonomous regions of Valencia (15.8 per cent), Catalonia (15.5 per cent) and Andalucia (14.1 per cent). In terms of dominant flows, the more central Valenciana region is the most prominent of these, accounting for five first order and five second order flows. Nationwide, many of the flows are again between neighbouring regions. The regionalized nature of Spanish domestic tourism is further emphasized by the large amount of intra-regional travel. For nine out of the seventeen regions the intra-regional flow is the largest, in some cases exceeding 50 per cent – the Canary Islands (69 per cent), Murcia (61.5 per cent), Andalucia (61.4 per cent) and Galicia (53.5 per cent). Location, tourist resources and regional wealth appear to be factors here. Landlocked regions such as Madrid, the Extremadura and Castilla-La Mancha, on the other hand, are more notable for their inter-regional movements.

Earlier analysis of French and Italian domestic travel

matrices (Secrétariat d'Etat au Tourisme 1977; Instituto Centrale di Statistica 1977; Pearce 1987c) also highlight the importance of inter-regional travel in Europe. Some 40 per cent of vacation days in Italy were spent in the home region in 1975, compared with 20 per cent of all summer holidays in France. This compares with only 13 per cent of domestic bednights in New Zealand. Other similarities and differences with the New Zealand patterns analysed here can also be identified. The pattern of origin and destination regions is more fragmented in New Zealand (Fig. 5.1) than France, where a neat division along the Le Havre–Marseille axis occurs, and Italy, which had only three regions with net negative balances. The prominence of three major markets indicated by the origin flows (Fig. 5.3c), places New Zealand behind Italy, which has two (Lombardy and Lazio), and France, which is dominated by the primacy of Paris. In each case the destination flows are more dispersed than those from the origins but strong regional biases were apparent in France. The structure of the urban hierarchy in each country and the distribution of resources appear to be factors contributing to the differences in the patterns observed but account also needs to be taken of differences in the units of measurement.

Canadian data on inter-provincial travel further highlight the importance of the measure used. For Canada as a whole inter-provincial travel in 1988 (third quarter) accounted for only 20 per cent of all domestic trips made but nearly half (49 per cent) of all expenditure on domestic travel (Statistics Canada 1987). Inter-provincial travellers spent proportionately more on transportation and in absolute terms four times as much on accommodation through a greater propensity to stay in commercial accommodation when travelling to other provinces.

Variations in inter-regional flows by purpose of travel were also reported by Perdue and Gustke (1985: 179) in their analysis of the 1977 American National Travel Survey (NTS) data:

> Although the South and Great Lakes regions were consistently the dominant attractor and generator, respectively, of inter-regional travel in the United States, differences are evident in the spatial patterns of travel by trip purpose. The theoretical implications of that conclusion are of importance to the development and specification of trip distribution models. The common practice of agglomerating leisure travel for several purposes into a single model may not be appropriate.

Noting the richness of the NTS data, Perdue and Gustke lament the cancellation of the 1982 survey and the lack of good longitudinal data. Without such surveys and analyses of the type outlined above, it will be difficult to track the evolution of domestic tourist travel and identify common patterns in its occurrence.

Regional Studies

Regional-level studies of domestic travel patterns can provide a useful complement to the national analyses outlined above. First, the regional focus may enable a more detailed analysis of inter-regional flows to specific regions. Second, studies at this level may enable other issues to be examined which will not be revealed at the larger scale of analysis, for example sub-regional variations in demand. In some cases national surveys can be disaggregated further, with the more detailed regional analysis being performed on subsets of the national data. This has the advantages of setting the regional data more readily in the national context and of comparing findings between regions. Frequently, however, regional studies are undertaken independently, particularly in the absence of national surveys. This may increase the detail and relevance of the results for the regions in question but limit the generalization of regional findings and thus the identification of common patterns at this scale. These issues are explored here by drawing on research undertaken by or for state tourist organizations in the United States. Emphasis is given to the range of approaches used and the different problems addressed.

Many of the principal aspects of regional domestic tourism studies are well illustrated by the annual Florida Visitor Study (Coggins 1990). Data for the study are drawn from an exit survey of approximately 9000 out-of-state visitors, approximately half with air visitors in airport departure lounges and half on highways near the Florida border. Quotas at each interview site are statistically weighted according to traffic volume at these airports and highways. Among other results, the Florida Visitor Study enables an analysis of state origins of air and auto visitors and their destinations within Florida at the county level.

Figure 5.5 shows a dominant north–south movement of air and auto visitors, particularly from the heavily populated states of the Mid-Atlantic and East North Central regions, states which also have considerably lower winter temperatures. In the case of auto visitors, these regions are complemented by a more immediately accessible hinterland, with Georgia (15 per cent) being the leading origin state compared with New York (15.4 per cent) for air visitors. Distance is clearly a factor affecting the patterns of the two sets of markets but it should be noted that Ontario (included in the Florida study as a 'domestic' origin) is ranked in the top ten markets by auto but not by air.

Tourism within Florida is strongly concentrated, with the ten leading destinations accounting for 94.3

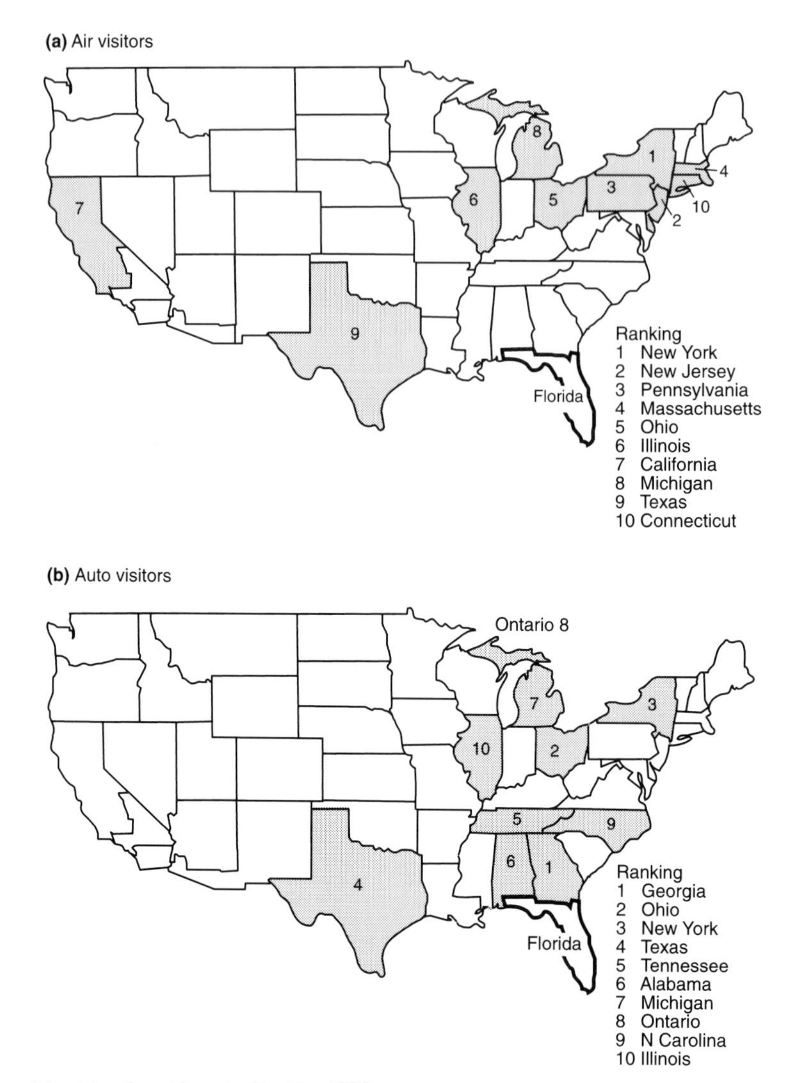

Figure 5.5 Top ten origin states for visitors to Florida, 1990.
Source: After Coggins (1990).

per cent of air visitors and 91.7 per cent of auto visitors (Fig. 5.6). In each case the Orange-Osceola-Walt Disney World region with its major theme parks was ranked number one, receiving approximately 22 per cent of visitors. However, there is a greater concentration of air visitors in the south of the state and auto visitors in the northern and north western countries of Duval, St Johns, Escambia and Bay, a further reflection of the different levels of accessibility by mode of transport.

Other American state studies elaborate on different aspects of their markets and within-state destinations of travellers. Analyses of pleasure travel to Texas have explored in detail the influence of proximity and market size (Texas Department of Commerce 1990, 1993; Hodge 1991). In terms of total out-of-state travellers to Texas, proximity is a significant factor but more populous distant ADIs (Areas of Dominant Influence, based on television broadcast markets), such as Los Angeles and New York are also very important (Fig. 5.7a). With 1.1 million person trips in 1992, Los Angeles ranked second to Oklahoma (1.2 million) while New York came sixth with 677,000.

For marketing purposes, it is argued, propensity to travel is a more significant factor than sheer market size (Texas Department of Commerce 1990: 8):

The inference is that it is easier to influence ad-

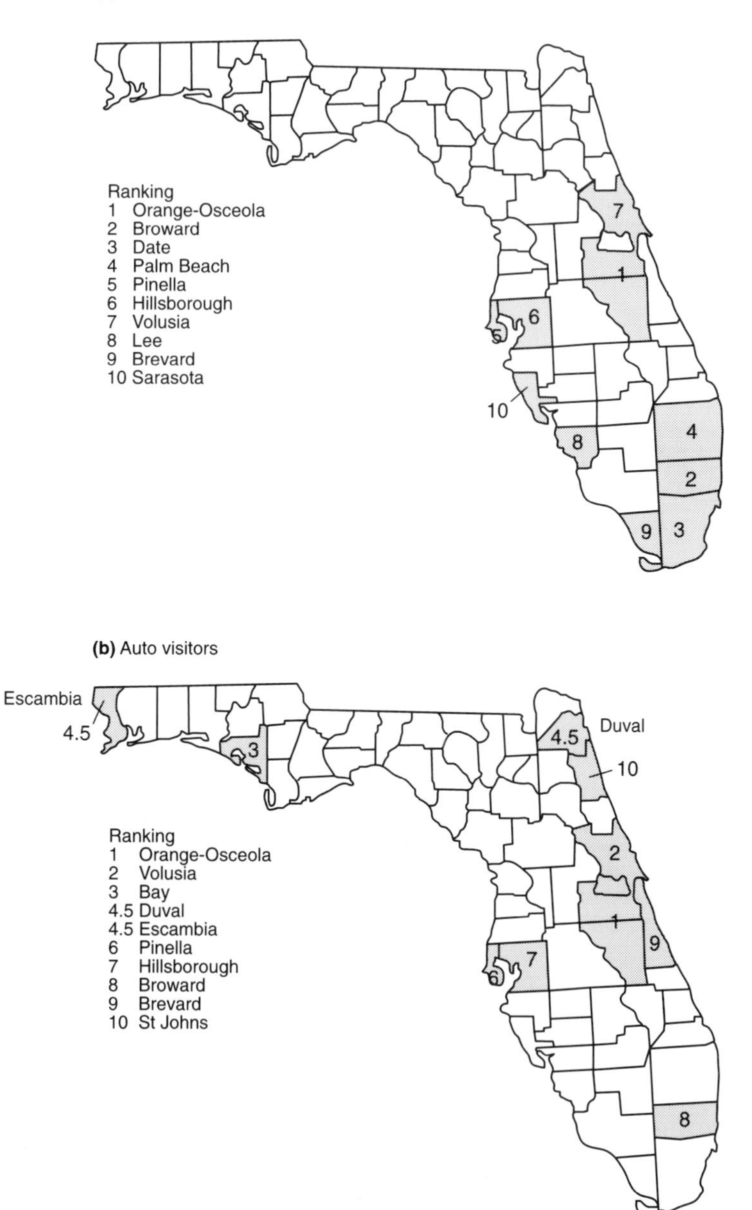

Figure 5.6 Top ten destinations within Florida, 1990.
Source: After Coggins (1990).

ditional people to visit Texas from a market that already has a high propensity to travel to this state than it is to increase the number of visitors from an area with a lower propensity – even though the high population of that area may be the source of a significant number of travellers to our state.

Propensity was determined by adjusting markets for population size by the use of a travel index. Travel volume for each ADI was divided by the area population and compared to the mean. With the mean set at 100, ADIs with indices greater than 100 are above average markets for pleasure travellers to Texas. When

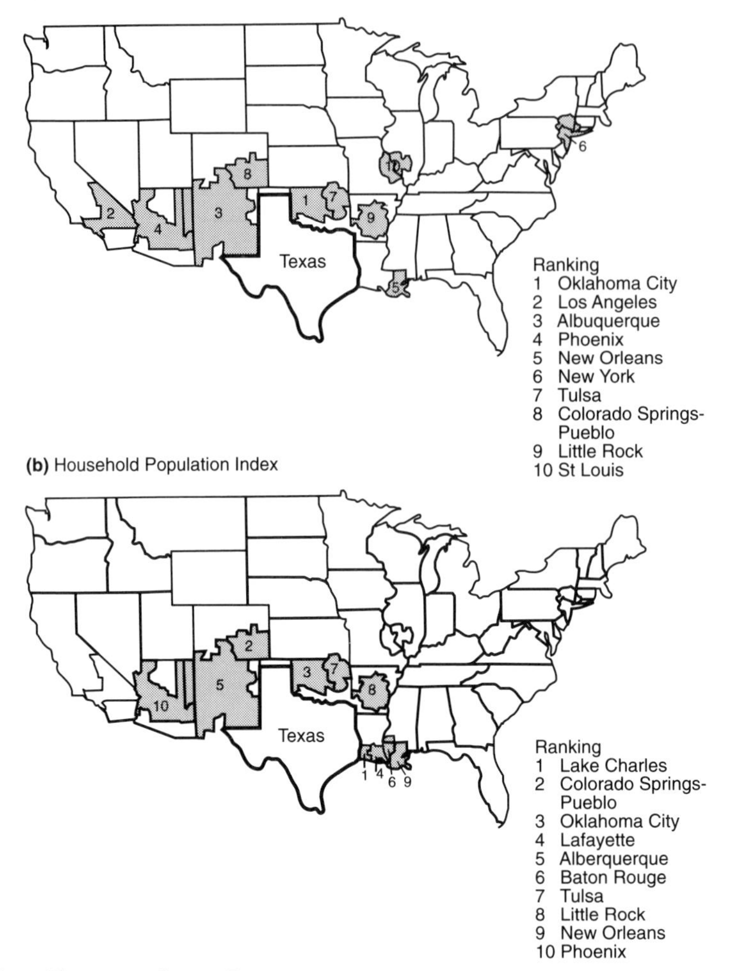

Figure 5.7 Origins of non-Texan travellers to Texas for leisure, 1992.
Source: After Texas Department of Commerce (1993).

population is taken into account in this way, propensity to travel to Texas is much more strongly related to propinquity (Fig. 5.7b), with the highest indices being registered in Lake Charles (363), Colorado Springs-Pueblo (235) and Lafayette (200). In contrast, Los Angeles has an index of 22, indicating 'the number of visitors to Texas are being driven by population size rather than a propensity to travel to Texas' (Texas Department of Commerce 1990: 8). Half of all the business and leisure travel within the state by Texans was generated by the two ADIs of Dallas–Fort Worth and Houston.

Elsewhere in the United States the domestic pleasure travel market has been disaggregated by purpose of visit. Detailed information on visitors to Colorado drawn from a syndicated nationwide survey (Longwoods International 1990) reveal marked variations in the geographic origins of different groups of visitors. Visitors to Colorado on a touring vacation ('a sightseeing trip through a region that can encompass any of the major tourism attractions that destination has to offer, and which match the specific interests of the travel party') or a ski vacation were drawn more widely from throughout the USA than those on an outdoors vacation or country resort holiday. The latter two markets were much more regionalized, with almost half of the outdoors vacationers, for instance, coming from Colorado itself. Denver alone provided more than one-third of the total. These spatial patterns reflect the drawing power of the state's scenery and natural resources, its abundant quality ski resorts and more general travel behaviour associated with vacation activities such as touring and outdoor recreation (Fig. 1.2). As the Colorado findings are drawn from a larger national survey, the profiles of state visitors and Colorado's position with regard to these segments can also be established in more detail *vis-à-vis* the overall United States' market.

More detailed data on within-state travel tends to be drawn most commonly from state-specific studies, for instance the Alaska Visitor Statistics Program (McDowell Group 1989). All visitors were included in this programme but 84 per cent of these in 1989 were Americans. Detailed travel pattern (and other) data collected by a series of visitor surveys enable a comprehensive picture to be built up of travel within Alaska, not only regions and places visited within the state, but also combinations of regions visited. In 1989, the more accessible and more populated South Central and South East regions were the most popular destinations, being visited respectively by 69 and 60 per cent of all visitors and 64 and 72 per cent of vacation/pleasure visitors, followed by Interior/Northern (35 and 39 per cent), Denali/McKinley (34 and 40 per cent) and South West (8 and 6 per cent). Anchorage (62 per cent of all visitors) and Juneau (48 per cent) were the two most frequently visited communities and the Portage Glacier (46 per cent of all visitors) and the Inside Passage (44 per cent) the most popular attractions.

Table 5.1 Alaska: regional visitor overlap, summer 1989

Also visited	Regions visited				
	South East (%)	South Central (%)	Interior/ Northern (%)	South West (%)	Denali/ McKinley (%)
South East	100	46	61	2	36
South Central	53	100	92	88	98
Interior/Northern	36	47	100	24	79
South West	15	10	6	100	3
Denali/McKinley	63	48	76	12	100

Source: McDowell Group (1989).

Analysis of travel itineraries indicates that a significant amount of 'regional overlap' occurs in Alaska, with most visitors to the state visiting more than one region on their trip (Table 5.1). About a half of visitors to South Central, for example, also visit the South East, Interior/Northern and Denali/McKinley. Visitors to this latter region travel widely throughout the state except to the South West region. South West visitors are much less mobile. Factors accounting for these variations in overlap include accessibility, the patterns of cruise ship tours and the nature of the attractions. One of the marketing implications of this overlap identified by the study is that 'Many businesses can reach potential customers by marketing in another region which may be visited by these customers' (McDowell Group 1989: 89).

Other insights into sub-regional patterns of travel are provided by earlier surveys of domestic visitors to the province of Quebec (Service de la Recherche Socio-Economique 1977) and to and between Australian states (Morgan Research Centre 1982). These two surveys indicate that large urban areas are the main focus for inter-regional travellers while areas bordering on to other regions may attract proportionately more visitors from the regions in question. In the first case, the strength of the flows appears to reflect the special character of the attractions of the big city, particularly where business and VFR travellers are concerned. Urban areas appear to be able to draw visitors from outside their region to a much greater extent than other domestic destinations which compete less successfully with intra-regional destinations offering similar features. Border areas, however, might be considered an extension of particular regional hinterlands. Propinquity is the major factor here for in the examples given the state or provincial boundaries are much less significant in practical terms than international ones. In Australia the major urban areas (Sydney, Melbourne, Brisbane, etc.) were also the prime focus for intra-regional (intra-state) travellers. As for the residents of the major urban areas, their visits tended to be rather dispersed but the most favoured destinations were in adjoining coastal areas.

Local Studies

Studies of tourist flows have been much more numerous at the local level, not only because of the greater range of possibilities there but also because data collection is generally much more feasible at this scale. Local studies normally focus either on flows emanating from a particular city or on the origins of visitors to a particular destination, the latter being of especial interest for marketing purposes by local tourist organizations (Market Opinion Research 1990). The flow data are usually collected as part of a larger visitor survey although a range of other data sources have been used, including mailing lists (Wolfe 1951), car licence plates (Sarramea 1979) and social columns (A. W. Carlson 1978). In the one case, details are sought on destinations visited so as to define the vacational hinterland of particular towns or cities (Cribier 1969; Greer and Wall 1979), in the second, place of origin or home town are used to delimit the market areas of particular resorts or destinations (Wolfe 1951; Deasy and Griess 1966). As the 1991 visitor profile for Reno-Sparks (Nevada), for example, indicated, 45.5 per cent of the resort's traffic originated in California, further attention was directed at analysing county-level variations of demand within that state (RRC Associates 1991).

Most studies of vacational hinterlands and market areas concentrate on aggregate origin/destination flows, either for the population at large or for specific groups of travellers, including: second home owners (Coppock 1977); campers (Gibson and Reeves 1972); honeymooners (A. W. Carlson 1978); and pilgrims (Chadefaud 1981). Other researchers have compared the hinterlands of different groups defined in terms of accommodation used, purpose of visit and activity patterns or intra-urban residential areas. A few have also considered changes in vacational hinterlands over time.

Vacational hinterlands or market areas are most frequently portrayed by dot distributions or some system of zoning. The general pattern to emerge is that of a decline in the intensity of flows with distance from the origin or destination. Various factors may affect this distance decay pattern, notably the distribution of recreational opportunities in the case of source areas or the distribution of population in the case of destinations. As noted in Chapter 1, Greer and Wall (1979) argued that theoretically visitation should peak some distance from the source area due to the interaction of supply and demand before declining (Fig. 1.3). They graphically illustrated their point with empirical evidence from Toronto. This pattern is also supported by the work of Fesenmaier and Lieber (1987) who specifically analysed the effects of the distribution of facilities in Oklahoma upon residents' recreational expenditure. They not only found a peaked spatial structure of supply but also showed (p. 33) that this influenced expenditure patterns: 'Expenditure generally increased as a function of distance travelled – but only to a point'. At greater distances a shift to multi-destination trips appeared to occur reducing the amount spent at or near each individual recreation facility. A. W. Carlson (1978) showed that aside from distance and financial resources, scenery, particularly that found in national parks, played a significant role in concentrating honeymooners from rural North Dakota and Wisconsin in certain localities.

Different quantitative techniques have been used by various writers to analyse further the role of distance and other factors. Gibson and Reeves (1972), for example, used regression analysis but their efforts were hampered by the size of their samples. A much more widely tested technique incorporating the friction of distance factor in relation to tourist flows is the gravity model. The basic gravity model proposed by Zipf (1946) expressed the interaction between an origin and a destination as a function of the population of each and the distance between them. Tourist flows, however, are not necessarily reciprocal, as Wolfe (1970) points out in the case of second homes where the traffic outward from an urban area to a cottage resort is frequently not complemented by a reverse flow to the city. Consequently, when applied to tourist flows, the gravity model has been modified in various ways. Such modifications include the incorporation of variables measuring the attractiveness of the destinations (the number of ski-lifts, the ratio of water to land area), intervening opportunities or the measurement of distance in terms of time or travel cost. Modified gravity models have been applied with some success to specific cases such as tourist travel to Las Vegas (Malamud 1973), the demand for second homes (Bell 1977) and travel to ski-fields (McAllister and Klett 1976). For a fuller review of methodological issues and for other applications of gravity models, see Archer and Shea (1973) and S. L. J. Smith (1983a, 1989).

In an innovative study, Murphy and Keller (1990) sought to examine the distance decay effect and the presence of hierarchical visitation patterns (Chapter 1) not in terms of trips from an origin but rather of travel away from gateways into a regional destination, namely Vancouver Island, British Columbia. The two major cities and gateways, Victoria and Nanaimo, were reported to receive the most visits, with visitation steadily declining as distance from these two points increased, thereby confirming a hierarchical distance decay function.

One of the earliest comparative studies was that by Cribier (1969) who delimited the vacational hinterlands of the major metropolitan centres in France then found, within each hinterland, spatial variations in the type of accommodation used. Non-commercial accommodation (second homes, visits to friends and relations) predominated close to home, while those travelling further afield tended to stay in hotels. Greer and Wall (1979) established the hinterlands of Toronto residents day tripping, owning cottages, camping and staying in commercial resorts. Different distance decay curves were found for each category but there was also a considerable degree of overlap in the zones frequented by each group.

In their survey of Canterbury (New Zealand) holidaymakers, Johnston, Pearce and Cant (1976) found that social visiting and sightseeing were the major objectives of those travelling to the North Island. Those pursuing water-based activities or seeking general relaxation usually found suitable holiday destinations closer to home. However, in a study of Dayton area households, Etzel and Woodside (1982) found no significant differences in terms of purpose of trip between distant and near-home travellers (defined as travellers within Ohio or to destinations in adjacent states). The two groups did differ though on a number of demographic and psychographic variables. Different activity preferences between local and extra-regional visitors have also been found in destination studies. E. L. Jackson and Schinkel (1981) examined the activity preferences of local and 'tourist' campers in the Yellowknife region of Canada and concluded (p. 361): 'Local residents expressed a stronger preference than tourists for activities such as resting and relaxing, swimming, boating and canoeing, while tourists more frequently preferred activities such as sightseeing, hiking, photography, visiting and meeting people'.

Wolfe (1951: 29) argued that it was necessary to disaggregate visitors from major urban areas, suggesting, in the case of Toronto, that 'the zonation of the city finds a rough extension in the zonation of summer resorts'. He later showed a significant directional bias among Toronto cottage-owners from different parts of the city, with residents tending to buy cottages in Ontario in locations that avoided cross-town travel (Wolfe 1966, cited by S. L. J. Smith 1983a). Despite the early recognition by Wolfe of the need to investigate variations in travel from different parts of the city, little serious work has subsequently been undertaken in this area. Washer (1977) showed a close association between the location of permanent residence in Christchurch and ownership of second homes on Banks Peninsula. Mercer (1971) examined the role of urban mental maps in producing directional bias in beach usage from different residential areas in Melbourne. The main handicap in developing this line of research with regard to vacation travel appears to be the need for an enlarged sample so that meaningful comparisons could be made among different parts of the city.

As yet, comparatively little is known about the extent to which vacational hinterlands and market areas expand or contract over time. Rajotte (1975) has shown, historically, how developments in transport technology have extended the vacational hinterland of Quebec, but inadequate time-series data and the absence of baseline studies mean little detailed information is available on how local tourist flows change through time. Chadefaud (1981) has used a series of different sources (diocesan records, hotel registers, surveys) to map long-term changes in the catchment area of French pilgrims to Lourdes. In 1880, a quarter of a century after the appearance of the Virgin Mary before Bernadette Soubirous, the majority of French pilgrims came from South West France, from a zone corresponding to that serviced by the regional railway. By 1925, with the development of a national rail network, pilgrims were arriving from all over the country, but principally from the North, the West and the southern Massif Central, that is, from rural and conservative France. Data from 1972 show an increasing dominance of the faithful from the North and West, with a noticeable absence of Parisians. There had also been a marked increase in foreign pilgrims who now constituted 60 per cent of all visitors to Lourdes.

A certain amount of interest, particularly in North America, was shown in changes induced by the increase in fuel prices and the shortage of petrol supplies experienced at particular times in the 1970s (Corsi and Harvey 1979; Kamp, Crompton and Hensarling 1979; P. W. Williams, Burke and Dalton 1979). These indicate some decrease in the distance travelled on holiday trips and even cancellation of vacation travel plans, but these effects appear short-lived and the overall picture is rather inconclusive. The surveys in question have relatively small samples, often low rates of return, and other methodological limitations (McCool 1980).

Flow patterns may also change throughout the year as, for example, the emphasis switches from the coast in summer to the mountains in winter. But the picture may be more complicated than this. Sarramea (1978) concluded from his studies of visitors to Frejus and Saint Raphael on the French Riviera that north–south movements predominate in summer, with visitors from neighbouring departments being more significant at other times of the year.

Travel Routes and Patterns

Despite the comparatively early models of Mariot and Campbell (Figs 1.1 and 1.2), few domestic tourism studies have examined itineraries or routes, with most surveys concentrating solely on aggregate flow patterns. However, the methodologies used in the analysis of intra-national travel by foreign tourists can in most cases be applied to studies of domestic travel patterns. Indeed, several of the examples cited in Chapter 4 embraced domestic as well as international tourists, for example the study of tourist movements in Scotland (Carter 1971) and the analyses using the Trip Index (Pearce and Elliott 1983).

Information from national travel surveys suggests that most domestic tourism does not involve multiple-stop touring but consists mainly of stay-put holidays at a single destination. Circuit tourism (as opposed to stays on the coast, in the mountains, the countryside or towns) accounted for only 5.5 per cent of French summer holidays in 1992, compared with 23 per cent of French holidays abroad (Observatoire National du Tourisme 1993). A similarly low proportion (4.2 per cent) of main holidays was also spent touring (*recorriendo varios lugares*) by the Spanish in 1990 (Dirección General de Politica Turística 1991). 'Touring' was reported to make up 8.8 per cent of the vacation mix for the United States as a whole in 1989. Some variation occurred from state to state; New Jersey reflected the national average while one trip in five to Colorado was a touring vacation (Longwoods International 1990,

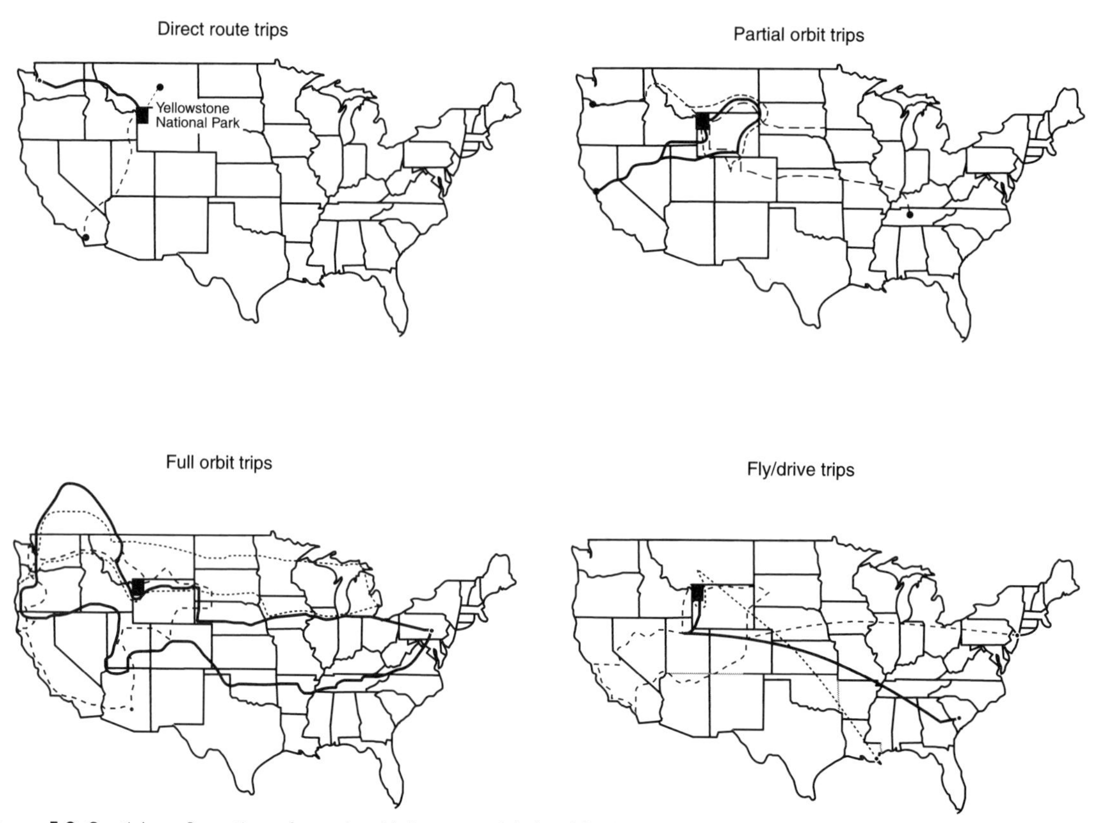

Figure 5.8 Spatial configuration of travel to Yellowstone National Park.
Source: Redrawn from Mings and McHugh (1992).

1991). Other state surveys indicate that three-quarters of air visitors to Arizona but only 38.5 per cent of highway tourists were limiting their visit to that state (Arizona Hospitality Research and Resource Center 1992) and that only 27 per cent of Las Vegas visitors were visiting other areas of Nevada or the Grand Canyon (Market Opinion Research 1990).

In a study of the vacation hinterlands of residents of Dunedin, New Zealand, Goldsmith and Forrest (1982) attempted to establish more directly the proportion of holidays corresponding to the different types of tourist and recreational travel postulated in Campbell's model (Fig. 1.2). Their survey showed that 61 per cent of the respondents had holidays involving trips to a recreational-vacational complex, 9 per cent of these being in the metropolitan hinterland and 52 per cent in the regional hinterland. A further 17 per cent involved tours of different regions, for example of Central Otago or of the North and South Islands. Goldsmith and Forrest also suggest a fourth category of travel not inherent in Campbell's model, the 'there and back' trip to a single urban centre. Such trips, however, correspond directly with the flows between major metropolitan areas depicted in Lundgren's model (Fig. 1.6). In the case of Dunedin holidaymakers, they constituted 21 per cent of all trips and were primarily associated with visiting friends and relations.

Goldsmith and Forrest do not comment on the nature of the vacation circuit nor on trips within the recreational-vacational regional complex. Although not conceived in such terms, patterns from a study of visitors to Albury–Wodonga in Australia (P.A. Management Consultants 1975) lend support to Campbell's concepts. Two distinct patterns emerged, one of linear highway oriented travel in the case of tourists originating in the Newcastle–Sydney–Wollongong area, the other of a regional vacation complex centred on Albury–Wodonga for visitors from Melbourne–Geelong. The Albury–Wodonga area appeared to be regarded more as a destination by the Victorians than by visitors from Sydney and surrounding areas who were to a greater extent on a round trip.

Mings and McHugh (1992) examined in detail the spatial configuration of travel to Yellowstone National Park based on 570 visitor route sketch maps and identified four principal trip types (Fig. 5.8). Although not couched in such terms, their categories correspond closely to some of the concepts expressed in Figs 1.1 and 1.2. Less than 10 per cent of respondents travelled to and from the park on a 'direct route'. Slightly more (11.2 per cent) followed a 'partial orbit', taking a direct route to the edge of the scenic mountain west where their route becomes circuitous as they 'orbit' a position of the Rocky Mountain region that includes examples of spectacular scenery before returning to their original route and proceeding directly back home. Almost half the respondents, the 'full orbiters', take a completely circular route, visiting 'a very wide array of scenic attractions in their grand tour of the American West, returning home via a completely different route to that taken on the outward journey'. Fly/drive visitors, who contributed almost a quarter of the total, exhibited similar travel behaviour to the 'partial orbiters', except they flew the direct legs to and from an air terminal city before renting a vehicle for the last circuitous stage. Mings and McHugh (1992: 46) also identify profile differences among the four groups (fly/drive visitors, for example, have higher incomes and are better educated; more 'direct route' travellers were making special trips to see the fire damage) and examine other trip attributes:

> Figures on duration of trip and other parks visited provide evidence that the overwhelming majority of Yellowstone visitors are, in essence, orbiting western landmarks. The Full Orbit represents the classic grand tour whereby travellers maximize their exposure to American landscapes.

Finally, the authors (p. 46) raise some pertinent questions regarding the apparent anomalous behaviour of the Yellowstone respondents:

> most Yellowstone visitors are involved in highly circuitous trips which, apparently, are not designed to minimize the time, distance or cost of their travel. . . . Why and how are Yellowstone visitors successfully overcoming these time-honoured obstacles to travel? . . .
>
> While most travel data and origin/destination studies assume some circuitous travel movement, few assume the degree outlined in this study. Perhaps we should re-examine our definitions of trip destination and route circuitry to determine the extent to which Yellowstone visitors are behaving differently from other travellers.

Conclusions

A variety of techniques have been applied to the analysis of domestic travel patterns at a range of scales, from inter-regional analyses at a national level to the localized case study. Studies at different scales can usefully complement one another. National studies, for example, may provide the broader context to assess the typicality or particularities of travel to or from a given place. The national survey data cited earlier, for instance, suggest that Yellowstone visitors are markedly more mobile than many other domestic travellers. At the same time, detailed local studies such as that on Yellowstone may go some way to answering questions of why particular travel behaviour occurs, even if much remains to be done in this domain.

While data and other constraints have limited the number of studies undertaken and the comparability of their findings, some general patterns are nevertheless starting to emerge. Domestic flows, like those of international tourists, are neither uniform nor random. Order is to be found at all levels, and finding that order is essential for understanding domestic tourism and for planning and managing it. The results of the studies at all scales generally confirm the existence of a distance decay function in domestic tourist travel and the importance of intra-regional flows. The effect of distance becomes particularly evident in those studies that have controlled for population size (e.g. Fig. 5.7). Studies that have disaggregated domestic travel by trip purpose or some other variable clearly indicate that the domestic market can no longer be treated as a homogenous entity. Rather, domestic travellers, like their international counterparts, can be segmented in various ways, with different segments exhibiting their own spatial preferences.

CHAPTER 6

Measuring Spatial Variations in Tourism

The preceding three chapters have focused on the interaction between origins and destinations through an analysis of tourist flows at various scales. The nature, intensity and direction of these flows can provide important insights into the spatial structure of tourism. Further understanding can be gained by examining the location and distribution of various tourist-related phenomena, for example hotels or ski-lifts. Analyses of such features commonly form part of general geographical studies of tourism in particular countries or regions. In other instances such analysis can constitute part of the national or regional scene setting for the examination and appreciation of particular places or specific problems. Pearce's (1991) discussion of the challenges facing the development of Split, for example, sets tourism in that city in the context of the Dalmatian coast, identifying differences between it and nearby resort areas. Appreciating the broader context in which tourist organizations operate was also shown to be crucial to understanding their goals, functions, structures and propensity for inter-organizational activities (Pearce 1992b). Planning and development, impact assessment and marketing are other areas which would benefit from further analysis of spatial variations in tourism (Pearce 1989b).

While some writers are adopting an increasingly critical approach to analysing spatial variations in tourism, in many cases there is little more than a cursory mention of why the measures and techniques used have been employed and what it is they are actually measuring. After addressing some general methodological issues this chapter reviews and evaluates a range of factors and approaches which have been used to define and describe spatial variations in tourism. Four major sets of factors are identified – those relating to accommodation, to attractions, to the economic impact of tourism and to the tourists themselves. Composite studies incorporating a range of different variables and emphasizing different types of tourism are then discussed. Not included in this section, however, are regional resource evaluation techniques used in planning and development as these have been reviewed elsewhere (Pearce 1989b). Subsequent chapters contain comparative studies of tourism at different scales as an attempt is made to identify common trends and patterns using a selection of these factors and techniques and so develop a more general understanding of the spatial structure of tourism.

General Issues

Whatever the specific application of a particular analysis of the spatial structure of tourism, three broad issues normally arise:

1. What elements or factors are to be analysed?
2. What spatial units are to be employed?
3. What analytical and presentation techniques are to be employed?

Given the multifaceted nature of tourism, a wide range of potential factors might be employed. As discussed in subsequent sections, the choice of those to be used may depend on the purpose of the study or, very commonly, be limited by available information and data. In either case the validity and reliability of the data should be carefully examined and limitations explicitly noted.

Some international studies have been undertaken, particularly of small island states (Rajotte 1977; Weaver 1992a), but most analyses of the spatial structure of tourism involve sub-national areal units. Care must be taken with the selection of the scale of analysis as, for example, regional units may conceal local area variations. Consideration must also be given to the use of formal or functional regions (G. W. S. Robinson 1953; Price 1981). Klaric (1992) suggests that there are three basic models for the spatial organization of tourism:

1. Tourist regions corresponding to administrative units (regions, provinces, etc.).
2. Tourist areas extracted and defined as special areas from the rest, i.e. the non-tourist parts of the country.
3. Tourist regions which may cover the whole of the country, but whose boundaries do not correspond with the actual administrative organization at the regional level.

While a case can be made for the definition of functional tourism regions for planning and development purposes (Town and Country Planning Directorate 1984; S. L. J. Smith 1989), in practice tourist areas or regions generally correspond to existing administrative regions or some amalgam of these. One reason for this is that most destination tourist organizations are funded substantially from the public sector and so reflect territories of local and regional government (Pearce 1992b). Where the concern is more with functional than formal regions, boundaries and divisions may be needed which differ from one activity to another. Sprincova (1968: 209) demonstrates this point well in her early study of the structure of tourism in the mountainous region of Hruby Jesenik in former Czechoslovakia:

> The investigation of the area has shown that the central ridge of the Hruby Jesenik ... had been a serious obstacle for a number of territorial and industrial relations and especially for the formation of a unified network of communications, as well as for the formation of a uniting centre. ... In the economic and administrative development of the examined territory it has always formed a natural dividing line and therefore a very constant one. Only in tourism it is different, here the ridge is no more a division but its very axis, its most attractive component part. All the other territorial units derive their function in tourism according to their relations to it.

Aggregating the smallest available administrative units into tourism regions rather than using larger administrative regions may be one way of identifying and retaining spatial variations in this sector which might otherwise be lost at the larger scale and of developing more coherent functional regions as in the case of the Meuse River in the Netherlands (Pearce 1992b).

One advantage of the use of administrative units is that tourism data compiled in this way might be related more readily to other statistics collected on the same areal basis, for example, population or employment figures. The ability to combine tourism with other data becomes particularly important when relative rather than absolute measures are used. Increasingly, especially in impact studies, researchers are acknowledging that absolute figures, be they of the number of visitors, hotel rooms or tourism jobs, provide only one measure of the importance of tourism. In particular, absolute figures may give a useful indication of where tourism is important in terms of a country or region as a whole, but they do not necessarily reflect the significance of tourism within particular areas. More and more attention is therefore being paid to relative measures of tourism intensity, variously defined by relating tourism indicators to other attributes of the areas in question through the derivation of some form of index. The choice of relative or absolute measures will depend in large part on the problem being addressed but in general a combination of both may be the most useful. Given the length of time over which tourism data are now available in many places, consideration might also be given to the dynamic nature of tourism, setting contemporary patterns against those of some earlier date to determine the extent of change and continuity (Pearce and Grimmeau 1985; López Palomeque 1988; Hinch 1990).

Accommodation

The distribution of accommodation is the most widely used measure of spatial variations in the tourist industry. A major reason for this is that accommodation is one of the more visible and tangible manifestations of tourism. Accommodation is generally readily inventoried, with statistics on commercial accommodation being compiled for fiscal and other reasons in many countries while data on second homes are often collected as part of a household census, although these are usually not without their flaws (Barke 1991). If necessary, the researcher can also go out and physically count the number of accommodation units in a given area, a task more readily accomplished than the enumeration of visitors to the same area, though one which is not without its problems.

Accommodation statistics tend to be used mainly to indicate spatial variations in the importance of tourism or to identify regions of different types of tourist activity; some studies incorporate elements of both. As a measure of the importance of the tourist industry, the use of accommodation is a logical one, a stay away from home being one of the defining characteristics of tourism. Except where a great deal of day tripping occurs from a tourist base, the distribution of accommodation does provide a fair measure of actual demand. As a measure of impact, it is also reasonable, given that expenditure on accommodation usually accounts for one-third to a half of a tourist's budget and that provision of accommodation plant contributes a similar or even greater proportion of tourism investment in most cases (Pearce 1989b).

Published accommodation statistics should not be used uncritically and note must be taken of the definitions employed and of the manner in which they are compiled. Conflicting figures from different sources are not uncommon, even for comparatively small regions or resorts (Barbier 1978; Vanhove 1980). Comprehensive studies are often complicated by variations in the availability and reliability of data from one sector to another. In particular, statistics on the less formal sector (camping, caravanning, second homes, etc.) tend to be less common and less reliable than those for the more commercial and structured hotel industry. Visits with friends and relations are also not recorded in accommodation inventories, yet surveys show a significant proportion of visitors may stay in private homes, particularly in urban areas (Hinch 1990). Where statistics are not available for all types of accommodation, some assessment should be made, for example through a visitor survey, of the significance of the missing sectors as these may vary widely from place to place and from market to market. Hotels, for example, may well account for virtually all the accommodation in some international resorts but play only a minor role in many areas of domestic tourism.

Accommodation statistics are usually expressed either in the number of units in each sector (e.g. the number of hotels or second homes) or some measure of capacity, commonly beds or rooms. Given that hotels in particular can be of widely varying size and that the mix of sizes may vary from one locality to another, significant differences may occur in the distribution of the number of hotels and the distribution of hotel rooms or beds. In most cases, the capacity figures will be more meaningful. Where comparisons are being made, capacity figures are also to be preferred, with the capacity of each sector being expressed in some standard unit, usually beds. Hotel and motel capacity is often so expressed in official statistics but for other sectors, such as holiday homes, camping grounds or caravan sites, a survey may be necessary or some more arbitrary decision may have to be made. Not all beds, however, are equal. Some will be filled virtually every night of the year while others will be used in certain seasons only. In general, occupancy rates will be higher in the hotels and lower in camping grounds, second homes and other less formal sectors. Occupancy rates also vary from one area to another, with metropolitan centres usually experiencing more consistent demand throughout the year than coastal or alpine resorts. Where the length of the season is known, adjustments might be made in terms of annual bed or room capacity (Keogh 1984) or occupancy rates shown (Gilg 1988). Actual use figures expressed in monthly or annual bednights are compiled in some countries, for example Belgium and Spain, and provide a valuable measure of demand, particularly where they are classified by origin (see pp. 90–1), and of impact.

Accommodation inventories may also provide data on the quality available. Information on the type and quality of the accommodation may suggest who the users might be, but inventories alone do not provide any direct information in this respect. It should be borne in mind, particularly in interpreting distribution patterns, that not all accommodation is directed at or used by tourists, in the sense of those travelling primarily for pleasure purposes. Indeed, in their study of the United States lodging industry, van Doren and Gustke (1982) report that the major market for hotels there in 1978 consisted of 43 per cent of business travellers, 32 per cent tourists and 17 per cent conference participants.

Spatial variations in the importance of tourism are commonly depicted by or inferred from maps portraying the distribution of one or more types of accommodation. The distribution of absolute numbers of units is usually portrayed by dot distribution maps (e.g. Vuoristo 1969; Bielckus 1977) or some form of shading by administrative area (Boyer 1972; Gutiérrez Ronco 1977). Capacity is usually depicted by proportional circles, squares or bar graphs with these being used at the scale of resorts, towns or administrative units (Bevilacqua 1984; King 1988; Schnell 1988; Berriane 1990; Lockhart and Ashton 1991). In other instances, the percentage of total national bed capacity has been mapped (Poncet 1976).

Various relative measures of tourist accommodation have been derived. That which has gained most acceptance is Defert's tourist function index (Defert 1967; Baretje and Defert 1972). The tourist function T(f) of an area is taken by Defert as a measure of 'tourist activity or intensity' as reflected in the juxtaposition of two populations – the visitors and the visited. It is derived by comparing the number of beds (N) available to tourists in the area with the resident population (P) of that area according to the formula:

$$T(f) = \frac{N \times 100}{P}$$

The theoretical limits of the index are T(f) = 0 where no tourist accommodation exists and T(f) = infinity where there is no resident population. A value of 100 indicates that the number of tourists would equal the number of local residents, assuming all beds available to tourists were being used.

Boyer (1972) proposes a six-fold classification of French communes based on their T(f) values:

- T(f) > 500 = recent 'hypertouristic' resort
- T(f) 100–500 = large tourist resort
- T(f) 40–100 = predominantly tourist commune
- T(f) 10–40 = communes with an important but not predominant tourist activity
- T(f) 4–10 = little tourist activity or tourist function 'submerged' in other urban functions
- T(f) <4 = practically no tourist activity.

While initially applied to towns and cities, the T(f) index has also been calculated for regions such as Colorado (Thompson 1971), Normandy (Clary 1977), São Paulo (Langenbuch 1977), New Brunswick (Keogh 1984), Ontario (Hinch 1990) and coastal South Carolina (Potts and Uysal 1992) and at a national level in New Zealand and Germany (Pearce 1979a, 1992b). Values tend to decline with the increase in the scale of the spatial unit as the role of tourism is diluted by other activities.

As a measure of tourist intensity, the T(f) index can be a useful discriminating variable. In the case of New Zealand, there was little correlation in the rankings of the 22 regions in terms of their absolute accommodation capacity and their T(f) value ($r_s = 0.27$). The region which ranked last in terms of capacity, Fiordland, recorded the highest T(f) value (103). Conversely, the major metropolitan regions which had a large number of beds tended to have low T(f) values. Differences in the population base of the regions account for much of the variation, but the T(f) index does indicate that whereas tourism is only one activity among many supporting a large population in places like Auckland and Wellington, it is a mainstay of the economy in Fiordland.

Similar conclusions were drawn by Hinch (1990: 252):

> although the two northern regions may be the least important regions in terms of their total inventory of bed spaces in Ontario, the bed spaces which they possess are relatively more significant to their regional economies than is the case of bed spaces in other regions. The opposite can be concluded in relation to Central Ontario which is the most important region in terms of its contribution to the total provincial bed space supply.

As a relative measure of tourist intensity or 'tourist ascendancy' (Ryan 1965), the T(f) index provides a useful complement to the traditional absolute capacity figures. Given that population statistics are normally readily available, the additional calculation should pose few problems. At the same time, the limitations of the capacity statistics should be borne in mind, particularly if there is much variation in the mix of accommodation types and the occupancy rates of these (Keogh 1984).

A number of variations on this approach have been offered. In his study of tourism in Quebec, Lundgren (1966) used a ratio exactly the opposite of Defert to obtain an index expressed simply in terms of the number of local residents per hotel unit. Places with a low numerical ratio were seen to have 'a more tourist-geared local economy' than those with a higher numerical ratio. In mapping spatial variations of tourism in the province, Lundgren effectively combined capacity and intensity by depicting the capacity of each place with proportional circles and then shading these according to the ratio of population to hotel units (see also Mirloup 1974). Bevilacqua (1984) simply presents bar graphs of guest beds and hotel nights for ski resorts in the European Alps. Plettner (1979) advocates the use of bednights rather than capacity and proposes an index of the 'intensity of tourism' which is the quotient of the number of bednights and the local population. A similar approach has been adopted by Jülg (1984) in Austria and Kulinat (1986) in Spain. In the study of second homes, maps are often prepared depicting the ratio of second homes to permanent residences (Breuer 1987; Barke 1991).

Measures of the density of tourist activity have also been suggested. Girard (1968) has mapped the number of hotel beds on Guernsey by half-mile squares and, in the Pacific, Rajotte (1977) has depicted the number of tourist days recorded at each destination in relation to their area as well as to the size of the local population. Marsden (1969) derived 'holiday homescape indices' for the Gold Coast of Australia based on the number, proportion and density of unoccupied dwellings as measured in unoccupied dwellings per square mile. In his study of second homes in former Czechoslovakia, Gardavsky (1977) suggests it is not total area but rather the potential recreation area available in each district which should be taken into account. This latter consists of (p. 67) 'areas of woodland and water, in addition to orchards, parks, meadows and pastures which contribute to the attractiveness of the environment for recreation'. He then proposes a modification to Defert's basic formula such that

$$Rp_1 = \frac{L \times 100}{0} \cdot \frac{1}{Rp}$$

where Rp_1 = recreation index
L = the number of second homes multiplied by four, representing the average number of occupants
0 = the number of permanent residents
Rp = the potential recreation area of the area surveyed.

Studies which employ variations in accommodation as a means of deriving regions or of identifying areas specializing in particular types of tourism are usually of two kinds. Some focus on a single accommodation sector and examine variables such as quality or comfort. Others are concerned with a variety of accommodation types but usually only one variable, normally capacity. In many of the single accommodation studies, particularly those dealing with second homes, the main objective may be more with analysing the distribution of that particular sector than with determining the spatial structure of tourism in general. Coppock (1977: 6), for example, notes that the main factors controlling the distribution of second homes appear to be 'the distance from major centres of population; the quality or character of the landscape; the presence of sea, rivers or lakes; the presence of other recreational resources; the availability of land; the climates of the importing and exporting region'.

In a number of countries, hotels, camping grounds and other types of accommodation are officially classified by grade or quality and this provides a first basis for categorization. In most cases, capacity is first represented by proportional circles, these then being subdivided and shaded according to the proportion of beds, rooms, hotels and so on in each category (Stang 1979; Chadefaud 1981; Fig. 9.6). Mirloup (1974) prefers a single comfort index for each locality based on the formula

$$\mathrm{RCI} = \frac{(0.25R_1 + 0.5R_2 + 0.75R_3 + R_4)}{TC} \times 100$$

where RCI = Room Comfort Index
R_1 to R_4 = the number of rooms in 1- to 4-star hotels
TC = total capacity of the resort or town

In using this index in his study of hotels in the Loire Valley, Mirloup distinguishes between low-quality rural hotels and high-quality hotels in towns near the chateaux or main roads, with the major cities falling between the two. He also proposes a diversification index based on the distribution of rooms in each category compared to the national average. Dewailly (1978) suggests a comfort index for second homes based on ten criteria such as the proportion of homes in each area being serviced with water and electricity or having a garage. He raises the question of weighting the variables but sees no consistent way of doing so. Use of this index showed that not all the second homes in the Nord-Pas de Calais region of France were equal, with those on the coast generally being of superior quality to those located in inland areas.

Types of accommodation have also been classified and mapped according to size distribution (Chadefaud 1981), type of location (Mankour 1980), national or foreign ownership (Berriane 1978), length of stay and seasonality (Jülg 1984) and change over time (López Palomeque 1988; Barke 1991).

Where a series of variables or indices is used, these may be presented in sequence, an attempt may be made to synthesize them through some form of classification, or both approaches may be used. Mirloup (1974) presents a series of maps of the Loire Valley depicting variations in the different indices he uses and then derives a more general classification based on threshold values for three measures – capacity, diversification and tourist function index. Graf (1984) derives a classification of tourism types in the Bavarian Alps based on average length of stay, tourism intensity and bed capacity, which he depicts via shading of proportional circles.

Studies which examine a range of accommodation types tend to do so either by examining the distribution of each in turn, or by combining different types of accommodation at different scales and analysing spatial variations in the composition of these. Typical of the first approach is A. S. Carlson's (1938) early study of the recreation industry of New Hampshire in which he presents a series of maps showing the number of summer residences and the capacity of hotels, lodgings, cabins and juvenile camps by county. Different spatial patterns emerge for each type of accommodation. Hotels, for example, tended to be located at or nearby railway junctions while cabins were found mainly in the lake and mountain districts. Carlson's emphasis is on the factors accounting for the distribution of each type, but some more general geographical statements are also made as a result of 'grouping various accommodations by recreation regions'.

Vuoristo (1969) too, in his examination of the geography of tourism in Finland, looks in turn at the distribution of different accommodation types and proposes a division of the country into six tourist regions based on a 'cartographic synthesis'. In concluding, he notes (p. 45):

> The differences between the regions are due to many factors, among which are a) differences with respect to the distribution of population and points of interest to tourists, b) differences with respect to the natural conditions needed to promote the tourist trade, c) the concentration of the points of entry into the country for foreign travellers in certain areas, d) the attraction of Lapland to foreigners, e) the formation of travel routes running north-south, f) differences in the distribution of various types of lodging and camping establishments, g) differences in the length of tourist seasons, etc. In a number of these cases, there can be observed as a background factor the elongated shape of Finland in a north–south direction.

Researchers who combine accommodation types usually do so by presenting the amount and proportion of accommodation, expressed in a standard unit of measurement such as beds, available in each resort or administrative unit by means of pie charts or bar graphs (Kulinat 1986; King 1988). This can be an effective means of visually presenting differences in capacity and composition from one area to another (Fig. 7.6). Resorts or regions can usually be identified which are composed of one or two dominant types of accommodation or even a balanced mix, but in most cases little attempt at synthesis is made, for example by combining like regions.

A notable exception is found in the ambitious, comprehensive and more analytical French project involving a detailed examination of 318 coastal resorts (Cribier and Kych 1977). Cribier and Kych adopted an original approach, that of estimating the actual number of holidaymakers in all types of accommodation on the coast during the peak summer period (the first two weeks of August 1974). This was achieved through mobilizing a substantial number of researchers who combined a comprehensive inventory of accommodation plant with field surveys. The latter, among other factors, ascertained, for a sample of the resorts, the number of holidaymakers staying in permanent residences and second homes. The proportions obtained were then used to calculate figures for the coast as a whole. The results of this exercise are instructive even if, as the authors stress, they are strictly applicable only to the survey period and do not reflect year-round usage or overall impact (Table 6.1). Nearly 60 per cent of all coastal holidaymakers were being accommodated in permanent or second homes: hotels accounted for only six per cent of the total.

Table 6.1 Distribution by accommodation of tourists during the peak summer season in 318 French coastal resorts, 1974

	% of tourists
Second homes	41
Camping and caravanning	27
Permanent residences	17
Hotels	6
Children and youth holiday camps	5.1
Other 'collective' accommodation[a]	3.4

Note: [a] Includes clubs, family holiday camps (VVF), tent villages etc.
Source: After Cribier and Kych (1977).

Seven categories of resorts were then identified on the basis of their accommodation profiles by means of factor analysis. Three categories were distinguished on the basis of the predominance of a single type of accommodation (second homes, permanent residences, camping); three according to a mix of two major types (camping and permanent residences; camping and second homes; hotels and second homes); while the seventh included those resorts whose profile approximated the average for the coast as a whole. Similar categories might also have been derived by a visual scanning of the profiles themselves but, given the large number of observations, the use of factor analysis should have led to a more consistent classification of the 318 individual resorts. Even so, the authors recognize that the classification of some resorts is rather marginal; some even appear surprising, such as the inclusion of La Grande Motte in the category typified by the 'omnipresence of camping'.

Each of these seven categories was found to correspond closely with a recognizable type of resort. A mix of mainly hotels and second homes, for example, was characteristic of well-established 'name' resorts such as Deauville, Biarritz, St Tropez and many of the resorts of the Côte d'Azur. Camping, on the other hand, was largely a feature of small rural communes.

Further investigation showed a marked degree of regional homogeneity, with large and distinct stretches of the coast being characterized by resorts of a single or contiguous categories. Camping, for example, predominated in southern Brittany and along much of the Atlantic coast. Concentrations of second homes were found along sectors of the Channel coast, around the mouth of the Loire and in parts of Languedoc. It should be remembered here that the categories of resorts are based on the relative composition of the accommodation types, not on their absolute distribution. In addition to these regions characterized by specific types of accommodation categories and associated resorts, there were also several regions where different types of resorts are found. In Languedoc, small blocks of communes characterized by camping alternated with others where second homes predominated; in the Var, hotels were interspersed with camping and, in the Cotentin, second homes and permanent residences were intermixed.

Cribier and Kych (1977: 124) concluded from this regional homogeneity that 'In France the vacation space in destination areas is not primarily a local resort level space, but a regional space. Each tourist region is perceived by the visitors in terms of its historic, climatic and cultural unity.' The emergence of this homogeneity is later attributed to the following factors (p. 158):

> The development of a particular mode of accommodation in each region depends on the size of the permanent population in each resort, on the volume of visitors . . . on physical factors (the size of the communes), but more importantly on the history of tourism, that is on the history of the relationships between supply and demand. We can thus distinguish between the established tourist resorts with important hotel industries, dependent sectors (in the orbit of a metropolitan area) characterized by second homes, sectors where camping has recently developed and those numerous but dispersed resorts where tourism is in the process of developing without a predominant accommodation type emerging or where communes are trying various paths of tourist development.

Cribier and Kych's methodology has much to commend it, combining as it does the regional insights typical of geography in France with a more systematic and analytical approach found less frequently among the French (Barbier and Pearce 1984). The authors also raise the question of whether the particular regions they define would have emerged if criteria other than accommodation types had been employed.

Attractions

A number of researchers, particularly Europeans, have examined various aspects of the spatial structure of the attractions sector. The discussion in most cases is limited to specific forms of tourism or types of recreational facilities and there is often little attempt to incorporate these analyses into a broader examination of tourism as a whole. Usually it is the more distinctive forms of tourism or types of facilities which have been treated in this way. Thus studies have been undertaken or maps prepared of the distribution of thermal resorts in Europe (Defert 1960), France (Ginier 1974; Clary 1993), and Germany (Dewailly 1990), ski resorts in France (Ginier 1974), Germany and Austria (Graf 1984), recreational boating facilities in France (Ginier 1974) and the Netherlands (Pinder 1988) and golf courses in Spain (Priestley 1987). Yamamura (1982) and Ishii (1982) deal with a range of recreational resources and facilities in Japan including hot springs, ski-fields and national parks. In general there has been little attempt to synthesize the different sorts of attractions although rather general maps of tourism resources have been prepared, for example by King (1988) in Italy and Valenzuela (1988) in Spain.

In his monograph on tourism in the Netherlands, Ashworth (1976) discusses the distribution of natural and human-made resources and includes maps of nature reserves and solitude, landscape evaluation, and two of 'attractions' in Western Europe. The latter are clearly biased towards urban historical and cultural attractions visited primarily by 'sightseers', with one, based on features cited in the Michelin guide books, offering a French perspective and the other depicting attractions recognized by the German tourist geographer, Ritter. Ashworth recognizes difficulties in the identification and assessment of attractions and notes that recorded patterns of use do not always correspond with the apparent distribution of attractions.

Maps of attractions in specific urban areas also tend to be biased towards historic monuments, museums, art galleries and other cultural features (Burnet and Valeix 1967; Burtenshaw, Bateman and Ashworth 1981). Some of the problems which arise in analysing the spatial structure of attractions can be seen in Burtenshaw, Bateman and Ashworth's map of major tourist attractions in Paris. They admit that it is not exhaustive but depict it as being based on 'those selected by the visitors themselves'. By this they appear to mean monuments and museums where an entrance fee is paid and a record is kept by the Secrétariats d'Etat au Tourisme and à la Culture. However, of the listed attractions, only three – the Eiffel Tower, the Louvre and the Musée de l'Armée – recorded more than half a million visits and of these some proportion would have been from Parisians. Even if the figure of 13 million visitors to Paris in 1977, cited by Chenery (1979), might also be viewed as imprecise and as containing a large proportion of business people and other travellers, the indications are that the majority of tourists are not visiting the 'major attractions'. The Eiffel Tower can, of course, be viewed without a visit being recorded if the tourist does not actually go up it, but clearly there are many attractions other than these official ones which are drawing visitors to *la ville lumière* (see Chapter 9).

The factors discussed here do provide a means of establishing the spatial structure of particular forms of tourism and have so far primarily been used in this way. Their value as indicators of the overall importance of tourism or of variations in its structure appears much more limited, however, unless more rigorous efforts are made to synthesize the many different types of attractions and to derive a composite measure of tourist attractiveness.

Economic Impact

With the growing interest in the economic impact of tourism, attention has been directed at establishing more directly spatial variations in the economic importance of the tourist industry. Such research can provide a very useful bridge between national studies of the contribution of tourism to a country's economy, with an emphasis on overseas exchange earnings, and detailed local impact studies, commonly based on expenditure surveys and calculations of tourist multipliers. The reliance in the literature on indirect measures of impact, notably accommodation statistics, reflects, however, the paucity of economic data on tourism at a sub-national scale. The studies of variations in the economic impact of tourism which have been undertaken are rarely comprehensive but relate to specific sectors of the tourism industry or to particular segments of the market.

One of the more detailed studies is that by S. G. Britton (1980b) who obtained, through a large-scale survey of tourist businesses, annual turnover figures for the different sectors of the tourist industry in Fiji. Although the majority of the tourist industry's gross turnover was concentrated in urban areas on Viti Levu, significant variations did occur between sectors. Turnover from accommodation was less concentrated than that from the travel and tour and the tourist shopping sectors due to the large numbers of resort hotels along the Coral Coast and on Mamanuca Island. Accommodation accounted for 96.5 per cent of tourist industry turnover in rural areas but only 31.4 per cent in urban areas. Britton's work shows the importance of considering all sectors of the tourist industry, for these figures clearly indicate that examination of the accommodation sector alone would have distorted the picture of the spatial structure of Fiji's tourist economy. Britton also shows that the degree of concentration of the tourist industry in Fiji depends to a certain extent on the economic indicator used. Rural areas generated only 17.5 per cent of gross tourist industry turnover but accounted for nearly 30 per cent of direct employment in tourism.

While S. G. Britton examined spatial variations in the economic impact of different sectors of the tourist industry, he did not assess how the relative importance of tourism in the economy as a whole varied throughout Fiji. As with the accommodation statistics, it is useful to obtain a measure of tourism dependency or the relative contribution of tourism to the economy of different areas by relating absolute economic measures of tourism to some other factor. Several relative measures have been proposed in the United States. Mings (1982) presents a map of travel-generated employment as a percentage of non-agricultural employment by state and suggests a number of 'dramatic' differences are evident when this is compared with a base map of person nights by state. Pearce (1992b) mapped variations in domestic travel generated expenditure to reveal different spatial patterns of absolute and relative (per capita) expenditure. In absolute terms California in 1987 was followed by Florida, New York, Texas, New Jersey and Pennsylvania – states with large populations. On a per capita basis, however, tourism was more important in Nevada, Hawaii and New England. Only two states – Florida and New Jersey – featured in the leading ten on both measures. Van Doren and Gustke (1982) analysed the evolution of the lodging industry in the United States from 1963 to 1977 by ranking states in terms of lodging receipts and then considering changes in state per capita income over the period. Unfortunately the two measures cannot be directly compared, but the general picture which emerges is that the pattern of growth on a per capita basis differed significantly from the changes in total revenue, with many of the Sunbelt states becoming more prominent on this relative measure.

Other United States writers have discussed the use of slightly more refined measures than simple per capita traveller expenditure (PCTE). Royer, McCool and Hunt (1974) proposed two such measures:

1. A Tourism Impact Factor (TIF), derived by dividing per capita traveller expenditure by per capita personal income and multiplying the result by 100.
2. A Tourism Proportion Factor (TPF), calculated by dividing the total travel expenditure in each state by its gross state product and multiplying the quotient by 100.

The traveller expenditure data in each case came from the 1972 National Travel Expenditure Survey and refer only to domestic tourism. Doering (1976) subsequently revised and expanded this study on the grounds that the authors had not fully considered the implications of the other data they had used. Doering found a high intercorrelation between the three measures and concluded (p. 15) that 'per capita traveller expenditure [is] at least as good a measure as the TIF or TPF and one that is certainly more communicable to both professional and lay audiences'. In a similar study in Australia, M. Cooper (1980) also found strong though not quite so significant relationships between these three measures.

Another perspective on expenditure has been adopted in Scandinavia where some innovative work is being done in terms of regional variations in the consumption patterns of tourists (Paajanen 1993). There it is recognized that regional variations in consumption may occur due to spatial differences in supply and demand. One Swedish study, for example, defines three sets of regions in that country:

- big city regions with higher levels of tourist expenditure
- mountain and forest regions with seasonal price rises
- other parts of the country with single-season tourism and higher than average spending on local attractions.

With the widespread concern for the growing ranks of the unemployed, the distribution of jobs in the tourism sector is becoming an increasingly relevant indicator, though one which is not without its methodological challenges (Pearce 1989b). Lewis and Williams (1988) found a highly concentrated pattern when they plotted the distribution of hotel and other accommodation employment in Portugal – four *distritos* accounted for three-quarters of all such jobs. Lewis and Williams were also able to depict the complex seasonal nature of tourism employment. By comparing the distribution of tourism-created jobs in the three major metropolitan regions of New Zealand to their share of the full-time labour force, Pearce (1990a) demonstrated that the international sector made no special contribution to regional development whereas domestic tourism did.

This work on the spatial patterns of the tourist economy, more than the research on the distribution of accommodation, emphasizes that the identification of these patterns is not an end in itself but rather a means of answering both specific and more general questions relating to society and the economy as a whole. Royer, McCool and Hunt (1974), for example, had derived their measures at the time of the 1973 energy crisis in order to identify states which might be hit the hardest by petrol rationing. Doering's revised figures suggest these would be the Mountain and South Atlantic states. He then goes on to examine other relationships between tourism and the economy using PCTE figures at the state level. Doering (1976: 14) found a negative correlation between state tourism dependency and both population ($\rho = -0.51$) and gross state product ($\rho = -0.54$), leading him to suggest that:

> as a state's population and economy increase in size its economy becomes more diversified and less dependent on any one sector including tourism.
>
> A corollary to this may be that in some states the establishment of a tourist sector is often a prelude to growth in other sectors. In California, Florida and Colorado, for example, there has been a large influx of 'footloose' activities (electronics firms, think tanks, etc.) which were attracted to these states by many of the amenities, real or imagined, associated with the earlier establishment of a tourist industry. More research is needed, however, to determine the exact linkages and causality of these relationships.

In the case of Fiji, S. G. Britton (1980b: 159) concluded that 'there seems little doubt that the spatial organization of tourism is directly related to pre-existing fixed capital originally developed to serve colonial interests'. Why this should come about is then explained in terms of the structural characteristics of peripheral economies as a whole (see Chapter 8). Because of this, Britton suggests that policy-makers' attempts at redistributing the impact of tourism will not be successful unless there is some change in these underlying structures.

Tourists

While studies of accommodation and economic impact may reveal the general spatial structure of the tourist industry, they rarely say much about the tourists themselves. Where possible, such studies should be complemented by direct analyses of the spatial preferences of different groups of tourists and of geographical variations in the composition of the tourist traffic. These analyses also complement the studies of tourist travel patterns discussed in earlier chapters.

Many countries, unfortunately, do not collect areal data on the composition of the tourist traffic. Where they are collected, the most common variable recorded is the origin or nationality of the visitor, although details on seasonality and length of stay may also be available. Information on nationality is usually expressed either in arrivals or bednights and sometimes both. Bednights generally give a better picture of total demand in a given region or for a given nationality. For instance in Portugal, differences in lengths of stay between the Lisbon region and the Algarve saw the former account for 27 per cent of all foreign arrivals but only 23 per cent of foreign bednights, whereas the latter's share of foreign tourism jumped from 25 per cent of arrivals to 36 per cent of bednights (Direccao Geral do Turismo 1980). Whatever the measure selected, the data have to be carefully considered in terms of their comprehensiveness and reliability.

Arrivals and bednight data are usually derived from one of two main sources – accommodation returns or visitor surveys. In many European countries, for example, the proprietors of different forms of accommodation have to return statistics on their guests for fiscal, police and other purposes. In some countries such data are limited to one sector only, notably hotels. In others, for example Belgium, they are more comprehensive but seldom cover all forms of accommodation. In particular, the more informal sectors involving visits with friends and relatives and the use of second homes are rarely included in such statistics. At a regional level, local authority records have been analysed to provide information on second home owners but, as Andrieux and Soulier (1980) note, ownership and use are not necessarily the same, particularly where second homes are bought primarily as an investment and for renting to others.

If the data used cover only one or a limited range of accommodation types, then this should be clearly stated and the representativeness of the patterns so derived explored. Since 1981 in Germany, for example, only enterprises with more than eight beds have been required to record visitor arrivals and bednights (Pearce 1992b). Schnell (1988) noted that private rooms rented on a cash basis accounted for 28 per cent of accommodation capacity and 19 per cent of bednights in 1979–80. Moreover, the effects of the changes in registration have not been experienced evenly, with up to a half of all beds in Schleswig-Holstein not being recorded subsequently. Another study (DRV 1989) estimated that nationwide there were 50 million bednights spent in private rooms or establishments with fewer than nine beds, from a total of 290 million.

Comprehensive visitor surveys – exit surveys for international visitors, household surveys for domestic tourists – have the potential to overcome some of the limitations of the accommodation data. In particular, such surveys can readily include the VFR and other informal sectors and collect information on such additional variables as expenditure and activities. Regional visitor data are collected in this way, for instance, by the International Passenger Survey and the British National Travel Survey in Great Britain (BTA 1992) and the International Visitor Survey and Domestic Travel Study in New Zealand (NZTP 1989a, 1989b). The main issues here relate to the logistical requirements of mounting sufficiently large surveys that they can be disaggregated with confidence at the regional level or below (see Chapters 4 and 5).

Regional data on the composition of the tourist traffic have been analysed in different ways. A basic approach is to distinguish between the domestic and foreign markets through the use of proportional circles and bar graphs, with the foreign traffic being further segmented by country of origin where the data allow (Pearce 1992b). Other more analytical approaches might also be employed. If, for purposes of synthesis, nationalities are to be grouped together, then this might be done on the basis of identified shared spatial preferences rather than on some basic assumption such as visitors from all European countries following the same pattern. Likewise, where the composition of the tourist traffic is comparable, administrative units might logically be amalgamated into larger formal regions. Where the matrix of origins and regions is reasonably large, principal components analysis has proved an effective means of identifying major patterns in the spatial preferences of tourists. Using this technique, Grimmeau (1980) identified three major groups of tourists to Belgium and nine distinct regions from a 9-by-30 matrix of origins and arrondissements. Similarly in a study of the structure of hotel demand in Spain (Pearce and Grimmeau 1985), principal components analysis suggested the nine nationality groups might be reduced to three, and that ten formal tourist regions might be identified among Spain's fifty-two provinces on the basis of similarities in their tourist traffic (Fig. 6.1). In each case, the principal components analysis was performed on the deviations of the recorded bednight values from a matrix of expected values constructed on the basis of an equal distribution of bednights by nationality throughout the country.

Other writers have focused on measures of spatial concentration. R. T. Jackson (1991), for example, identifies those destinations within Australia where particular groups of tourists are concentrated or absent in terms of whether they lie more than one standard deviation above or below the mean proportion of all nights at those destinations. Ashworth (1976) prepared a detailed series of maps showing the distribution of foreign visitors in the Netherlands based on location quotients. Location quotients provide a measure of concentration and in this case were derived by relating the percentage of bednights to the percentage of land area in each region. Values greater than unity indicate a region is receiving more tourists from a given market than the national average; values less than 1 show that it is below the national average.

Derivation of location quotients also constitutes an initial step in the construction of localization or Lorenz curves, which can be an effective means of depicting concentration or dispersion. Lorenz curves are constructed by ranking the areal units in descending order of their location quotients and calculating, then plotting, cumulative percentage values of the target variable (e.g. foreign tourist bednights) against the base variable (e.g. area). If there were a perfect correspondence between the distribution of the target variable and the base variable, then a straight diagonal line would result. The greater the divergence between the localization curve and this diagonal, the greater the degree of localization or concentration present in the distribution of the target variable. Thus if localization curves are plotted for different periods, it can be determined whether tourists are becoming more or less concentrated over time or whether dispersion patterns are relatively constant. The distribution of different groups of tourists within a country can also be compared in this way, as can the extent of concentration between countries (Ashworth 1976). Only general comparisons can be made, however, where the number of areal units varies from country to country, as the form the localization curves take depends in part on the number of units used. It should also be borne in mind that particular geographic areas can no longer be identified from the localization curves. It may be, for instance, that two groups of tourists show a comparable degree of concentration but are concentrated in different regions. The use of localization curves for comparative purposes will be explored more fully in Chapter 7 (Figs 7.1 to 7.4).

Composite Studies

The majority of the studies cited so far have dealt primarily with specific aspects of tourism, particularly with types of accommodation and groups of visitors. In other cases, authors have considered in turn several different facets of the tourist industry. Ashworth's (1976) maps showing the distribution of foreign tourists in the Netherlands, for example, are complemented by others dealing with the distribution of accommodation and attractions, and Ishii (1982) deals with a range of recreational resources and facilities in Japan. Herbin (1982) has produced maps showing the distribution of resorts in the Austro-German Alps classified by accommodation capacity, the importance of winter tourism and visitor origin. These are but parts of a larger series which also includes maps of tourist intensity, ski-lifts, communications, occupancy rates and length of stay.

Such maps can, individually, provide a lot of useful information. They could also contribute to a broader picture of the spatial structure of tourism but, in the cases cited, there is little or no attempt at synthesis. Other writers have endeavoured to integrate a variety of tourism characteristics to produce such a synthesis in

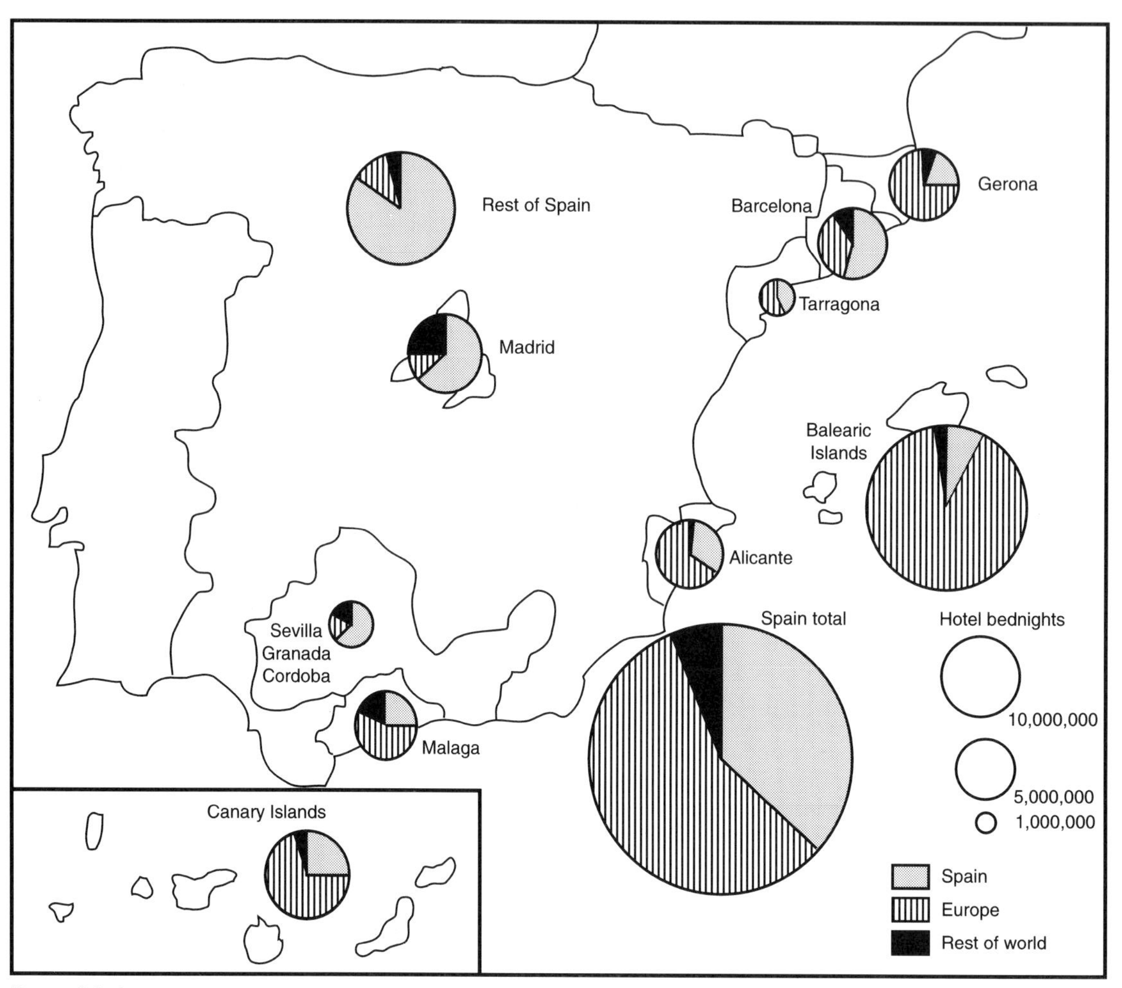

Figure 6.1 Spain: regional variations in hotel demand by major national groups.
Source: Pearce and Grimmeau (1985).

one of three ways: classification schemes or typologies, quantitative analyses and models of tourist space.

Typologies

Plettner (1979) classified resorts in Ireland according to the following factors: prevailing kind of tourism, intensity of tourism, type of accommodation, bednights, seasonality, average length of stay, and visitor origin. A composite symbol, featuring proportional pie charts embellished with other graphics, was derived for each resort. Unfortunately, the key to the resultant map is a little obscure. No attempt was made to group similar resorts although Plettner did identify general patterns. A similar approach was followed by Molnar, Mihail and Maier (1976) who derived a typology of Romanian tourist centres based on the main types of tourism and variations on these. Geographic position, size and composition of the tourist traffic were other criteria used. A map of tourist centres in Romania was then presented but, like Plettner's map of Ireland, its utility is limited by an inadequate key. And, like Plettner, the authors attempted to identify the salient features of the distribution of the different centres. They noted, for instance, the role of the Black Sea coast, particularly for foreign tourists, and observed how the tourist towns were to be found mainly along the major arteries. A second map of tourism in Romania was prepared by Swizewski and Oancea (1978) who defined three categories of tourism types: structural, dynamic and 'stay-put'. The structural category was based on an area's tourist resources while different travel patterns determined the dynamic types. The 'stay-put' types are said to be based on length of stay, but statistical difficulties apparently occurred and different functional types (e.g. winter sports, water sports) are depicted. The resultant map is rather detailed and not readily assimilated: moreover, no interpretation is given by the authors. The lack of a measure of capacity or intensity also makes it difficult to appreciate variations in the scale and importance of tourism throughout Romania. In particular, the importance of the Black Sea coast is lost on Swizewski and Oancea's map.

In Belgium, Piavaux (1977) classified and mapped tourist units in the province of Luxembourg by combining data from a visitor survey with a detailed examination of physical and socio-economic features. Although the details on how this was achieved are not too clear, the dominant forms of tourism practised appear to have played a significant role. Piavaux identified three main forms of tourism: transit tourism, weekend tourism and destination tourism (*le tourisme de séjour*). This latter category was subdivided into two classes depending on whether the stay was spent in one locality (*le séjour fixe*) or whether that locality was used as a base to visit other sites (*le séjour mobile*). Piavaux then goes on to propose specific measures to promote tourism in the units he has identified.

Despite their shortcomings, the Romanian examples and that of Piavaux are interesting in that they attempt to incorporate types of tourism defined in terms of travel patterns as discussed in earlier chapters. The notions of circuit and destination travel, for example, are embodied in each of these examples. In spite of their different origins, Piavaux's *séjour mobile* is almost identical to Campbell's recreational vacational regional complex' (Fig. 1.2). What is required now is work integrating the studies of travel patterns with research on the distribution of tourism features. It may be that those concerned with the distribution of different types of tourism would find that some concise measure of circuit/destination travel, such as the Trip Index, could be incorporated into their classification. Conversely, researchers examining travel patterns should be encouraged to develop their work further to identify the function of different places in their networks and circuits, as Forer and Pearce (1984) have endeavoured to do in their study on package tourism (Table 4.4).

Quantitative analyses

Quantitative techniques have been favoured by the few North American researchers who have attempted regional analyses of tourism resources, notably S. L. J. Smith (1987) in Ontario and Lovingood and Mitchell (1989) and Backman, Uysal and Backman (1991) in South Carolina. In an approach similar to that used by Grimmeau (1980) and Pearce and Grimmeau (1985) on tourist data, principal components analysis (factor analysis in the case of Backman *et al.*) was used to aggregate a set of quantifiable resource data (number of hotels and hotel rooms, campgrounds, boat ramps, historical sites, etc.) into a smaller number of statistically defined components. County-level variations in the indices so obtained were then plotted, after which clustering algorithms were used to identify clusters of similar regions based on their component scores.

The regions identified in this way were shown to differ markedly from existing 'tourist regions' used for organizational and marketing purposes. As is acknowledged by the writers themselves, their data are resource based, essentially measures of accommodation and attractions, with no account being taken of spatial variations in demand, a factor which might significantly affect promotional campaigns and regional images. This supply-side focus leaves many unanswered questions as Lovingood and Mitchell (1989: 315) note:

> what is the value of a clean, white, sandy and unbroken beach? What role does the moderate climate play? What is the influence of location in reference to the various points of origin of tourists.

Moreover, the clustering techniques used have not produced sets of counties which are always contiguous, a factor which also raises questions for functional regionalization.

Despite these limitations, an important original contribution by these analyses in the context of composite studies is their facility to link spatial variations in tourism resources directly to local or regional level impact. This can be achieved by regressing the components identified against measures of economic impact. S. L. J. Smith (1987), for example used local retail receipts attributable to tourist expenditure as a measure of 'local' importance and the percentage contribution of tourism expenditure in a county to total provincial receipts as an indicator of 'provincial importance'. He found that the most important component in terms of local impact was cottaging/boating whereas 'urban' tourism and 'urban fringe' tourism were the greatest contributors at the provincial level.

As Smith and the others point out, on their various measures tourism is not identified as being important at all in some counties or regions. Few studies, however, have systematically examined spatial variations in tourism *vis-à-vis* other activities for as Jackowski (1980: 86) observes: 'In geographical-tourist studies ... the typology of tourist localities was usually carried out independently of the whole settlement network of a given region.' Jackowski thus sought to discriminate not only between different types of tourist regions but also between tourist and other socio-economic regions. This has involved extending the number and range of variables examined and also analysing them quantitatively.

Jackowski selected 18 variables in his analysis of 453 settlements in the Polish vovoidship of Nowy Sacz. These he divided into two groups, one 'characterising the level of economic development' (agricultural population and area of agricultural land) and the other 'covering features of social development'. This latter group covers an intriguing range of 16 variables, some of which might be considered direct measures of tourism, e.g. restaurants and sleeping facilities, while others include shops, cinemas, length of paved roads, total telephone subscribers and private telephone subscribers. Eight types of localities were then identified by Jackowski using nearest neighbour analysis. Over half of the 453 settlements were classified as agricultural localities or farming-service localities. The remaining six types are characterized by a varying mix of agricultural and tourism functions.

In their study of 74 coastal cantons in Brittany, Bonnieux and Rainelli (1979) chose 28 variables – 5 demographic, 15 covering employment in different sectors and 8 relating to accommodation – and subjected the resultant matrix to factor analysis. Four major types of cantons were identified in this way: agricultural, urban and peri-urban, tourist and maritime. Two or three subgroups were found in each type, e.g. summer tourism, and summer tourism and retirement. The authors then list and map the cantons belonging to each category. However, neither Bonnieux and Rainelli nor Jackowski go on to explore and account for the patterns they have identified.

Models

Both the typological and quantitative studies have proved useful in drawing together and synthesizing a range of tourism variables and identifying spatial variations in their occurrence. Most such studies, however, tend to result in rather static depictions of tourism regions. What is generally missing is a sense of the dynamic relationships which characterize tourism, a feeling for the functional linkages between regions and between different sectors. In particular, the transport component is absent from the studies cited above as is the notion of flows elaborated on in earlier chapters. As classification and quantification of these elements and linkages remains difficult, but not impossible, a more conceptual approach to the problem might usefully complement those techniques just outlined. Some interesting and innovative approaches of this type have been developed by French geographers (Rognant 1990; Clary 1993). Although no direct bibliographic references are made to it, this work might derive from the earlier more general concepts of national spatial structure originated by Brunet (1973). Links might also be seen between the French models of *'l'espace touristique'* outlined in Chapter 1 and these later attempts to portray the tourist space of particular countries and regions.

Rognant (1990) uses a systems approach incorporating flows, nodes, networks and surfaces to develop first of all a general model of the economic spatial structure of Italy, one which incorporates notions of the North–South imbalance, of contrasts and gradients. This general structure established, Rognant progressively builds up his tourism model (Fig. 6.2) based on an analysis of domestic and international flows, networks of different types of destinations (coastal, thermal and alpine resorts, historic cities) and the dispersion of other forms of tourism (agritourism and second homes). Again a systems terminology is employed as Rognant refers to sources, gradients, axes and poles, of a core, peripheries and pioneer zones. The sources (reservoirs), for example, are domestic and international; the domestic flows are shown to radiate out from the major urban centres, decreasing in intensity with distance in the manner discussed earlier (Chapters 1 and 5). The gradient is north–south; flows follow a 'trident' of motorways, two along the coast, the major one inland linking urban and cultural centres along the Bologna–Florence–Rome axis. A tourism core is defined, consisting of a mix of densely developed coastal zones (the Italian Riviera and the Rimini coast) and four major crossroads (Bologna, Florence, Verona–Padua and Genoa), with the Rome region constituting a secondary southern core. The southern periphery of the peninsula and islands is complemented by 'interior peripheries' in the north and areas such as the Po Valley which are passed through quickly (transit zones in the terminology of Chapter 1). Rognant also links these three zones to the broader economic structures of Italy, noting 'the divergence between these in dynamic terms' (Table 6.2). Finally, with this overall national model established, Rognant develops various regional tourism models, ranging from the 'complete' model of Tuscany (including 'metropolitan', 'coastal', 'mountain' components), to the dispersed nodal model of southern Italy, Sicily and Sardinia.

Clary (1993) develops a similar approach, at lesser length and on a smaller scale, for the French département of Maine-et-Loire. Whether depicted in 'structural' (attractions and accommodation) or 'functional' (summer flows) terms, Clary suggests the department's tourist space is essentially limited to a central axis, with some extensions in the south. Tourism in this fashion is set against the more traditional agricultural structure of the Maine-et-Loire. Clary observes (p. 281):

> Over the disorderly territory of the rural residents is drawn a pattern of flows and poles. The only points of concordance between these different spatial enti-

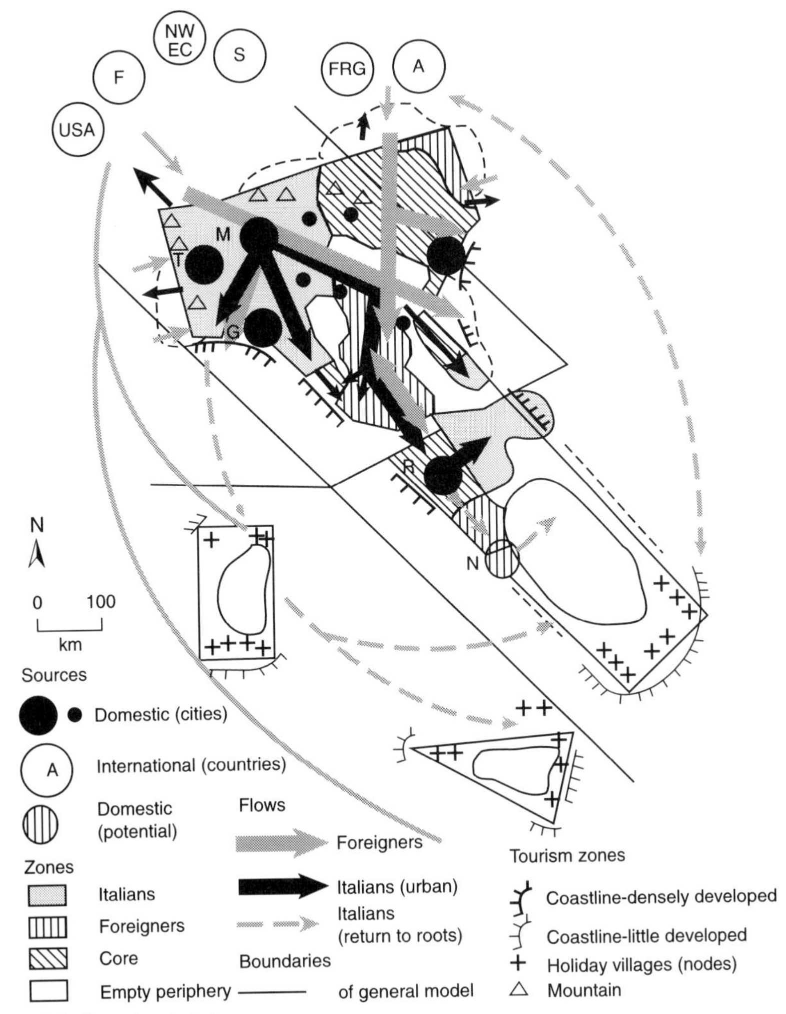

Figure 6.2 Basic model of tourism in Italy.
Source: Redrawn from Rognant (1990).

ties are these nodes, small centres and lively market towns, which are both poles of life for the rural areas and poles of activities and services for the tourists.

It is around such poles, he argues, that coherent rural development strategies must be centred, signalling again how such spatial analysis can have practical application.

Conclusions

Geographers and others have used a wide range of data sources, techniques and approaches to examine spatial variations in tourism. This diversity is in part due to the multifaceted nature of tourism with the potential that it offers to select one of a number of components for study. While in some respects healthy, this diversity also appears to reflect a general lack of direction in the literature and common purpose among researchers, rather than a conscious striving to select and develop the most appropriate techniques for particular problems. Only in a few cases have researchers adopted a critical approach, for example evaluations of the tourist function index, and only recently have deliberate attempts been made to replicate earlier studies to compare techniques and findings, notably the quantitative regional analyses of tourism in North America. Nevertheless, the literature, diverse and fragmented though it is, does provide the base on which to develop a more systematic approach to the spatial structure of tourism. Such an approach is attempted in the following three

Table 6.2 Economic and tourism zones in Italy

	Economy	Tourism
North West	Industrial core I • megapoles • large-scale agriculture • high travel propensities	Tourism periphery • nuclei of lakes • ski resorts • peri-urban recreation (holiday homes)
Centre and North East	Industrial core II • central Italian model ('dilatation') • family agriculture • industry and small and medium enterprises • isolated metropolises • alpine depopulation	Tourism core • domestic and international tourism • cultural tourism • large-scale coastal
South and Islands	Poor periphery • emigration • industrial nuclei • archaic agriculture • economically non-integrated	Tourism periphery • tourism 'rich' but reduced • tourist nuclei (villages)

Source: Rognant (1990).

chapters, which consider in turn national and regional patterns (Chapter 7), the spatial structure of island tourism (Chapter 8) and then of coastal resorts and urban areas (Chapter 9).

In each case an attempt is made to identify basic patterns through comparative studies so as to develop a more general understanding of the processes at work (Pearce 1993b). Two types of comparisons are used. Where suitable data are available for a range of countries, regions or other areas, these are analysed using the same technique or approach. Thus in Chapter 7, localization curves are used to examine the concentration of domestic and international tourism in selected European countries, and in Chapter 8 the distribution of accommodation provides a means of establishing regularities in the spatial structure of tourism on islands. In other instances, case studies which use different techniques or data sources but which address common problems are systematically compared, as in the discussion in Chapter 7 of the functional structure of coastal regions. Particularly with this latter approach, there are limitations in the extent to which generalizations can be made.

CHAPTER 7

The National and Regional Structure of Tourism

In this chapter the spatial relationships between domestic and international tourism are explored. This is a field which has attracted relatively little attention (Pearce 1989a). As noted in Chapter 6, attempts have been made to differentiate spatially between different groups of tourists, but researchers have often focused on just one sector, particularly international tourism. Moreover, most have confined themselves to the study of one country and there have been few attempts to examine systematically variations from country to country in order to establish whether any general patterns exist.

In an early report, the United Nations (1970) observed that in many instances domestic and international tourism might best be seen as complementary:

> in some countries development of domestic tourism might lead to a development of foreign tourism, whilst in others, as yet undeveloped, but well endowed with tourist attractions, the encouragement of foreign tourism would lead in due course to growth in domestic tourism.

How does this complementarity, if it does exist, manifest itself geographically? Do domestic and international tourists share the same preferences or do certain tourists favour particular parts of a country or region while others take their holidays elsewhere?

The basic spatial relationships between domestic and international tourism are implied in Fig. 1.5. There, Thurot (1980) distinguishes between supply and demand and between demand generated domestically and that originating externally. The question now on the supply side is whether, and to what extent, these two markets seek and use the same resources or whether they exhibit distinct spatial preferences. If spatial variations do occur, is this essentially at a regional level or are different structures also to be found within regions?

These questions are examined at a national level in the first part of this chapter with reference to a range of European countries for which adequate and recent comparable data on demand were available. The focus in the second part then switches to the regional level and is broadened to include a discussion of functional structures generated by different forms of tourism. The regional examples are again European and are drawn primarily from coastal regions along the Mediterranean.

The Spatial Structure of Tourism at a National Level

The basic features of domestic and international tourism in the countries of Europe to be studied in this section are depicted in Table 7.1. This shows that the countries to be considered range from the major destinations such as France, Italy and Spain, large countries with a variety of tourist resources, through to smaller countries (Austria and Switzerland) and secondary destinations (Ireland, Norway and Finland). With the exception of Ireland, where statistics are collected on visitors to each region, demand is expressed in bed-nights. In four cases these relate to all accommodation (noting the qualification of 'all' outlined in Chapter 6), but the data for the remainder relate to specific sectors, notably hotels and camping. In Spain, Pearce and Grimmeau (1985) showed that hotels accounted for about half of the recorded bedspace but regional differences occurred in the distribution of different types of accommodation there. Hotel demand in France is much less representative of domestic tourism in general than international tourism. These differences in comprehensiveness limit and complicate cross-national comparisons but, through the use of relative measures such as localization curves, it is possible to identify general patterns.

Table 7.1 shows considerable variation in the relative importance of domestic and international tourism. On the measures used, domestic demand exceeds that from international visitors in all countries shown except Austria, Spain and Switzerland, while in Ireland the two sectors are in balance. The extremes are found in neighbouring Germany and Austria. Interrelated factors contributing to these differences include population size and the propensity for travel, which together influence the size of the domestic market, and the nature, extent and degree of development of the tourist resources which affects the volume of foreign visitors.

Table 7.1 International and domestic tourism in selected European countries

	Accommodation included	Year	International bednights	Domestic bednights	International bednights / Domestic bednights
Austria	All	1992	99,757,595	30,658,674	3.25
Finland	All	1992	3,096,383	8,086,132	0.38
France	Classified hotels	1991	53,046,000	88,016,000	0.60
	Camping	1990	29,603,000	39,141,000	0.44
Germany	All (12 beds plus)	1988	29,779,408	204,216,204	0.15
Ireland	Visitors	1988	4,790,000[a]	4,377,000[a]	1.09
Italy	Hotels and parahotelry	1991	87,150,400	174,542,800	0.60
Netherlands	All	1991	17,206,000	39,141,000	0.44
Norway	Camping	1990	2,189,260	2,880,416	0.76
Spain	Hotels	1992	76,911,908	54,321,322	1.42
Switzerland	Hotels and health resorts	1987	19,907,605	15,686,671	1.27
Yugoslavia	All	1991	17,206,000	39,141,000	0.44

Note: [a] Visitors.
Source: National tourist organizations.

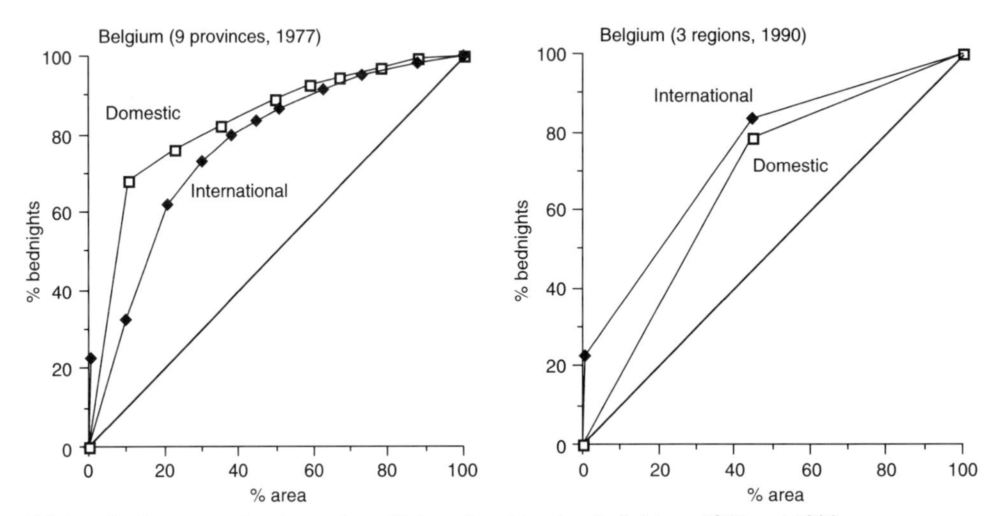

Figure 7.1 Localization curves for domestic and international tourism in Belgium, 1977 and 1990.
Data sources: *Annuaire de Statistiques Regionales*; Tourist Organization of Flanders.

The first question considered with regard to the spatial structure of tourism within these countries is to what extent is domestic and international demand concentrated or dispersed?

As discussed in Chapter 6, localization curves provide a useful means of analysing such concentration and making cross-country comparisons. However, in addition to the data limitations outlined above, it should also be noted that the number of spatial units may vary from country to country as a result of different administrative structures. Variations in the number and scale of the areal units can have a marked impact on the form of localization curves and their interpretation. This is graphically illustrated by the case of Belgium shown in Fig. 7.1. With the moves towards federalism in Belgium has come a tendency for data to be presented at the levels of the three regions, Brussels, Flanders and Wallonia, rather than in terms of the nine provinces as was the case previously. A nine unit or three unit solution produces vastly different results as is illustrated by the curves for 1977 and 1990. In 1977 domestic tourism is shown to be more concentrated than international tourism, a pattern which is reversed in 1990. This is a direct consequence of the reduction in units from nine to three. In 1977, two-thirds of domestic and one-third of international bednights were spent in the province of West Flanders, a function of the concentration of demand along the 67 km of the Belgian coastline (Fig. 7.7). In the 1990 figures, the province is incorporated in the larger region of

Flanders, thereby reducing markedly its location quotient. In contrast, Brussels, previously incorporated in the province of Brabant, is now separated out, the concentration of international visitors in the capital region being emphasized by its small base area (0.5 per cent of the country's total). While this is an extreme case, care obviously has to be taken in making cross-national comparisons when the number and size of the spatial units varies, especially where smaller metropolitan regions are involved. However, for each of the countries depicted in Fig. 7.2, the number of units is equal for both domestic and international tourism, thus, enabling general tendencies to be established.

Two general patterns can be identified in Fig. 7.2. First, the tendency is for both forms of tourism, domestic and international, to be spatially concentrated. Only in the cases of Switzerland, Germany and Ireland do the two curves approach the diagonal, indicating that tourism there is spread fairly evenly throughout the country. Elsewhere, the curves diverge much more, revealing a greater degree of concentration, particularly in Spain, Finland, former Yugoslavia and Austria. Second, especially in these four countries, international tourism is more concentrated than domestic tourism. The only exceptions here, and then rather marginally, are camping in Norway and France.

Tourism, of course, is not the sole social and economic activity to be concentrated in certain areas more than others. As with agriculture or manufacturing, a variety of factors such as resource endowment, demand, accessibility and government policies combine to favour the development of tourism in particular parts of a country rather than an even distribution throughout the nation. Of particular interest in this respect is why international tourism is generally more concentrated than domestic demand.

Various interrelated supply and demand factors play a role here. The nature, distribution and development of tourist resources appear to be decisive factors. In several instances, much of the international tourist traffic is directed to coastal regions with several favoured destinations accounting for a significant share of demand but only a small proportion of surface area. In Spain, the small Balearic and Canary Islands accounted for almost two-thirds of international hotel demand in 1992 but only 2.5 per cent of the country's surface area. Similarly, in former Yugoslavia, the Croatian coastline attracted the majority of the republic's international visitors (Gosar 1989), some 80 per cent of the country's total.

Other countries, such as Austria and Switzerland, have no coastline and depend on other natural or cultural attractions. International tourism in Austria, for example, is mainly concentrated on the Alps proper in the west, particularly in the Tyrol. This is not only the winter sports traffic, for the summer influx is greater than that of the winter months. Austrian domestic tourism, particularly in summer, extends to the areas of gentler relief in the east, which no doubt receives much of the demand from Vienna. Alpine attractions are found throughout much of Switzerland, with the main urban centres, several of which have a major international role being located in the lowland and valley areas. As a result, both domestic and international tourism are fairly evenly dispersed.

In other instances, the major urban centres are a prime attraction for holidaymakers. Conversely, for the domestic market, major metropolitan centres tend to be sources rather than destinations (Chapter 5) and much of the traffic they do attract often contains a large VFR element which is not included in the accommodation statistics used here. North Holland, with Amsterdam and the Hague, attracted over one-third of the foreign bednights in the Netherlands while in Finland, Uudenmaan, with Helsinki, catered for 60 per cent of that country's international demand. The Ile de France region (Paris) accounted for 40 per cent of hotel demand in France in 1991, compared with 14 per cent of domestic hotel demand. Unfortunately the region's share of camping was not reported, but is likely to rank well behind that of the leading coastal provinces of Corsica, Aquitaine, Languedoc-Roussillon and Brittany in terms of location quotients for foreign campers.

Italy, like France, offers a range of natural and cultural attractions as reflected in the three regions with the largest location quotients: Trentino Alto Adige (alpine), Veneto (with the cultural attractions of Venice) and Liguria (a small region containing the 'Italian Riviera').

Regional variations in resource endowment are reinforced by different patterns of demand and accessibility. The quality of the resources or experiences sought by international tourists will normally be higher and perhaps more specific than that of many domestic tourists. In leaving their own country, international tourists are generally investing more time, money and effort to get to their destination. The sites and sights they seek are thus likely to be much less common and therefore more localized than those sought by, or at least able to be afforded by, many domestic tourists in the destination country. Non-vacation travel by international visitors, for example those attending conferences or on official or private business, is also likely to be more concentrated than that of their domestic counterparts, notably in capital cities and other major urban areas. In Chapter 5 it was shown that much domestic tourist travel is intra-regional in nature and that the volume of travel decreases with distance from each market area (Fig. 1.3). Although patterns of

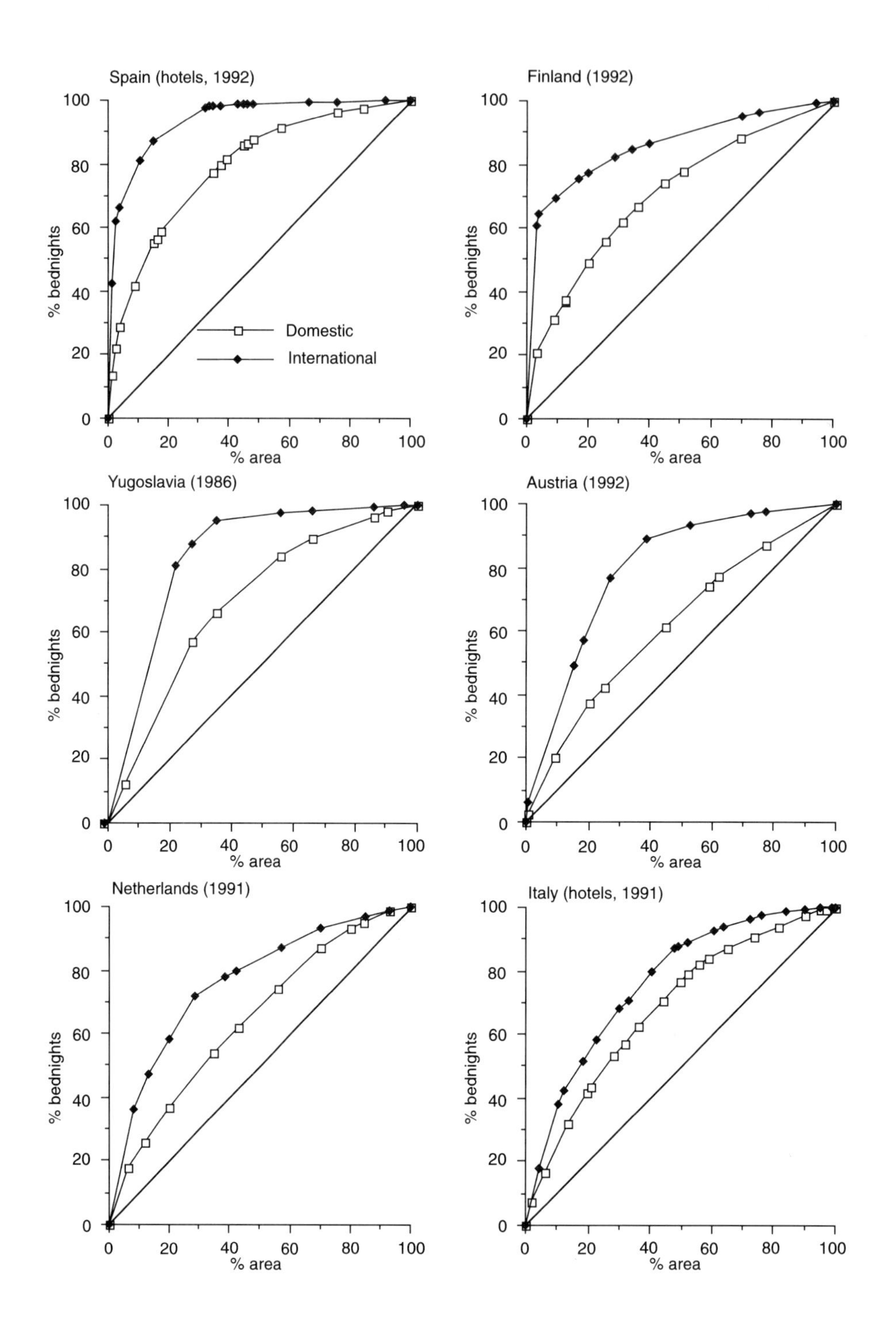

Figure 7.2 Localization curves for domestic and international tourism in selected European countries.
Data sources: National tourist organizations.

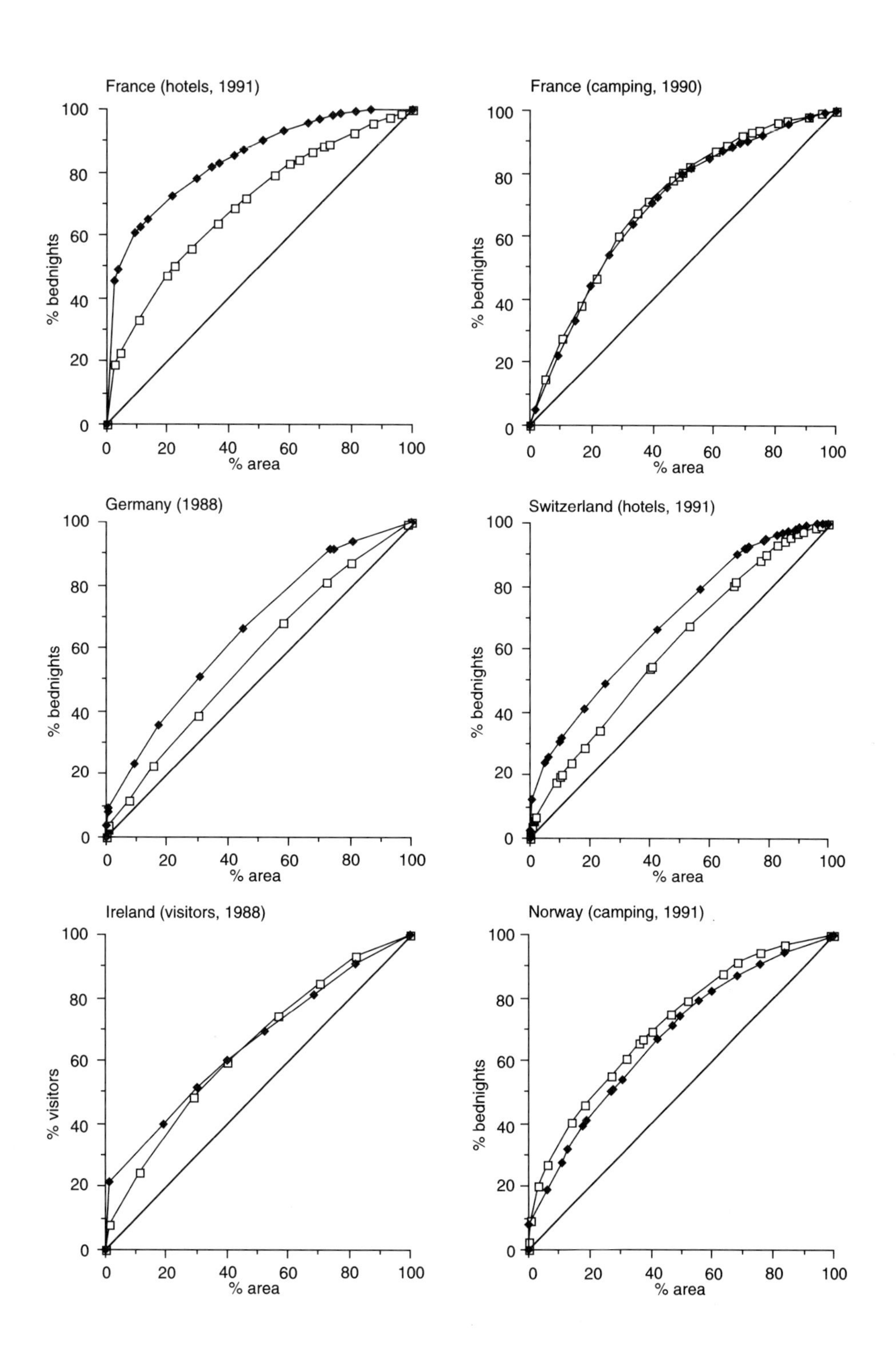

Figure 7.2 *(cont.)*

settlement and urban hierarchies vary from country to country, in most of the countries studied here there is a reasonable number of these markets, each giving rise to a regional pattern of demand. Germany, in particular, has a reasonably well distributed population. In contrast, international gateways, which were shown in Chapter 4 to influence the intra-national travel patterns of international visitors, are generally much fewer in number, particularly where arrivals are by air. Major air gateways are usually associated with the principal urban centres, for example Schipol between Amsterdam and the Hague, although large resort areas may also develop their own gateways, as with the charter traffic to Majorca (Pearce 1987a). Moreover, economies of scale in development and operations are achieved by focusing on a limited number of development areas. This is particularly the case with international tourism, whether in terms of constructing international airports, developing mass tourism or of selectively promoting particular areas once a country's image has been established in the international market-place. For these and other reasons suggested by Miossec (1976), the international tourist is also less likely to be aware of the full range of vacation opportunities than the domestic holidaymaker (Fig. 1.4).

These factors may also contribute to the continuing concentration of the tourist industry in already developed parts of the country. The payback period on tourist plant, e.g. airports and hotels, is relatively long, encouraging continuing efforts to promote these areas (Burkart and Medlik 1974). This is not true to the same extent with condominium construction where the developer is often more interested in the initial sale of the apartment. But even here, the desire to reduce investment risks will frequently lead to new projects being located in or near areas which are already well known and which have an established servicing and infrastructural base. An informal feedback system will also develop, with many marketing surveys showing the 'influence of friends and relations' as a major factor affecting the selection of a holiday destination. As the market grows, so the amount of information feedback will increase, especially important where the quality of the experience is satisfactory. McEachern and Towle (1974) suggest that even where there is a decline in environmental quality a decrease in the tourist traffic is not inevitable:

> As the character of the tourist island changes from 'unspoiled' to 'spoiled' (e.g. more urbanized) the influx of tourists does not stop. Due to various stimuli, such as man's advertising and other publicity, it continues and may even accelerate according to a positive feedback loop that reinforces the degree of urbanization and the influx of new tourists with different tastes.

The concentration of tourism may thus develop from a process of circular cumulative growth similar to that observed in industrialization and economic development (Myrdal 1957; Pred 1965) or in agricultural specialization (Moran 1979). Certain types of tourism, unlike these other industries, however, are characterized by a search for novelty. Developing or maintaining the more allocentric or perhaps fashionable segment of the market may depend in part on opening up new areas. Given that the measures of concentration used here are relative ones, the development of new areas does not necessarily mean an increase in dispersion. In absolute numbers, the traffic to the new resorts may be offset nationally by an expansion in the volume of visitors to or on the fringe of the more traditional areas. Where, however, the expansion of the tourist traffic has come about through the concerted development of a specific attraction or a new area, a marked increase in the degree of concentration might be expected. In Romania, the Black Sea coast's share of advertised hotel accommodation increased from 33 per cent in 1967 to 76 per cent in 1975 as a result of 'vigorous efforts to develop tour business with Western operators' (Turnock 1977: 53).

Table 7.2 Rank correlation coefficients of the regional distribution of international and domestic tourism in selected countries

	Measure of regional distribution	
	% of bednights	Location quotient
Yugoslavia (1991)	0.9286	0.9524
Spain (1992)	0.9142	0.9632
Italy (1991)	0.9083	0.8541
Germany (1988)	0.8909	0.7
Switzerland (1987)	0.8345	0.6191
France (camping) (1990)	0.8012	0.7685
Finland (1992)	0.7972	0.6737
France (hotels) (1991)	0.6737	0.7624
Norway (1990)	0.3316	0.5982
Netherlands (1991)	0.3273	0.3091
Ireland (1988)	0.119[a]	0.8571[a]
Austria (1992)	0.0833	0.3333

Note: [a] Visitors.

Although the localization curves depicted in Fig. 7.2 enable some conclusions to be drawn regarding the concentration of domestic and international tourism, they do not show directly the extent to which the two groups of tourists favour the same regions or not. In other words, are domestic tourists concentrated in the same regions as international visitors? Table 7.2 provides some answers to this question, depicting the rank correlation coefficients of the two sets of regions – domestic and international – in each country ranked in terms of their location quotient and share of each market measured in bednights. Table 7.2 indicates that in most cases there is a reasonable degree of correspondence on both measures, especially for former Yugoslavia, Spain, Italy and Germany. In these countries domestic and international visitors generally prefer the same regions, though not necessarily to the same extent. The rankings may also vary with the measure used (location quotient or percentage of bednights). In Spain, for example, coastal tourism predominates with both groups although the small island destinations of the Balearics and the Canaries have a higher concentration of bednights while the domestic tourists favour more Andalucia and the Valenciana region. Conversely, regions such as Navarra and Rioja experience the fewest bednights for both groups while the more arid, inland regions of Castille and the Extremadura record the smallest location quotients.

In contrast, domestic and international visitors in Norway, the Netherlands and Austria exhibit markedly different spatial preferences as shown by the much weaker correlations in Table 7.2. In the Netherlands, for example, the heavily urbanized provinces of North and South Holland attract half of the international bednights but only 12 per cent of domestic demand. In contrast, Limburg (18 per cent) and Gelderland (17 per cent) are the two leading provinces for Dutch visitors, having lower densities and containing what relief this generally low-lying country has to offer. In Austria the Tyrol alone accounts for 43 per cent of international bednights while ranking only sixth with almost 10 per cent of the domestic bednights. Domestic demand there is more evenly spread amongst the top five domestic regions, ranging from Steiermark (18.9 per cent) to Lower Austria (13.1 per cent). The disparity in Ireland between the degree of correlation on the two measures – very low for visitors, relatively strong for location quotients – reflects the position of Dublin, Ireland's capital and primate city. Its large location quotient (5.9 for domestic, 15.9 for international) is due in part to its small areal extent which contributes to its leading rank on both measures. In terms of bednights, however, it is primarily a source of domestic tourists, ranking seventh of the eight Irish regions (7.8 per cent), particularly for the second-ranked and adjacent East region, but holds first place with international visitors (21.3 per cent), ahead of Cork/Kerry (18.5 per cent).

With these general patterns of domestic and international tourism identified, the international component might be examined more closely to see whether any systematic variations exist in its spatial structure. Are specific international markets more spatially concentrated than others and do they manifest distinct regional preferences? Again, localization curves can be plotted, in this case of the leading international markets – the USA, Germany, Japan, the United Kingdom and France – together with other markets which might dominate or be prominent in particular cases, for example the Swedish market in Norway and Finland and the Dutch in Germany (Fig. 7.3). For the destinations considered, this also involves a mix of the shorter-haul European markets and the longer-haul origins of the United States and Japan, although in several cases traffic from the latter is so small as not to warrant separate designation in the national statistics.

Figure 7.3 clearly shows that in most cases these different markets exhibit different preferences and in terms of their spatial behaviour they are by no means homogenous. There is a general tendency for those destinations that showed a greater degree of concentration of the overall international component in Fig. 7.2 to experience a wider spread among the constituent markets, for example Finland and Austria, while those with a more even distribution, such as Ireland and Germany, also show greater uniformity in the distribution of the individual markets. The United Kingdom, not shown in Fig. 7.2 due to a lack of comparable domestic data, also exhibits quite a spread in its different markets.

When the behaviour of individual markets is examined a fairly consistent pattern emerges in which Japanese and American visitors are more concentrated than those from Europe. To a certain extent this appears to reflect a basic split between wanderlust and sunlust tourism, the former associated with longer-haul visitors and focused on urban, cultural features, the latter with shorter-haul traffic and coastal and alpine areas. A degree of complementarity is apparent both in terms of cultural tourism (Old World/New World, West/East) and sunlust tourism where the destinations' natural resources complement or extend those found in the originating countries. Accessibility is also a factor. Longer-haul visitors arriving by air are more likely to be concentrated due to the limited number of gateways whereas overland travel in continental Europe may be more widely dispersed through multiple points of entry.

In France the Ile de France attracted 86 per cent of Japanese hotel demand, 61 per cent of American but only 31 per cent and 27 per cent of the British and German respectively. Approximately 80 per cent of American and Japanese demand in the Netherlands is concentrated in the provinces of North and South Holland. These provinces also attracted a significant amount of French (72 per cent) and British demand (66 per cent) but a lesser share of the neighbouring German traffic (35 per cent). The Germans are more widely

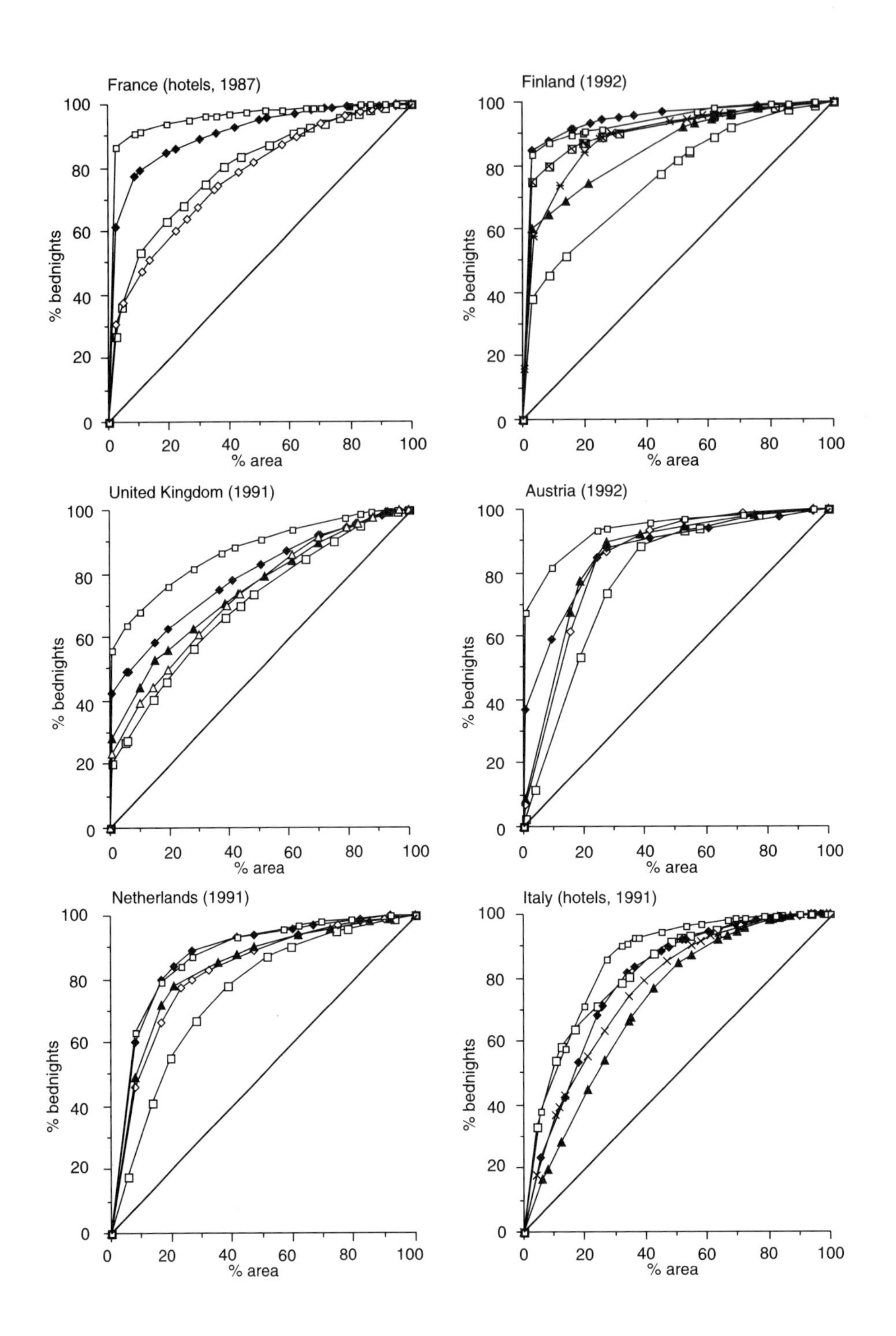

Figure 7.3 Localization curves for major international markets in selected European countries.
Data sources: National tourist organizations.

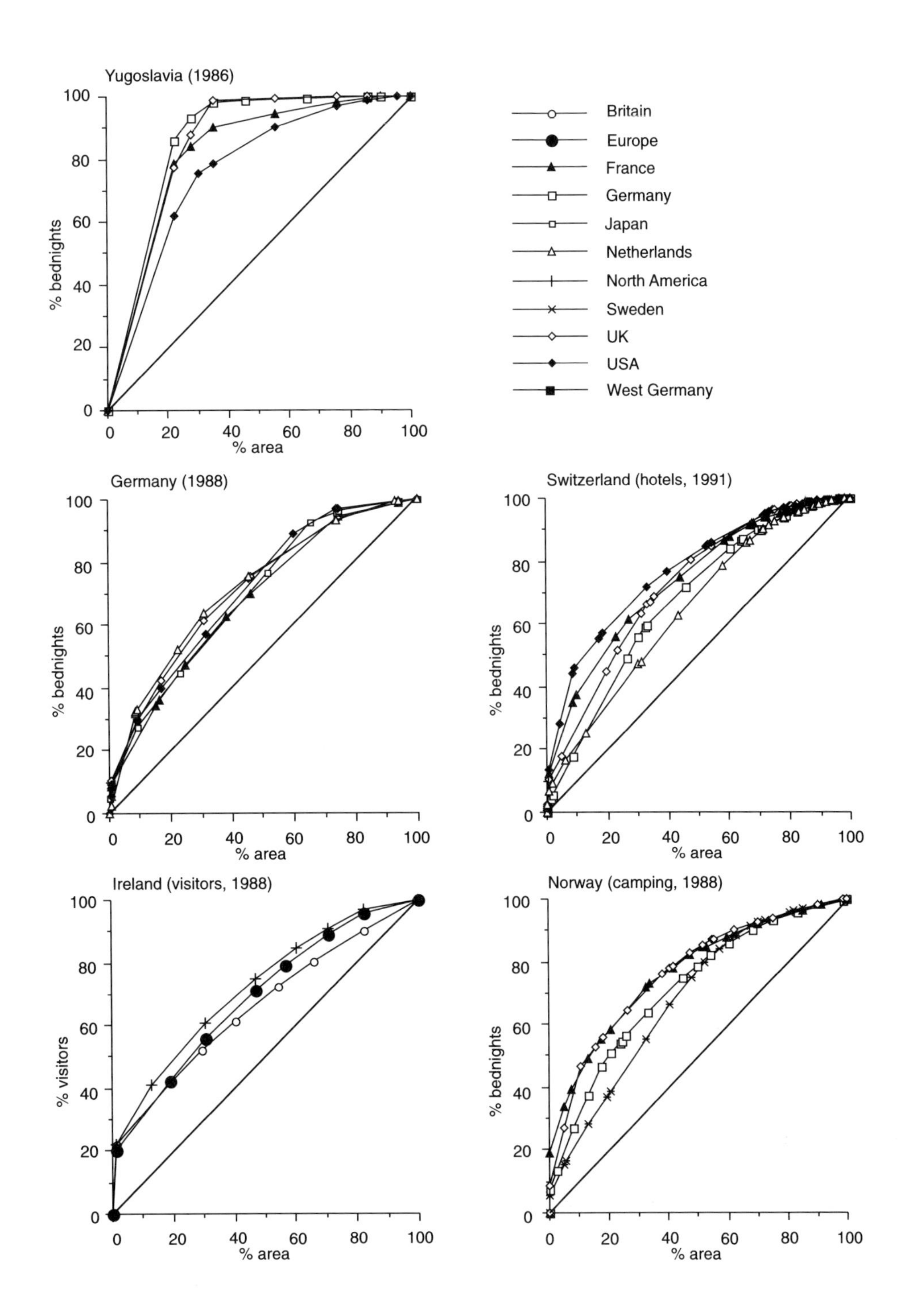

Figure 7.3 *(cont.)*

dispersed, favouring in particular the coastal and water-based resources of Zeeland, Friesland and Limburg. A similar pattern is apparent in Austria where Vienna accounts for two-thirds of Japanese bednights and one-third of Americans but only 2.4 per cent of German bednights while the Tyrol is by far the leading destination for each of the three European markets (France 59 per cent; UK 55 per cent; Germany 42 per cent). Again, the neighbouring Germans, the largest market, are the most widely dispersed of the three European groups, also being over-represented in Voralberg, Salzburg and Corinthia.

In Italy, the mix of resources and markets produces variations on this general pattern. The Japanese are heavily concentrated in the urban and cultural regions of Lazio (38 per cent), Lombardy (20 per cent), Veneto (13 per cent) and Tuscany (14 per cent). Americans follow a similar pattern to the Japanese, adding Campania (with Naples) to the list of favoured regions to produce a slightly more widespread pattern. Campania had the largest location quotient for the British as a consequence of the popularity of Capri, but Veneto had marginally more bednights (19 and 17.8 per cent respectively). Veneto and Campania are also popular with the French but this market is distinguished by its preference for Sicily (16 per cent of bednights). German demand in Italy is more concentrated; one-third of German bednights were spent in the adjoining alpine region of Trentino Alto Adige, a function of propinquity and complementarity, while the Venice region attracted one-fifth of the demand.

The reversal of the concentrated wanderlust/dispersed sunlust pattern in the former Yugoslavia reflects the geography of the cultural and coastal features. German, French and British demand is heavily concentrated in Croatia, respectively 86, 78 and 77 per cent of bednights, particularly in the Dalmatian coastal resorts. Croatia also attracted 62 per cent of American demand, but much of this was associated with sightseeing circuits which took in not only Dubrovnik and Split, but also other parts of the country (Slovenia 13 per cent; Serbia 11 per cent).

In other instances, cultural or wanderlust tourism might dominate but the difference between the European and Japanese/American market still persists. Such is the case in England where Greater London accounted for 54 per cent of Japanese and 50 per cent of American demand but only 32 per cent and 25 per cent of bednights from French and German visitors respectively. The Helsinki region is the prime destination of American (85 per cent of bednights) and Japanese (83 per cent) visitors to Finland. The region also captures a significant share of British (75 per cent), French (60 per cent) and German (38 per cent) bednights. Visitors from the latter two markets are also particularly attracted to Lapland. Swedes, who constitute Finland's largest market, are drawn to the Helsinki region (42 per cent), but are also over-represented in the neighbouring regions of Vaasan, Turun ja Porin and Ahvenanmaa.

Germany represents an interesting case, for although little divergence between the markets is apparent from the graph, more detailed analysis of the regions reveals some quite striking differences due in part to propinquity. In general, the three city states of West Berlin, Hamburg and Bremen record the largest location quotients but this is effectively due to their small size relative to the other *Länder* as together they accounted for less than 10 per cent of the bednights for each of the markets. In contrast, Bavaria, the largest state, attracted one-quarter of the total demand while recording a location quotient of 0.88. Much of the foreign demand within Bavaria is concentrated in Munich. Bavaria drew one-third of the Japanese and American bednights, but demand from these markets was also spread to Hesse, the Rhineland-Palatinate and Baden-Württemberg, the four *Länder* constituting about 80 per cent of their total nights in Germany. The second-ranking of Hesse can be attributed to Frankfurt's prominence as the German financial centre, the gateway role of Frankfurt airport and, for the American market, the state, like Bavaria, constituted part of the US zone of occupation. The British market is spread fairly evenly among the above four *Länder* and North Rhine Westphalia (19 per cent), the latter again a function in part of occupational ties. Propinquity is a factor in the cases of France and Baden-Württemberg (25 per cent of bednights) and of the Netherlands (Germany's largest foreign market) and the Rhineland-Palatinate (29 per cent) and North Rhine Westphalia (19 per cent). These different national preferences reinforce the general tendency towards dispersal to produce the fairly flat curve for all foreigners shown in Fig. 7.2.

Similar patterns are evident in Switzerland. Basel has the highest location quotient in all markets, again in part due to its small areal extent. Americans, and to a lesser extent the British, are prominent in the urban cantons of Basel, Geneva, Luzern, Zurich and Berne. Propinquity and a common language lead to a disproportionate share of Germans in the Tessin and Graubuenden and of the French in Geneva, Waadt and Wallis while the Dutch are the most widely dispersed.

The use of localization curves in Figs 7.2 and 7.3 thus allows a large amount of data to be brought together and cross-sectional comparisons to be made. In interpreting the patterns so identified, however, closer examination of the situation in each country is required. While one level of explanation has been provided here there is also scope for more in-depth treatment of individual destinations using other analytical techniques, such as the analyses by principal components of tourism in Belgium (Grimmeau 1980) and Spain (Pearce and Grimmeau 1985) noted in Chapter 6.

Grimmeau identified three major groups of visitors on the basis of their observed spatial preferences: the Dutch (26.8 per cent of all foreign bednights in 1977), other neighbouring countries (former West Germany, Luxembourg, France and Great Britain, totalling 46.7 per cent) and more distant markets (Italy, Spain, USA and others, 26.5 per cent). Grimmeau suggested that the distribution of tourists from neighbouring countries could be explained in terms of proximity and complementarity. Proximity is a major factor in the predominance of the Dutch in the Campine, the over-representation of the French in certain frontier arrondissements such as Furnes and Ypres, and the greater share of British tourists in Ostend, the Channel ferry port. Complementarity in physical resources gives rise to the Dutch concentration in the Ardennes and the German and Luxembourgeois preference for the Belgian coast. Longer-haul visitors are much more significant, both proportionately and numerically, in Brussels and Antwerp. While much of this can be attributed to a *villes d'art* form of tourism, other travel is associated with the commercial function of these cities and with visits to friends and relations. Special factors are also evident, such as the location of SHAPE headquarters which increases significantly the American traffic to Mons-Soignies.

In the case of hotel demand in Spain, Pearce and Grimmeau's (1985) results suggested three categories of tourists, domestic tourists, visitors from Spain and those from the rest of the world (Fig. 6.1). Domestic demand predominated in the urban and historical centres, notably Madrid, Barcelona and the Moorish cities of Seville, Granada and Cordoba, as well as in the inland and Atlantic coastal provinces. Much of the urban-centred domestic demand appeared to be related to non-vacation travel.

The distribution of European and rest of the world bednights again typifies to a large extent patterns of sunlust and wanderlust. Demand from European visitors was highly concentrated in the islands and major Mediterranean coastal provinces where charter tourism plays a significant role. Visitors from beyond Europe focused primarily on the urban and historical centres which are visited by many as part of a cultural circuit. Madrid's traffic is also augmented by its gateway role and central place functions. The intra-national travel patterns which Gray (1970) associated with sunlust and wanderlust (Table 2.2) are evident in other research results. Over 40 per cent of respondents in a survey of 1500 foreign visitors to Spain indicated that the locality in which they were interviewed constituted the sole place visited on their trip (Equipo Investigador del IET 1981). The proportion varied from insular and coastal zones (Canary Islands, 75 per cent; Costa Blanca, 65 per cent) to urban and historical centres (Madrid, 24 per cent; Barcelona, 11 per cent; Seville, Cordoba and Granada, 6 per cent). North Americans were much more mobile than Germans, with only 9 per cent of the former group making a single stop compared with 63 per cent of the latter.

Pearce and Grimmeau (1985) also found that over the period 1965–80, the spatial preferences of each nationality remained remarkably constant. However, as growth rates differed significantly from market to market during these years, marked regional differences in demand occurred. The number of hotel bednights increased much more rapidly in the coastal and insular provinces favoured by the Europeans, than in the regions more heavily dependent on the 'rest of the world' and domestic markets.

The results of these individual studies generally confirm the conclusions based on the comparative analysis presented earlier with regard to the spatial structure of domestic and international tourism and variations in the latter. To test their universality, however, further examination of these issues is now required in parts of the world other than Europe. Wide-ranging analysis is hampered by the paucity of appropriate data elsewhere but Fig. 7.4 depicts two examples from Malaysia and New Zealand. In each case a marked degree of concentration occurs, particularly in Malaysia due to the popularity of Kuala Lumpur and Penang, and international bednights are more concentrated than domestic ones. In terms of individual market behaviour in New Zealand, the Japanese again exhibit the most concentrated pattern but, in contrast to some of the discussion of intra-national patterns in Chapter 4, the localization curves do not discriminate markedly between the spatial behaviour of the other long-haul markets (Great Britain, the USA and Germany) and the short-haul Australians. This lack of divergence may reflect the general emphasis in all markets on sightseeing, with very little sunlust activity occurring. The inclusion of all visitors – the VFR traffic being especially important from the United Kingdom and Australia – may also have a levelling effect.

While it is tempting to emphasize the consistently highly concentrated nature of Japanese tourism in New Zealand as well as Europe, it should be recalled that this too is a long-haul destination for this market. What happens in Asia where distances from Japan are much less? Likewise, what is the pattern in Canada, Mexico and the Caribbean where the Americans constitute the short-haul markets and the Europeans the travellers from further afield? Are their respective roles and patterns reversed there? What will happen in developing countries as their potential to generate large volumes of domestic tourism is realized over coming decades? In

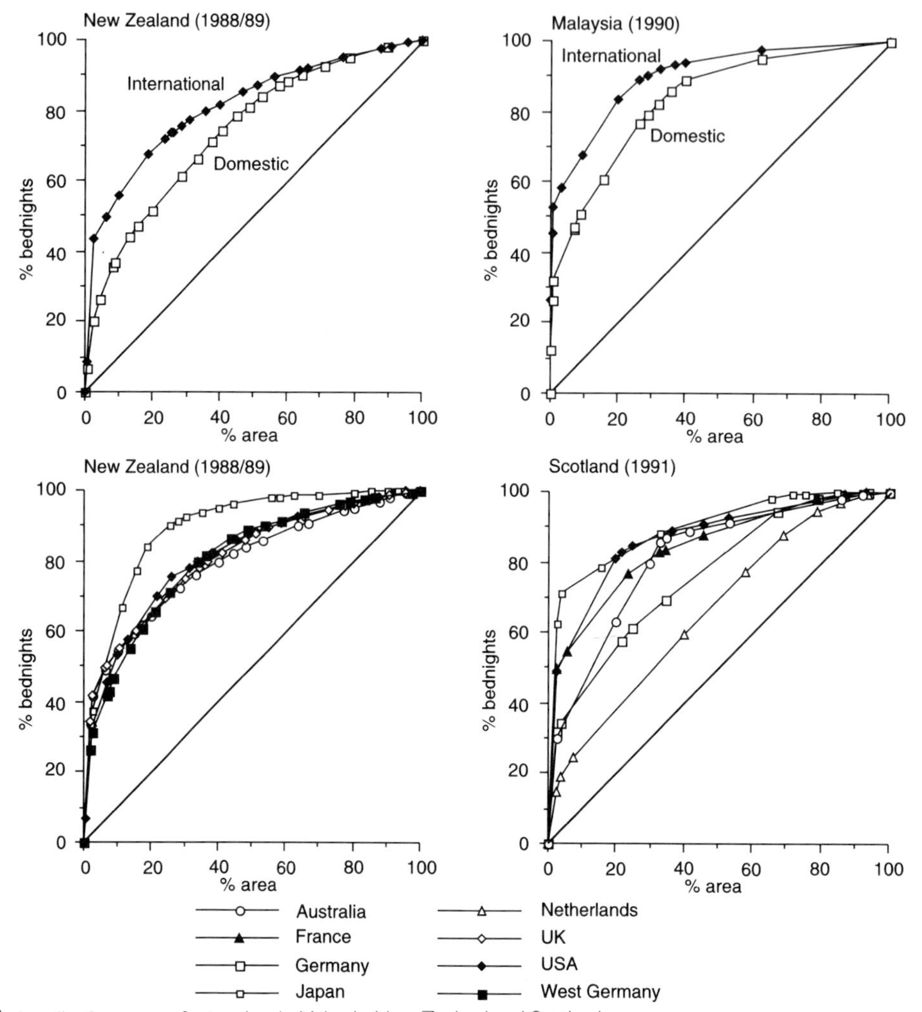

Figure 7.4 Localization curves for tourism in Malaysia, New Zealand and Scotland.
Data sources: National tourist organizations.

one of the few such studies undertaken so far, Berriane (1992) shows that the rapidly emerging domestic market in Morocco exhibits a similar degree of concentration to international tourism and that Moroccan tourists favour coastal destinations, and not just major cities located along the coast.

The Spatial Structure of Tourism in Regions

There have been comparatively few studies at the regional level of the extent to which differences occur in the structure of domestic and international tourism. In part this appears to be a function of the data available. Although the national studies are often based on bednight returns made at a local level, the data are seldom published or made available at this scale. As a result, the regional studies of variations in market composition have usually been based on specific surveys or analysis of second home ownership records. Studies at this scale have tended to be concerned more with the functional structure of tourism, giving attention to the nature and location of tourist facilities and their relation to landscape features and other forms of land use. Such an approach is in keeping with the regional focus of much European geography in general (Barbier and Pearce 1984) but it is one which has been adopted by few Anglo-American tourist geographers. This is no doubt a reflection of general trends in their discipline in recent decades which have favoured more systematic and quantitative methods, though (as noted in Chapter 6), there have been some recent attempts to apply these to regional problems (S. L. J. Smith 1987; Lovingood and

Mitchell 1989; Backman, Uysal and Backman 1991). Studies of the spatial structure of tourism in regions can provide a useful bridge between national analyses of tourism and the many more detailed case studies of individual resorts and cities (Chapter 9).

This section reviews examples of different types of regional studies in Europe. The examples from Scotland and France focus on spatial variations in the composition of different geographic markets, those from Spain emphasize the structure of varioius forms of tourism while that of the Belgian coast constitutes an intermediate case, containing elements of both supply and demand. The functional structure of each of the coastal examples is then examined in terms of the various factors outlined in Miossec's model of the development of a tourist region (Fig. 1.12) to provide some basis for general comments on the spatial structure of coastal tourism at a regional level.

Scotland

Spatial variations in international demand within Scotland can be determined from the International Passenger Survey data which are made available for the ten Scottish Tourist Board districts (Fig. 7.4). Comparison of Figs 7.3 and 7.4 indicates many of the patterns identified at the national level are replicated at the regional scale. Within Scotland, a marked concentration of international demand occurs, with almost 70 per cent of all foreign bednights being spent in the two major urban districts of Lothian (Edinburgh) and Strathclyde (Glasgow). The Americans and the Japanese are the most concentrated, these two districts accounting respectively for 81 and 72 per cent of their bednights. Differences exist between these markets, however, as almost two-thirds of all Japanese demand is generated by Edinburgh whereas Glasgow attracts one-third of American bednights in Scotland. Europeans are again more dispersed although the two cities continue to attract a significant share of the demand, particularly from the French (71 per cent) and German (55 per cent) markets. In relative terms the Dutch are the most widespread travellers, the Highlands attracting one-third of all Dutch bednights. While the Dutch have a greater propensity to visit the Highlands – a further illustration of complementarity – in absolute terms Americans and Germans spend more time in the region as they are the two largest markets overall for Scotland.

Provence and Languedoc-Roussillon

Tourism has developed in different ways and at different rates along the Mediterranean coast of France. Provence, to the east, has a well-established tourist industry which grew up gradually in the nineteenth century, and has since experienced substantial spontaneous expansion to become the country's foremost tourist region. Languedoc-Roussillon, however, has been the object of a major, highly planned programme of development initiated in the early 1960s which has been responsible for lifting the area from one of regional importance to national and even international significance (Pearce 1989b). Given these and other differences, for example in the urban network, it is not surprising that variations occur in the patterns found in each region. At the same time, however, certain common tendencies exist.

In Provence, significant differences occur in the origins of second home owners along the coast and between the coast and the interior (Barbier 1975). The highest rates of foreign ownership are found from St Tropez to the Italian border, a zone where extra-regional ownership, particularly by Parisians, is also the most significant (Fig. 7.5). Classified hotel capacity, both in absolute and relative terms, is also generally much greater along this part of the coast. Further west, notably in Cassis and in the lower reaches of the Rhône Valley, most of the second home owners are from the region itself, particularly from Marseille. History and geography contribute to this division. International tourism has been important on the Côte d'Azur proper (from Cannes to Menton) since the area first began attracting tourists during the winter months in the late eighteenth and early nineteenth centuries. Cannes's origins as a winter resort date from 1839 when an Englishman, Lord Brougham, was obliged to spend some time in the village on his way back from Italy due to an outbreak of cholera in Provence. He found the locale and climate agreeable and returned the following year to build his villa; his English friends soon followed suit. Cassis owes its development to its proximity to the largest metropolitan area in the south of France as well as to its charming port and *calanques*. Most of the second homes in interior Provence are owned by coastal dwellers from within the region, with foreign and extra-regional ownership being noticeable in the immediate hinterland of Nice and Cannes and in the more accessible Rhône Valley. Proximity, complementarity and sunlust are factors contributing to this pattern.

A different pattern of second home ownership west of the Rhône is suggested by Andrieux and Soulier's (1980) study of Languedoc-Roussillon. Their figures indicate that about half of the second homes belong to owners living within the region, another 40 per cent from owners residing elsewhere in France, with about 8 per cent being in foreign ownership. Proportionately more of the second homes on the coast belong to local (within the same commune) and regional residents, many of whom have purchased them as an investment and for rental to other users. As a result, differences in

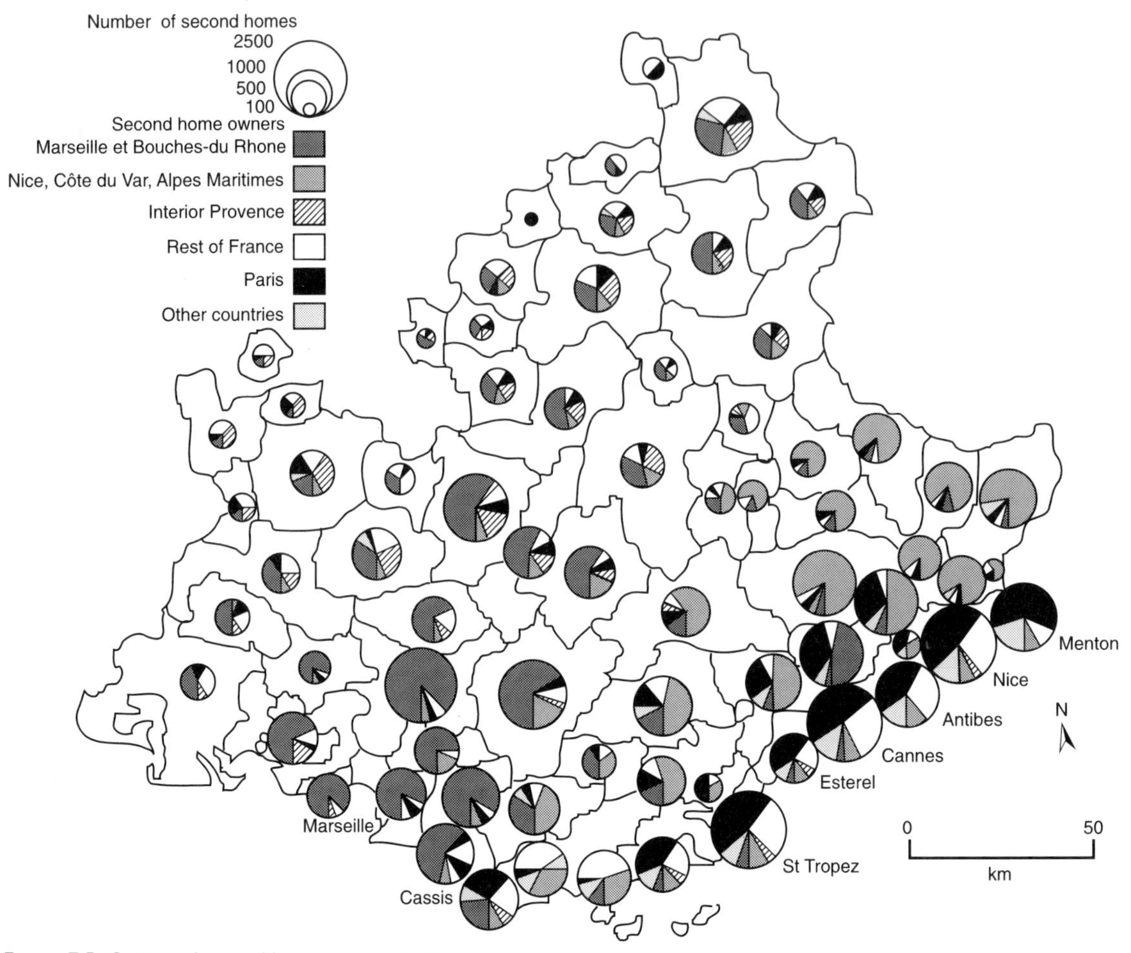

Figure 7.5 Origins of second home owners in Provence.
Source: After Barbier (1975).

the origins of owners and users appears to be greater here than on the Côte d'Azur. Further inland, foreign ownership is about twice the regional average in the Garrigues-Soubergues district, 'an area which is not important agriculturally, is not densely populated but which is still relatively close to the coast'. Differences are also found throughout the region in the composition of the international sector. Germans, for example, are over-represented on the coast, while Belgians are relatively more numerous along the margins of the Rhône Valley and Massif Central, with Dutch second home owners being dominant in other upland areas. In the absence of absolute figutres, these trends must be regarded with caution, but it does seem that the complementarity of resources plays a role in the distribution of international demand within regions as well as between them.

Other research suggests that significant differences occur in the structure of demand along the Languedoc-Roussillon coast. Fornairon (1978) carried out an extensive survey of the origin of cars at different beaches throughout the region during the summer of 1977. Such an approach is likely to overestimate regional demand, through a greater likelihood of nearby beach users travelling by private car rather than some other mode of transport and through the inclusion of regional day trippers, but the general trends are nevertheless interesting. These indicate that the pattern of usage reflects the history of tourist development in the region. Regional demand is proportionately greater in the older resorts, many of which, through access and proximity, have long-established ties to cities located further inland, for example Grau du Roi to Nîmes, Palavas and Carnon to Montpellier. Foreigners, who accounted on average for 12 per cent of the car registrations, are over-represented both in the large modern resorts and in the

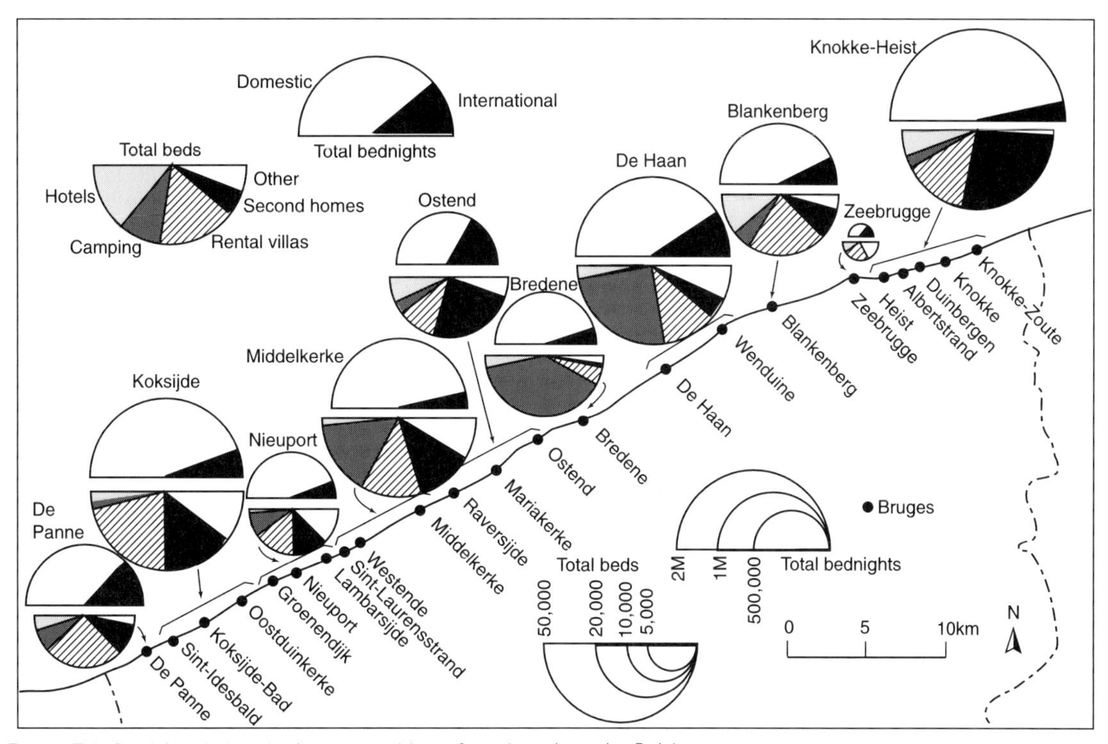

Figure 7.6 Spatial variations in the composition of tourism along the Belgian coast.
Data source: Vanhove (1980).

underdeveloped areas, the *plages sauvages* (L'Espiguette, 15.6 per cent; Les Aresquiers, 16.8 per cent). Overall, there was little variation in demand from Parisians throughout the region and no systematic variation in the demand from elsewhere in France was apparent.

Similar variations in demand between older and newer resorts were also reported by Ashworth and de Haan (1987) in a shoulder-season survey of resorts from Port Camargue to Gruisson. When images of these resorts were considered in an innovative aspect of this study, few marked differences were identified. However in terms of the functioning of the coast, Ashworth and de Haan noted (1987: 71): 'Most holiday visitors stay on their trips within a certain sector along the coast and its hinterland, this indicates an "activity region" of interrelated facilities'. They later conclude that the key functional unit is the resort not the region or the subregion of the physical planners, asserting (p. 73):

> It is the resort not the cluster that largely circumscribes the behaviour pattern of holidaymakers, that holds their loyalty in subsequent years, attracts day visitors and second homes owners from particular inland towns and is publicized and conceptualized by the visitor.

The Belgian coast

Despite its limited length, significant differences occur in the structure of tourism along the Belgian coast, both in terms of supply and demand. With over twenty resorts located along its 67 km, virtually the entire length of the coastline has been developed to some degree for tourism. Rarely, however, does the tourist zone extend for more than a kilometre inland, the urbanized area being abruptly terminated in many instances where the polders are reached. The oldest resorts – Ostend, Blankenberg and Knokke – were already well established by the end of the nineteenth century (Vanhove 1980). While these resorts benefited from early rail links, the construction of a coastal tramway later opened up access to other parts of the littoral. This tramway, now paralleled by the road system, continues to play a significant role in unifying the region. Many of the new resorts were coastwards extensions of existing communities which have been separated from the sea by large dune systems, e.g. Koksijde-Bad and Koksijde, Oostduinkeerke-Bad and Oostduinkerke. These dune systems are still responsible today for breaking up the ribbon development of the coast, separating, for instance, De Panne from St Idesbald and St Idesbald from Koksijde-Bad.

Figure 7.6 shows that for the coast as a whole, camping, rental accommodation (villas and apartments) and second homes each account for about a quarter of the total bedspace along the coast, with the remainder being shared by hotels and other forms of accommodation (essentially holiday villages and various forms of 'social' tourist centres). However, as Fig. 7.6 shows, the composition of the accommodation varies from commune to commune. According to Flament (1973), the coast has traditionally been divided into three distinct sectors on the basis of the accommodation available and the history of tourist development. Hotels are relatively more important in the eastern zone with its old, established aristocratic resorts of Knokke, Heist and Blankenberg. But even here hotels have long been surpassed by the bedspace available in villas and second homes, the latter often taking the form of apartments in high-rise buildings along the beachfront. With the exception of Ostend, which shares a similar structure to Knokke and Blankenberg, the central zone (from De Haan to Middelkirke) has attracted tourists of more modest means. Camping is the dominant form of accommodation here, particularly at Bredene where tents were first pitched in 1912 (Vanhove 1980). The western zone, where tourism expanded in the inter-war period, is characterized by its broad beaches, developed beachfront, villas, holiday villages and social tourism centres. Here the twinning of resorts with established settlements is particularly noticeable (Flament 1973).

Figure 7.6 also emphasizes the role of domestic tourism along the coast, with Belgian holidaymakers accounting for about 90 per cent of all estimated bednights in 1978 (Vanhove 1980). However, the domestic share fell to 68 per cent at Ostend and 77 per cent at De Panne, largely as a result of the concentration of British tourists in the former, the ferry port, and of French holidaymakers in the latter, a border resort. Flament (1973), commenting on the results of a more detailed survey in 1969, showed that visitors from the north of France, the largest group, tended to favour De Panne and Koksijde, while Parisians were relatively more numerous in the renowned resorts of Knokke and Ostend. The Parisians favoured hotels and pensions while the majority of holidaymakers from the north rented villas or apartments. This latter tendency has attributed to the greater facility which those in the north had for establishing contact directly with rental agencies and individuals. Flament also suggests that the presence of the French along the western part of the coast is in part responsible for a greater concentration of French-speaking Walloons in this zone. Flemish holidaymakers are particularly dominant in the east, while vacationers from the Brabant (Brussels) are more equally distributed along the length of the coast. Proximity, communications (e.g. the rail links from Bruges to Blankenberg and Knokke) and habit are factors contributing to these variations in domestic tourism.

Finally, it should be recalled that the coast accounts for two-thirds of domestic bednights and over half of all tourist bednights in Belgium. It is this tourist function which sets the Belgian coast apart from the adjoining industrialized Channel coast of France whose physical attributes are, however, very similar. As Dewailly (1979) notes, these differences must be seen in a national context. France's long coastline provides a range of much better opportunities elsewhere for tourist development as well as space for other forms of economic activities. In Belgium, the choice is strictly limited. Although some industrial development has occurred at Zeebrugge, it is Antwerp on the Schelde estuary which has assumed the major port-industrial role, thus fortunately reducing the potential for conflict with tourism along the coast itself. That one can stand on the promenade at St Idesbald and look out to the industrial smoke stacks at Dunkirk is in large part due to the intervening political frontier and to decisions taken in two inland capitals.

The Costa Brava, Costa Blanca and Costa del Sol

Some of the world's most intensively developed coastal tourism is to be found along the Mediterranean coast of Spain (Fig. 6.1). Much of this development occurred from the early 1960s and was often spontaneous and unplanned. The scale and extent of coastal tourism in Spain has generated a number of in-depth regional and local studies, notably by Spanish and French geographers. These have generally focused on physical forms and functional structures rather than on the origins and other characteristics of visitors. Different types of structures and forms of development do, however, reflect different types of tourism which are discernible, if not always quantifiable, along various stretches of the Mediterranean coast. Regional studies have tended to focus on individual *costas*. The three main ones are examined in this section: the Costa Brava, Costa Blanca and Costa del Sol. Here the development of mass tourism has resulted in a major re-orientation of the regions' economies and structures towards a narrow coastal fringe (Barbaza 1970; Vera Rebollo 1987; Jurdao 1990). As detailed fieldwork has shown, however, this development has been far from uniform with distinctive differences being detected from resort to resort and from one stretch of the coastline to another.

Barbaza (1970) notes that prior to the development of tourism, the Costa Brava was rather isolated and had not been dominated by any large city, neither Gerona, further inland, nor Barcelona, further along the coast. The rapid and massive development of tourism in the 1960s ended this isolation, linking the coast, by means

Plate 4 The colonization of hillsides by tourism: villas sprawl across Punta Ifach, Costa Blanca, Spain.

of flows of workers and capital as well as the new transport infrastructure (new motorways and the Gerona–Costa Brava airport), to the rest of Catalonia, to other parts of Spain and even to neighbouring countries. At that time (1970), a fully developed hierarchy of urban areas along the coast had not yet emerged but, as Barbaza observed (1970: 451), 'tourism introduces a powerful unifying element and leads to the development of a truly functional region, *l'espace touristique*'. Cals, Esteban and Teixidor (1977: 200) identify two major markets on the Costa Brava: 'in several centres, Lloret, l'Estartit, Tossa, Rosas et Blanes, etc, the dominant form of tourism, which has expanded rapidly since 1965, is that of the tour operators. Elsewhere, on the contrary, the tourist clientele, continues to be basically private and individual'. The first form of tourism is typically reflected in large densely settled resorts, comprised of high-rise hotels and apartments, the second by sprawling subdivisions of villas known as *urbanizaciones*.

Further south, in the region of Alicante (see Fig. 6.1), Dumas (1976: 45) observed: 'One of the major characteristics of the Costa Blanca is the juxtaposition of very varied types of tourism: family style villas as at certain French beaches, or organised, industrial tourism linked to the tour-operators'. Benidorm is the classic example of the latter. Popular originally with well-to-do Madrileños, being the closest resort to the capital, Benidorm during the 1960s, 'became part of the tourist space of the North European tour operators, in particular the British and Germans'. By the mid-1970s, with a resident population of only 12,000, Benidorm was able to accommodate 120,000 visitors at the one time. Factors accounting for its development include a splendid physical site (two long curving sandy beaches separated by a promontory), access via the N332 and later the Valencia–Alicante motorway and proximity to the Alicante airport (40 km away) which was modernized in 1967.

North of Benidorm, *urbanizaciones* sprawl over the broken relief of the Cap de la Nao, as hillsides have been progressively colonized by villas, many owned by German families (Plate 4). Subdivisions here tend to be smaller in scale and larger in number than those found to the south of Alicante. Construction of apartments and villas has also occurred around existing coastal communities such as Denia and Moraira.

The coast south of Alicante to Cabo de Palos is more sparsely populated. Development there has been more recent and taken the form of a series of large Centros de Interes Turistico Nacional (CITN), resort enclaves in which the activities of the tour operators have been linked with those of the real estate promoters. As CITNs, these developments benefited from generous credit facilities and fiscal assistance (Vila Fradera 1966). With the exception of Santa Pola del Este which has been grafted on to a fishing port with some established holiday homes, the others – La Zenia, Dehesa de Campoamor and La manga del Mar Menor – have been developed *ex-nihilo* on large agricultural holdings at the initiative of the landowner. Dumas (1976: 48) describes the development of La Zenia and Campoamor thus:

> The property along the coast (between the road and the sea) is subdivided and several apartment buildings and bungalows and villas are built. Other serviced sections are put up for sale and to rapidly attract clients [and to give some life to the resort] a large hotel is built and filled by tour operators. In the same way, a large number of apartments are rented through German, Dutch and French agencies.

The process at La Manga del Mar Menor has been similar but on a larger scale and with more tour operator involvement (Dumas 1975). Development there has occurred along a sandy bar, 30 m wide and 22 km long, which separates the Mar Menor from the Mediterranean (Plate 11). Both in its physical setting and its scale of development, La Manga recalls other recent tourist projects, notably the new resorts of Languedoc-Roussillon and Cancun in Mexico (Pearce 1983b).

Vera Rebollo (1987) concludes a later, detailed study of the region with various syntheses of the different forms of tourist development along the Costa Blanca. Stressing the need to integrate a range of factors, in the typologies he derives, Vera Rebollo lists the following as critical elements:

1. Forms of land use, particularly the process of urban development, and the model of '*asentamiento*' (settlement). Types of urban development identified include
   - redevelopment of existing centres, incorporating new forms of dwellings
   - *ensanches*, or zones of expansion adjacent to the traditional centre
   - *rurbanización* – unplanned, dispersed, single family dwellings in rural areas
   - suburban dwellings (around Alicante) where the distinction may be lost between primary and secondary residences
   - planned *urbanizaciones* – housing projects of various sizes based on a *plan parcial*.
2. Types of building, notably
   - high-rise constructions
   - single family dwellings *entre medianerías*
   - detached single family dwellings.
3. The extent to which the development is successfully integrated into its physical setting.
4. Infrastructure.
5. Recreational facilities of different types: marine, sporting etc.

At the same time, Vera Rebollo emphasizes that these physical attributes of development must be seen in terms of the mechanisms which have transformed the littoral, including: planning, capital gain (*el plusvalor*), land values (*el mercado del suelo*), demand and the road network. After analysing the different sectors of the coast in terms of the above typology, Vera Rebollo also delimits functional hinterlands based on commercial and service sector functions, derives a classification of the region's municipalities according to population size and economic base and distinguishes between areas transformed by foreign, national and regional agents of development. While underlining in these ways internal differences in the structure of tourism along the Costa Blanca, Vera Rebollo also situates the region in its international context, noting (1987: 371): 'In summary, the transformation of the coastal landscapes of Alicante, as with other parts of the Mediterranean basin, is a product of its new economic function as the leisure region of industrialized Europe.'

These detailed supply-side studies are complemented by some demand data which reinforce the spatial variations identified. Iribas Sánchez (1991) distinguishes between *turistas*, in his terms the classic pre-paid, stay-put, package tourist travelling by charter and *veraneantes*, longer-stay summer visitors, travelling by car, making all their own arrangements, staying in accommodation not subject to 'industrial management' and whose real expenditure is much more uncertain. Visitors to Benidorm typify the first group, those to other northern parts of Alicante the second. Each has distinctive characteristics. In 1986, *turistas* in Benidorm on average stayed 14 days, travelled in groups of 2 and spent 3250 pesetas per person per day; 60 per cent of them had their trip organized through a travel agent. *Veraneantes* in other parts of Alicante averaged 25 days, had a group size of 3.5, spent less than 2000 pesetas a day each and had no need of travel agents. Irribas Sánchez then explores the development implications of these differences, particularly in terms of the greater demand for services other than accommodation which the *turistas* generate. Table 7.3 highlights the dominant role of Benidorm which accounted for about 80 per cent of all hotel bednights in the province of Alicante in 1991. Benidorm also has a much greater percentage of international bednights than other parts of the province, although even there Spanish visitors generate the greatest demand in the high season. Half of all the

Table 7.3 Distribution of hotel demand in the province of Alicante, 1991

	Average monthly bednights							
	Benidorm		Other coastal areas		Inland areas		Total Alicante province	
	(no.)	(%)	(no.)	(%)	(no.)	(%)	(no.)	(%)
High season								
International	298,522	39.2	32,535	18	1,587	12.9	332,644	34.8
Domestic	463,548	60.8	148,545	82	10,748	87.1	622,841	65.2
All visitors	762,070	100	181,080	100	12,335	100	955,485	100
% of prov. total		79.7		19		1.3		100
Low season								
International	301,921	54.4	14,358	18.4	1,489	13.7	317,768	49.3
Domestic	253,313	45.6	63,625	81.6	9,393	86.3	326,331	50.7
All visitors	555,234	100	77,983	100	10,882	100	644,099	100
% of prov. total		86.2		12.1		1.7		100

Source: After Generalitat Valenciana (1992).

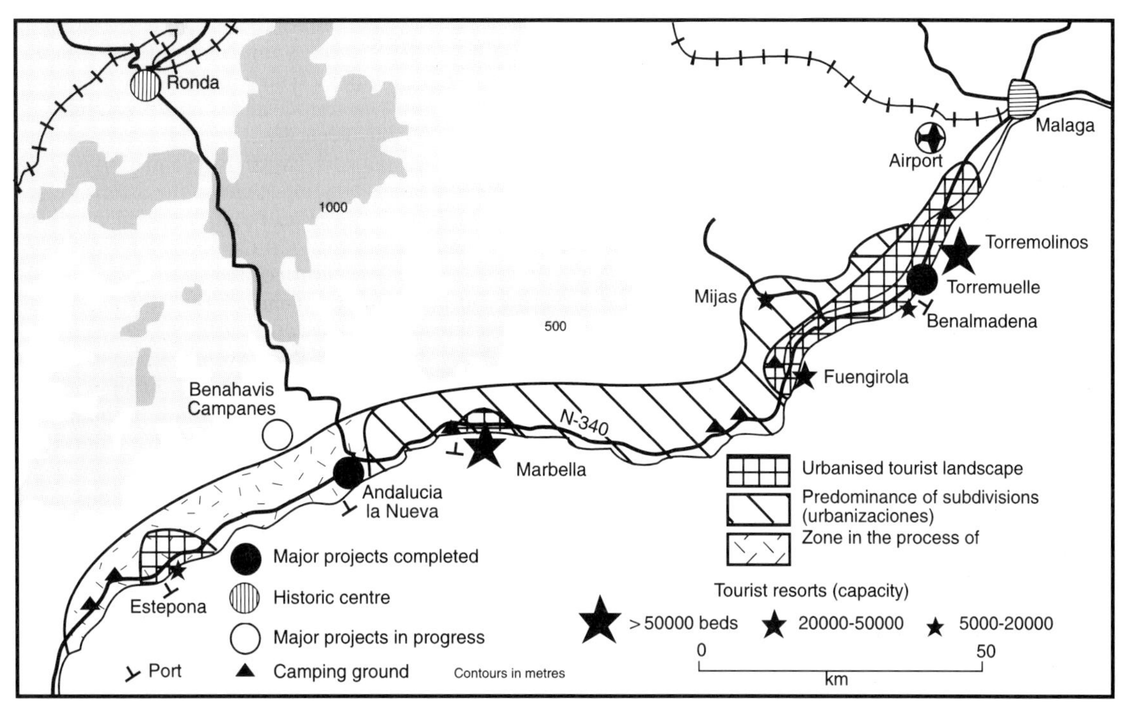

Figure 7.7 The structure of tourism along the Costa del Sol.
Source: Adapted from Mignon and Heran (1979).

international hotel bednights in Benidorm come from British visitors but German, Dutch and other continental European visitors may be more significant in other resorts and outside the hotel sector (Generalitat Valenciana 1992).

A similar mix of resorts filled by package tourists and *urbanizaciones* is found along the Costa del Sol where Malaga is the major urban centre and Torremolinos the counterpart of Benidorm (Fig. 7.7). Villegas Molina (1975) shows Torremolinos had almost twice the hotel bedspace of Malaga in 1970, with the capacity of Marbella exceeding that of the provincial capital as well. Three major structural elements are identified by Mignon and Heran (1979) along the Costa del Sol:

1. the *front de mer* featuring a promenade and dominated by a wall of high-rise hotels and apartment buildings
2. the commercial axis bordering the N340 and containing shops, travel and real estate agencies, and restaurants
3. villas and *urbanizaciones* which cover the lower slopes beyond the resorts.

Between the first two zones, residential quarters and more modest hotels mix with boutiques and occasionally the remnants of the original residential quarters. In most cases, however, local residential areas have been pushed further inland, aside from the resort itself.

These three elements do not assume the same importance along the coast with the density of development decreasing westwards, away from Malaga and its airport to Andalucia La Nueva and Estepona where new *urbanizaciones* dominate (Fig. 7.7). In some cases, inland towns have been incorporated into the regional structure, as in the case of the hilltop town of Migas, and Ronda, and historic town famed for its ravine and bridge, and now, after the construction of a new sealed road from the coast, the focus of day tours. Other excursions are organized to the major Andalucian cities of Sevilla, Granada and Cordoba.

Jurdao (1990) provides a very detailed account of the social, economic and physical transformation of Mijas, particularly the structural reorientation of the commune away from the inland municipal centre to the coastal *urbanizaciones*. Estimates based on comprehensive fieldwork outline the scale of the changes which have occurred: Mijas had a resident population of 8000 in 1965, by 1989 Jurdao estimates its 'real' population at 67,000, of whom about three-quarters were foreigners spending extended periods there, not simply vacation visits. Jurdao too sets this local change in a wider context, denouncing (1990: 16) the broader national policies and international processes by which Spain, in his terms, is being sold: 'The Spanish supermarket is open twenty-four hours a day [to foreign investors]'.

Regional structure

The limited number of examples examined show that differences occur in the composition of the tourist traffic not only between but also within regions. In the cases of the Belgian coast, Provence and, to a lesser extent, Languedoc-Roussillon, variations were found in the distribution of domestic and international tourists. Distinct spatial preferences may also exist within these two markets. Flemish and Walloon vacationers, for example, tend to take their holidays along different parts of the Belgian coast while French and British tourists there favour the resorts most accessible to them. Within Scotland varying patterns of concentration were found among international visitors that generally replicated those observed at the national scale. Different forms also reflect different types of tourism along the Mediterranean coast of Spain. These variations are less quantifiable, but certain resorts or zones are dominated by hotels and apartments oriented towards package tourists with *urbanizaciones* and camping grounds catering more for domestic tourists and independent foreign holidaymakers.

These regions are clearly at different stages of maturity in terms of Miossec's model (Fig. 1.12). The coasts of Belgium and of Provence have undoubtedly reached a very mature stage, being well developed or even saturated for virtually their entire length, with a dense and well-integrated transport network being established and tourism coming to shape and even dominate the urban and economic structure of each region. Market specialization occurs in both cases, distinct variations in types of accommodation are found and a hierarchy of resorts and urban centres is apparent, especially on the Côte d'Azur. Languedoc-Roussillon, the Costa Brava, Costa Blanca and Costa del Sol are also maturing. A range of resorts has been developed in each case, the communications network has been expanded to include regional motorways and, especially in the Spanish case, international airports and an overall regional structure has emerged. In Languedoc-Roussillon, a formal hierarchy of resorts was imposed by the official development plan while in Spain major resorts such as Torremolinos and Benidorm, which started to expand dramatically in the 1960s, have been complemented by smaller resorts. The development of excursion circuits is illustrated in Fig. 7.7 by the road link to the inland town of Ronda. Research in Languedoc-Roussillon suggests that such circuits there are rather localized.

In each of these cases, major resorts and indeed tourist regions have grown up in relatively undeveloped areas, often some distance away from the large urban centres. This appears to be the case especially with those areas catering for international visitors and non-local domestic tourists. In Provence, for example, international and inter-regional domestic tourism has developed over a long period along the easternmost part of the coast, essentially outside Marseille's zone of influence and recreational hinterland. Likewise the Costa Brava, where tourism has developed much more recently, fell outside the zone of influence of both Gerona and Barcelona. In many cases, small fishing and other villages formed the nucleus of resorts, but from the 1960s a number of large projects were created *ex-nihilo* on sparsely populated parts of the coast. Factors contributing to this pattern include historical accident (as at Cannes), the non-compatibility of coastal tourism and industrial and other urban uses, variations in the resource base, the availability of relatively cheap land and the development of package tours to self-contained resorts. Land appears to have been a particularly important factor in Spain where property speculation fuelled much of the expansion of tourism, particularly in the form of *urbanizaciones*. Even in the state-planned project to develop the coast of Languedoc-Roussillon, the existence of large blocks of undeveloped land was a major criterion in the selection of sites for the new resorts. Given the relatively isolated and undeveloped nature of these regions, the willingness of the state to provide infrastructural assistance was a significant development factor, not only in Languedoc-Roussillon but also in Spain. At times in Spain, major problems emerged when the provision of infrastructure did not keep pace with the demand-led tourist expansion. This new infrastructure and the scale of tourist development have clearly had a marked impact on the geographical structure and character of each of the regions studied. In Provence and along the Belgian coast, this character is now well defined. Elsewhere the process of regional restructuring and reorientation along the coast is still taking place.

Conclusions

Significant spatial variations occur between and within countries and regions in terms of the composition, volume and relative importance of domestic and international tourism. While the structure of tourism in each country and region reflects local conditions and unique features, certain general patterns have emerged from the comparative approach adopted here. These include: greater spatial concentration of international compared to domestic tourism and of longer-haul visitors from the USA and Japan *vis-à-vis* those from shorter-haul European markets, the preference exhibited by long-haul tourists for wanderlust destinations and short-haul tourists for sunlust regions, the growth of some major tourist destinations in comparatively isolated and undeveloped coastal regions and a consequent reorientation of regional structures and economies.

Without this systematic comparison, the definition of specific objectives and questions at the outset (prompted in part by existing models) and the use of a common method or framework of analysis, it would not have been possible to distinguish the general patterns and trends from situations peculiar to a particular country or region. An attempt has also been made to explain and account for the patterns and trends identified in this way. Although the patterns described and explanations given must be seen in the light of the range of examples used, they perhaps advance our understanding of the processes involved in tourist development and our application of the structure of tourism in general more than the isolated, ideographic studies which have characterized this field of research in the past.

CHAPTER 8

The Spatial Structure of Tourism on Islands

In Chapter 3 it was shown that in absolute terms the largest international tourist flows are within continental Europe, in North America and between these two regions. The developed countries on both sides of the Atlantic also experience large-scale domestic tourism. However, in relative terms international tourist flows are also very significant in many islands in the Caribbean, the Mediterranean, the Pacific, the Indian Ocean and in South East Asia. Cazes (1989) estimates that islands received about half the international tourist traffic in the inter-tropical zone in 1985, some 19 million arrivals. On many small islands, the ratio of tourist arrivals to the host population, and particularly to the land area, is much greater than in North America and many parts of Europe (Huetz de Lemps 1989a; McElroy, de Albuquerque and Dioguardi 1993). In comparing the development of tourism in the Caribbean and the Pacific on these measures, McElroy *et al.* also highlight the diversity which exists between islands with greater densities and a higher level of development generally being recorded in the Caribbean.

As with tourism at other scales, research on the spatial structure of tourism on islands has largely been confined to case studies of particular islands, in which the spatial aspects, depicted in the main by the distribution of tourist accommodation, form part of a general overview of tourism. Exceptions to this include the comparative study of Corsica and Mallorca by Richez and Richez-Battesti (1982) and the more conceptual work of the structure of tourism in peripheral economies (noted in Chapter 1) by Hills and Lundgren (1977), and S. G. Britton (1982) who exemplify their work by reference respectively to islands in the Caribbean and in the Pacific.

This chapter identifies systematically the various interrelated factors which influence the spatial structure of tourism on islands, particularly those which distinguish tourism on islands from that in other areas. Examples are drawn from a range of islands, which differ in their location, size, resource base, political status and their degree of tourist development (Table 8.1). In particular, Table 8.1 provides some perspective on accommodation capacity, highlighting the massive amount of development on Mallorca and in the Canary Islands compared to other regional leaders such as Bali and Oahu.

Basic Patterns

Many of the major features of the spatial structure of island tourism are shown in Figs 8.1 to 8.8 which depict the distribution of tourist accommodation on selected islands in the Mediterranean, the Caribbean, the Pacific, the Atlantic and South East Asia. The general pattern is one where tourist accommodation is concentrated in a small number of coastal localities in or close to the major urban centres. Metropolitan San Juan accounted for over two-thirds of the 8400 tourist rooms in San Juan in 1992, marginally down on the situation in 1980. In other instances, clusters of hotels are located adjacent to or outside the major urban centres. In Guadeloupe most of the hotels are situated along the Riviera Sud, with little tourist accommodation being found either in Pointe-à-Pitre itself or the capital, Basse Terre. Over half the accommodation on Mallorca in 1986 was located in the Bay of Palma. While many of these hotels are in the city itself, much of the accommodation occurs along intensive coastal strips further around the bay, for example at El Arenal and Magaluf (Fig. 8.1). Other tourist accommodation is spread along the east coast in ports such as Alcudia and Pollensa. Such concentration reaches an extreme on Oahu in Hawaii where 90 per cent of the tourist accommodation is concentrated just outside Honolulu in Waikiki – a massive 32,800 rooms on a one square mile site (Pearce 1992b). On the other Hawaiian islands, there are two to four major resort areas. For example, 60 per cent of the units on Maui are found along the Kaanapali coast between Lahaina and Napili, with 36 per cent along the stretch from Kikei to Wailea (Fig. 8.2). Very little accommodation on the Neighbour Islands is to be found in the urban centres such as Kapalua or Hilo.

Elsewhere, hotels are shared in varying degrees between the major urban centre and coastal resort clusters. Over one-fifth of the 13,500 rooms on Phuket, southern Thailand were located in Phuket Town in 1991 while almost 60 per cent occurred in the three resort areas of Patong Beach (32 per cent), Karon Beach

Table 8.1 Major features of selected islands

Island or island group	Political status	Area (km²)	Accommodation capacity (rooms)	Accommodation density (rooms/km²)
Bali	Province of Indonesia	5,561	24,223 (1991)	4.4
Canary Islands	Autonomous region of Spain	7,270	95,500 (1986)	13.1
Dominican Republic	Independent	48,734	10,334 (1987)	0.2
Guadeloupe	French overseas department	1,373	4,740[h] (1988)	3.5
Hawaii	State of USA	16,641	68,034[h,a] (1989)	4.0
Hawaii (the Big Island)		10,414	8,161	0.8
Kauai		1,427	7,398	5.2
Maui		1,886	15,439	8.2
Oahu		1,526	36,467	23.9
Mallorca	Part of autonomous region of Balearics, Spain	3,639	102,500[h,a] (1986)	56.4
Phuket	Changwat (admin. div.) of Thailand	800	13,500[h] (1991)	16.9
Puerto Rico	Commonwealth associated with USA	8,897	8,415 (1992)	0.9

Notes: [a] Apartments.
[h] Hotels.
The room figures for the Canary Islands and Mallorca are approximate only, and have been derived by halving the number of beds (*plazas*) recorded.
Sources: As for Figures 8.1 to 8.8.

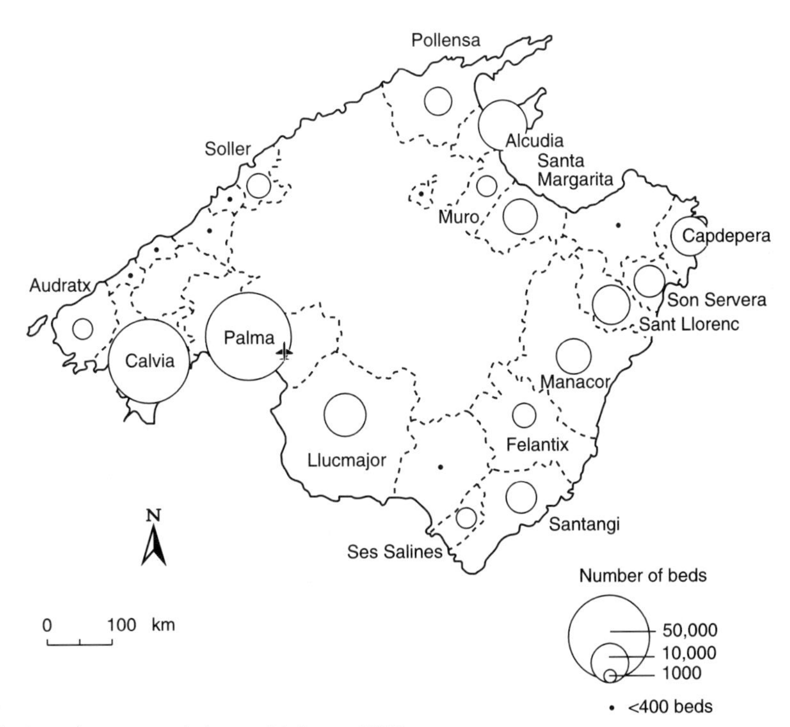

Figure 8.1 Distribution of accommodation on Mallorca, 1986.
Data source: Socias Fuster (1986).

(16 per cent) and Kata Beach (11 per cent) (Fig. 8.3). On the other hand, relatively few of Bali's 24,000 rooms are located in Denpasar. Three-quarters of the island's tourist capacity and 98 per cent of classified hotel rooms have been built in the three resort areas of Kuta, Sanur and Nusa Dua. In the Canary Islands much recent growth has occurred on the south of Tenerife and Gran Canaria as Puerto de la Cruz and Las Palmas have lost

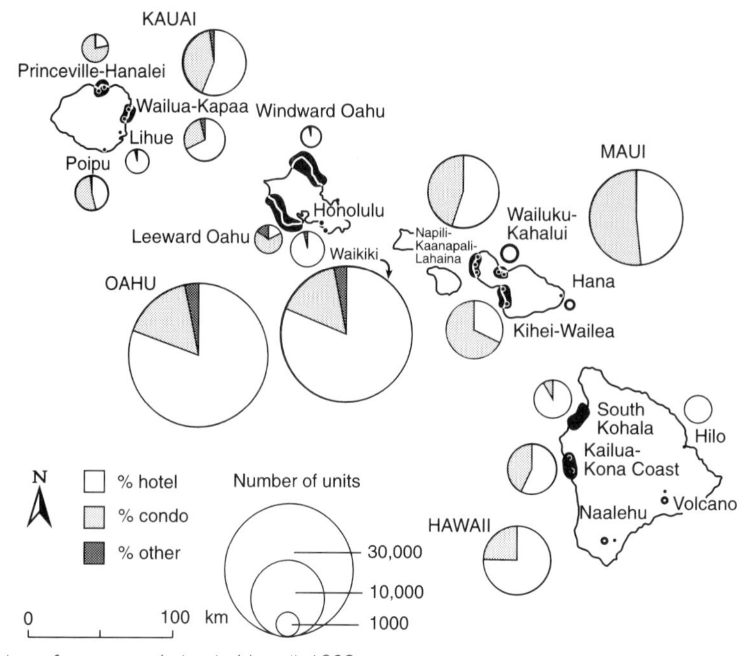

Figure 8.2 Distribution of accommodation in Hawaii, 1989.
Source: After Pearce (1992b).

market share (Odouard 1989). In the Dominican Republic the concentration of hotels in Santo Domingo has weakened (27 per cent of rooms in 1987) in the face of new coastal resort developments, particularly along the northern coast at Puerto Plata and Sosua.

Other tropical and subtropical islands also exhibit similar general patterns to those shown in Figs 8.1 to 8.8. Hotels in and around Funchal account for 80 per cent of Madeira's capacity (Huetz de Lemps 1989b). One-third of the hotel beds on Mauritius in 1985 were found in fourteen hotels clustered along the sandy beaches to the north of Port Louis with a similar proportion in the hotel complexes on the south-west coast between Flic en Flac and Morne Brabant (Singaravelou 1989). On Mahé in the Seychelles, hotels are clustered along the three kilometres of sandy beaches at Beauvallon, just to the north-east of Victoria, with several large hotels also being found along the western coast (Doumenge 1989). Virtually all tourist accommodation in Barbados is on the western coast of the island, notably immediately to the north and south of the capital, Bridgetown, especially along the road to the airport (Potter 1981; Burac 1989). Similarly in Antigua, many of the hotels are located on the beachfront between St Johns and the airport although other hotels are scattered elsewhere around the coastline (Weaver 1988, 1992b). Most hotels in Tahiti share a similar position, being located in Papeete itself, along the airport road or in a specially developed resort enclave adjacent to the airport (Blanchet 1989). The bulk of tourist accommodation in Fiji is to be found in Nadi, Suva or along the Coral Coast (S. G. Britton 1980b). On Penang (Malaysia), hotels are clustered in Georgetown, the major urban centre, or along the north coast at Batu Ferringhi.

It should be borne in mind, however, that the use of accommodation statistics, the only comparable figures readily available at this scale, may accentuate or underestimate the degree of tourist concentration on particular islands (see Chapter 6). In Guadeloupe, tourist-related employment at the level of the commune is more widely dispersed than accommodation capacity. In this case, the hotels lie outside the main urban area, Le Raizet airport in another commune generates a large number of jobs and restaurant-related employment is scattered throughout the island. In Hawaii, the island of Oahu accounted for only 54 per cent of accommodation in 1989, but 63 per cent of visitor expenditure due to higher occupancy rates and a larger share of higher spending Japanese visitors (Pearce 1992b).

Economic Structures

The basic pattern shown in Figs 8.1 and 8.8 and evident elsewhere in the literature generally supports the conceptual models of Hills and Lundgren (1977) and S. G. Britton (1980a, 1982) which highlight the role of the main urban centre and a limited number of resort enclaves (Fig. 1.8). The emphasis in each case, however, is on the resort enclaves, with the urban centre seen as providing a point of arrival, and associated tourist facilities. Hills and Lundgren write (p. 259) of 'a structured descent within the tourist product, via the arrival mechanism ... to the resort-accommodation facilities where the dispersal tends to stabilize'. Interaction at this level is limited and 'illustrates an enclavic, institutionalized tourist circulation between resort facilities, a process often actively promoted by hotel management, or a certain type of resort hotel'. In a similar vein, S. G. Britton (1982: 341–3) writes:

> In physical, commercial and socio-psychological terms, ... tourism in a peripheral economy can be conceptualized as an enclave industry. ... Tourist arrival points in the periphery are typically the primary urban centres of ex-colonies, now functioning as political and economic centres of independent countries. Within these towns are located the national headquarters of foreign and local tourism companies and retail outlets of travel, tour, accommodation, airline, bank and shopping enterprises. If on package tours, tourists will be transported from international transport terminals to hotels and resort enclaves. ... Tourists will then travel between resort clusters and return to the primary urban areas for departure. While resident in the resort enclaves, tourists will make brief excursions from their 'environmental bubbles' into artisan and subsistence sectors of the economy for the purchase of shopping items, entertainment, and sightseeing.

Goonatilake (1978: 7), discussing tourism in Sri Lanka, sees tourist enclaves as 'islands of affluence within the country, walled in and separate from the rest of the population'. A tourist enclave then is not just a physical entity but also a social and economic structure.

Weaver (1988, 1993) also interprets the spatial structure of tourism in Antigua in terms of dependency theory, as well as Butler's life-cycle model (Fig. 1.10), developing a three stage 'plantation model' of the emergence of the island's tourism landscape. The emphasis here is not on enclaves but, in its abstracted form, on a circular zonal model of decreasing intensity of tourism land use away from the coastal fringe (Weaver 1988: 330):

> In the plantation model, the resort cycle commences with the peripheralization of the destination, and tourism initially assumes a passive and minor role within the infrastructure of an economy based upon plantation agriculture. A transitional stage follows when tourism emerges as a viable alternative to agriculture, and becomes an active though still limited agent of spatial change. In the mature phase of the tourism-dominant stage, the island landscape is characterized by a series of Von Thunen rings, with the relationship to tourism decreasing with distance from the coast. Despite the major transformations which occur in the cultural landscape during this process, the basic structural relationships between the core and the periphery remain essentially unchanged, and the tourism landscape like its agricultural predecessor reflects profound spatial inequality. This is evident in the juxtaposition of a privileged resort-oriented coast with an underprivileged labour reserve in the interior. Such a spatial duality reflects the overall failure of tourism to promote the economic, social and environmental development of Antigua.

Weaver presents (1988: 330) his 'pure' plantation model, as represented by Antigua, as 'a standard from which other destinations might be measured' but it does not yet appear to have been applied elsewhere, leaving its wider applicability untested in the literature. However, Figs 8.1 to 8.8 provide little evidence of continuous zones of tourist development around the coast; rather the pattern is one of concentrated clusters.

Although Weaver does comment on some of the characteristics of Antigua, these writers have focused primarily on the economic structure of tourism. Accordingly they have sought their explanations essentially in economic terms, emphasizing the broader economic and political structures of peripheral or Third World countries rather than the insular character of the examples they have chosen. Such a perspective does provide valuable insights into the spatial structure of tourism but by itself is incomplete, as it fails to take into account other basic features which arise out of the size and insularity of small islands.

Smallness

Islands, although cited frequently as examples, usually have not been treated separately from other small territories in the general literature on smallness, where tourism is usually considered only in passing if at all (Knox 1967; Ward 1975; Shand 1980). Nevertheless, the main characteristics of smallness identified by these writers have an important bearing on the structure of tourism on islands. In particular, a small land area usually implies a less diverse resource base while a small population means a limited domestic market. This in turn gives rise to a heavy reliance on foreign trade based on a limited number of products and a narrow range of markets. Diseconomies of scale, which may also extend to the public sector, are also apparent.

Most islands lack the diversity of resources to attract a broad range of international tourists and depend overwhelmingly on the three S's – sun, sand and sea. These may be augmented by other attractions such as gambling (the Bahamas), duty-free shopping (Fiji), historic features (the Dominican Republic), natural attractions (the volcanoes of Hawaii) and the ubiquitous folklore concert and day or half-day circle island tour. However, except in the cases of gambling and where the islands form convenient stopovers on a larger circuit, such attractions will usually generate relatively little demand by themselves. Singapore, with its shopping, Easter Island, with its archaeological interest (Porteous 1981) and the Galapagos Islands with their enormous ecological appeal (Collin-Delavaud 1989), are exceptions to this pattern. Cultural differences may distinguish one island from others within a region, for example the French influence in Guadeloupe and Martinique, but they rarely lead to internal differentiation on individual islands except where political divisions occur (Haiti/Dominican Republic, St Martin). Moreover, the number of attractive coastal sites may be limited and sites suitable for development few. Despite the images conveyed by brochures and posters and a ratio of coastline to surface area greater than most other places, few islands are encircled by continuous stretches of white sand. On volcanic and other mountainous islands, land suitable for building may be restricted to a narrow coastal strip. For ease of access, a circle island road will often be located along this same strip. In other instances, broken relief will isolate parts of the island.

These factors, stemming from the limited resource base, contribute significantly to the coastal location and concentration of tourist accommodation characteristic of islands. Potter (1981: 47) notes in the case of Barbados that 'The principal beaches and hence tourist accommodation zones are all located on the sheltered leeward . . . coast of the island'. On Phuket the major resorts on the south-west coast of the island each have good beaches; in contrast those beaches to the south are narrow while on many parts of the east coast tidal flats predominate (Fig. 8.3). In Tahiti, Donehower (1969: 85) observes 'as all roads closely follow the coastline, hotel location on the coast or beach implies also location on the main roadway'. Attractions and access, two of the main locating factors for tourism (Pearce 1989b), thus go hand in hand here and on many other islands. In Mallorca, Richez and Richez-Battesti (1982) see the coastal range 'protecting' the north-west coast from tourist development and the small surface area limiting the island to a single international airport whose location has favoured development in the Bay of La Palma. In Corsica they attribute the emergence of three major tourist zones rather than a single concentration to the *compartimentage du relief* and to the well-established multiple gateways which have in part arisen out of the difficulties of overland travel on the island.

On many islands, a small land area and a small population contribute to limited domestic tourism. Where distances are not great, overnight stays are not warranted and beach usage, fishing and other recreational activities are accomplished on day trips. In island groups, domestic tourism may, however, involve inter-island travel but there informal accommodation is often used. On some of the larger islands with a mountainous interior, movement inland to hill stations may occur in the hot season, particularly where there is or was an important expatriate community. Larger islands, particularly when under colonial rule, may also see a certain amount of business travel, both official and private, as in Fiji (S. G. Britton 1980b). In any event, the numbers involved in these forms of domestic tourism are usually not very great given the size of the population and the levels of economic development of many islands.

This lack of domestic tourism makes the concentration of foreign tourism more pronounced and contrasts markedly with the situation in Europe discussed in Chapter 7. There foreign and domestic tourists share, in varying degrees, the major tourist regions. Moreover, the domestic demand, which may exceed foreign demand (Table 7.1), also tends to be distributed to a range of minor regions. On many islands this recreational interaction may be limited to local day use of hotel beaches. Even this may be discouraged or prohibited in some places. Where the island is not a nation state in itself, domestic tourists from the mainland may play a more significant role, as in Hawaii where Americans make up about two-thirds of all visitors (Pearce 1992b).

In general, through the dependence on foreign tourists attracted primarily by a 'sun-sand-sea' holiday, tourism is not dissimilar to other sectors of island economies which also rely on an external market and a limited range of products. The implications of this situation are stressed by Ritchie (1993: 305):

> the limited capacity of an island tourism destination means that the managerial task is a much more difficult one. In effect the number of visitors that can be accommodated both comfortably and profitably falls into a very narrow range. . . . Because the 'window of success' is a very narrow one, it is relatively easy to under-shoot or over-shoot the number of visitors required for success. This is in contrast to larger more integrated destinations where the visitor is absorbed, often invisibly, into the fabric of the local community and economy.

Large-scale foreign or metropolitan involvement in the provision of tourist plant is in part a function of the

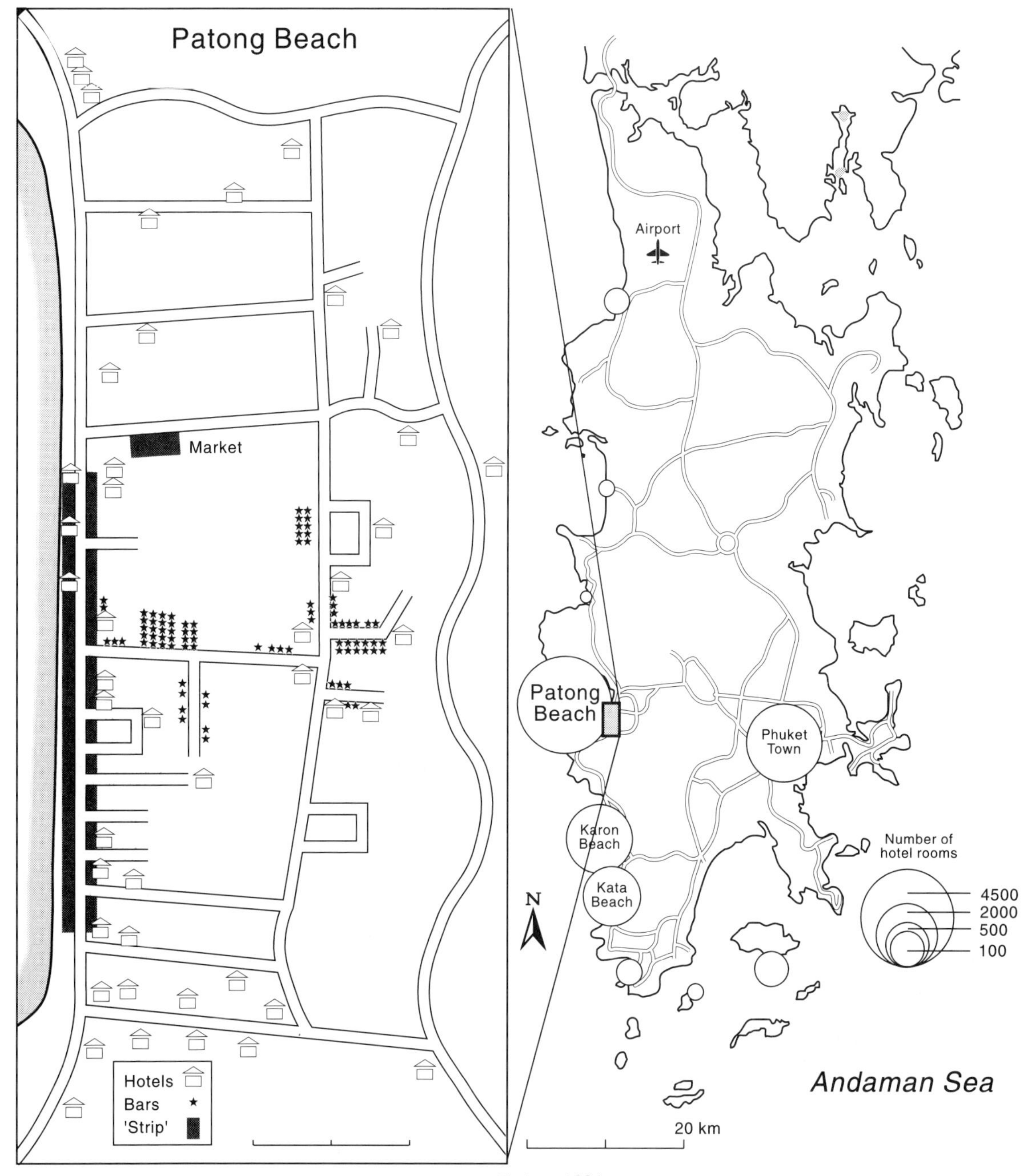

Figure 8.3 Distribution of hotel and bungalow capacity on Phuket, 1991.
Data source: Tourism Authority of Thailand.

small size of island economies. A lack of experience in catering for a domestic tourist market has also meant a more difficult entry by local entrepreneurs into the international tourist industry, particularly in the face of competition from large, well-established foreign or metropolitan developers. However, the activities of this latter group have also been influenced by the small size of the public sector. In particular, the provision (or lack

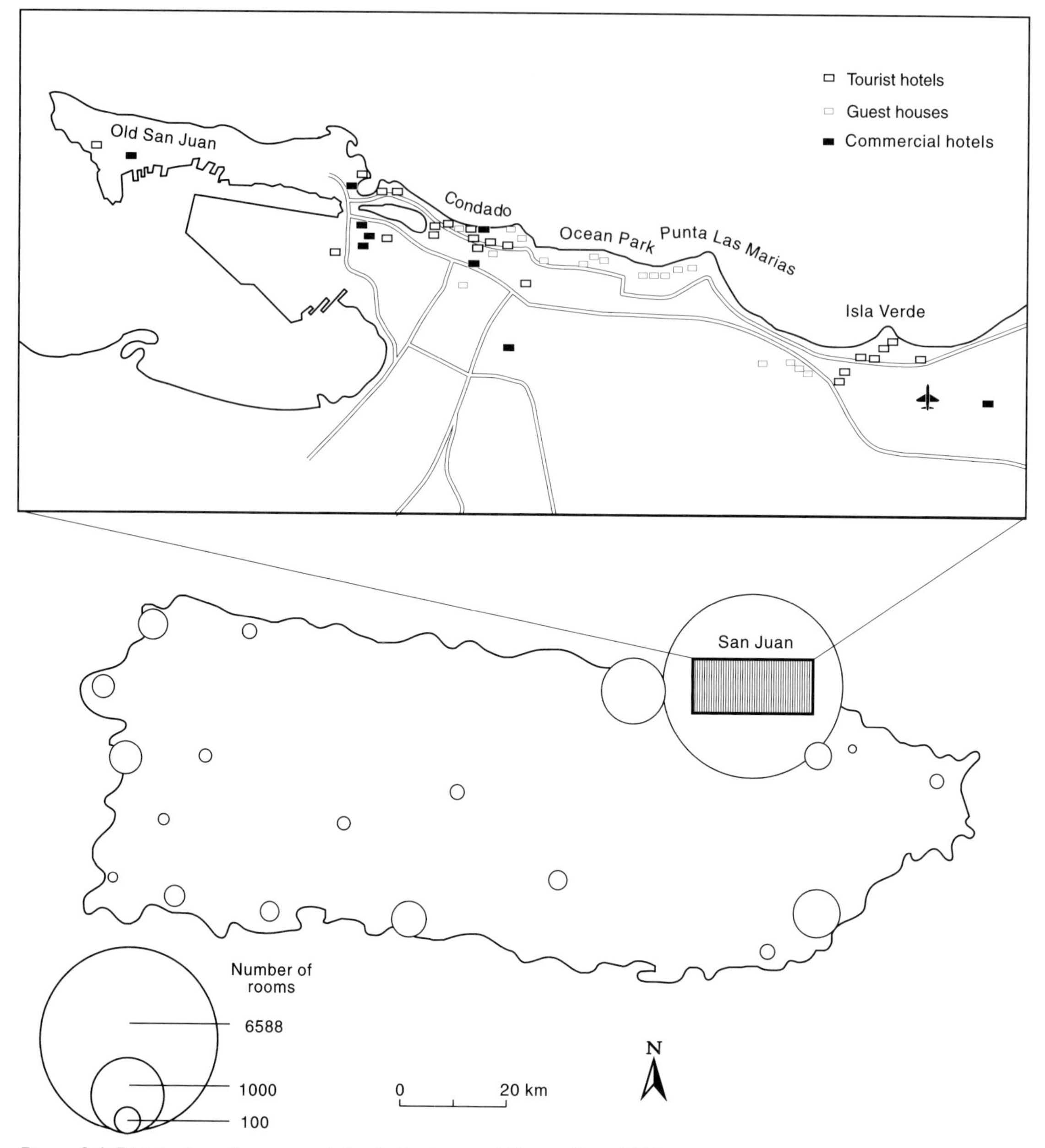

Figure 8.4 Distribution of accommodation in San Juan and Puerto Rico, 1980.
Data source: Tourism Company of Puerto Rico.

of provision) of sealed roading, electricity and adequate water supply has often furthered the trends towards concentration.

The lack of public services on the island at the time of the development of mass tourism to Puerto Rico in the 1960s was a major reason for the concentration of hotels in metropolitan San Juan (Fig. 8.4). Hotel construction there was part of a more general process of urban expansion which saw the city's share of Puerto Rico's population increase dramatically in the 1960s while 'the urban fringe exploded by 213% in one decade' (Cross 1979). New hotels were located on beachfront sites in the Condado, Ocean Park and Punta Las Marias among the high-rise quality residential accommodation. There guests enjoy the advantages of both an urban and coastal location. Further out, lower land values, proximity to the airport and a good beach site have led to another cluster of hotels at Isla Verde. A similar process was later evident in Santo Domingo, while in Haiti the general lack of development has restricted hotels to the capital. Comparable conditions constrained the early development of tourism in the Canaries and led to the concentration of hotels near Las Palmas and Puerto de la Cruz (Odouard 1989). In Fiji, an unwillingness to extend public utilities has effectively discouraged the construction of hotels on more isolated sites. In Sicily, the general lack of services has led, paradoxically, to a juxtaposition of new tourist areas and industrial zones (Ciaccio Campagnoli 1979: 137):

> the dominance of private interests in the tourist industry led to the planning of new tourist facilities in areas which were already developed and were already furnished with auxiliary services which private investors were not willing or able to finance. Therefore, tourist installations were often concentrated by their planners in areas adjoining large industrial establishments producing high degrees of pollution, and the major part of financial aid was granted with a total lack of interest in the existing socio-economic structure.

A similar situation occurs in the south-east coast of Barbados where the Holiday Inn is separated from the Hilton Hotel by an oil refinery and power and light company. Elsewhere, Ciaccio Campagnoli (1975: 86) cites the example of the Sicilian motorway network. Rather than facilitating the dispersion of tourism by linking Messina, the bridgehead to continental Italy, with five designated tourist development zones, the first motorway ran between Catania and Palermo 'in order to facilitate contacts between the regional capital and the domain of Montedison'.

On other islands the public sector has attempted to foster tourist development through infrastructural provision. Often supported by overseas assistance, such projects also tend to concentrate development through the economic and technical advantages this brings. Examples of this strategy in the Caribbean include the Popeshead Coast road north of St John's in Antigua (Weaver 1988), Frigate Bay in St Kitts, Rodney Bay in St Lucia, Trois Ilets in Martinique and Gosier in Guadeloupe (Burac 1989).

Baptistide (1979) and Larroque-Chounet (1989) show that in Guadeloupe the administration favoured large international class hotels aimed at the American market and sought to establish economies of scale through concentration:

> As experience has shown that the success of hotels depends on their concentration, the need to achieve a satisfactory return on infrastructural investment, implies that the southern coast of Grande-Terre must be the sole focus for their development. (Extract from the Sixth National Plan 1971–75, cited by Baptistide 1979: 18).

This policy was in part a reaction to the difficulties experienced by three relatively isolated hotels built in the 1960s, at Sainte Anne, Deshaies and at Moule (Fig. 8.6). The first two were later taken over by Club Méditerranée and the third was converted into apartments. Isolation was not solely responsible for their failure, however, other contributing factors cited by Baptistide being poor management, the emphasis on luxury hotels in lower quality sites and the fact that Guadeloupe was still a relatively unknown destination. Local political interest and the availability of land at St François were important factors which led to the concentration of hotels on the Riviera Sud. The more extensive development at Le Gosier (eight 3-star hotels) is a function of its proximity to Point-à-Pitre and to the airport at Le Raizet rather than the physical attributes of the site. Extensive infrastructural works involving the creation of artificial beaches and funded with the assistance of the European Regional Development Fund were required here.

Similar processes can be observed in French Polynesia where official policy in the 1960s favoured the construction of large, luxurious *hôtels d'impact* on Tahiti with 90 per cent of public credit to the tourist industry in the period 1960–9 going to that island (Blanchet 1981, 1989). A significant sum went to the construction of the Maeva Beach Hotel on a reclaimed site near Faaa airport. The following decade saw a shift in emphasis to bungalow-type accommodation on other islands, notably Moorea and Bora Bora, so that by 1977, Tahiti's share of public investment in tourism had dropped to two-thirds. In 1981 the island still accounted for half of all classified hotel rooms in French Polynesia.

The resort enclave at Nusa Dua in Bali has resulted from institutionalized development, first proposed in a plan in the early 1970s and subsequently developed by the Bali Tourist Development Corporation with a loan from an international aid agency for infrastructural development (Noronha 1979; Hussey 1982, 1989; Lihou-Perry 1991). Located on a peninsula to the south of Bali (Fig. 8.7), Nusa Dua was chosen because the land was of limited agricultural value, it was thought its relative isolation would enable contact between tourists and the Balinese to be regulated thereby reducing the disruptive effects on the local culture, while construction of a major resort on the site would permit economies of scale in the provision of infrastructure. In the event, the

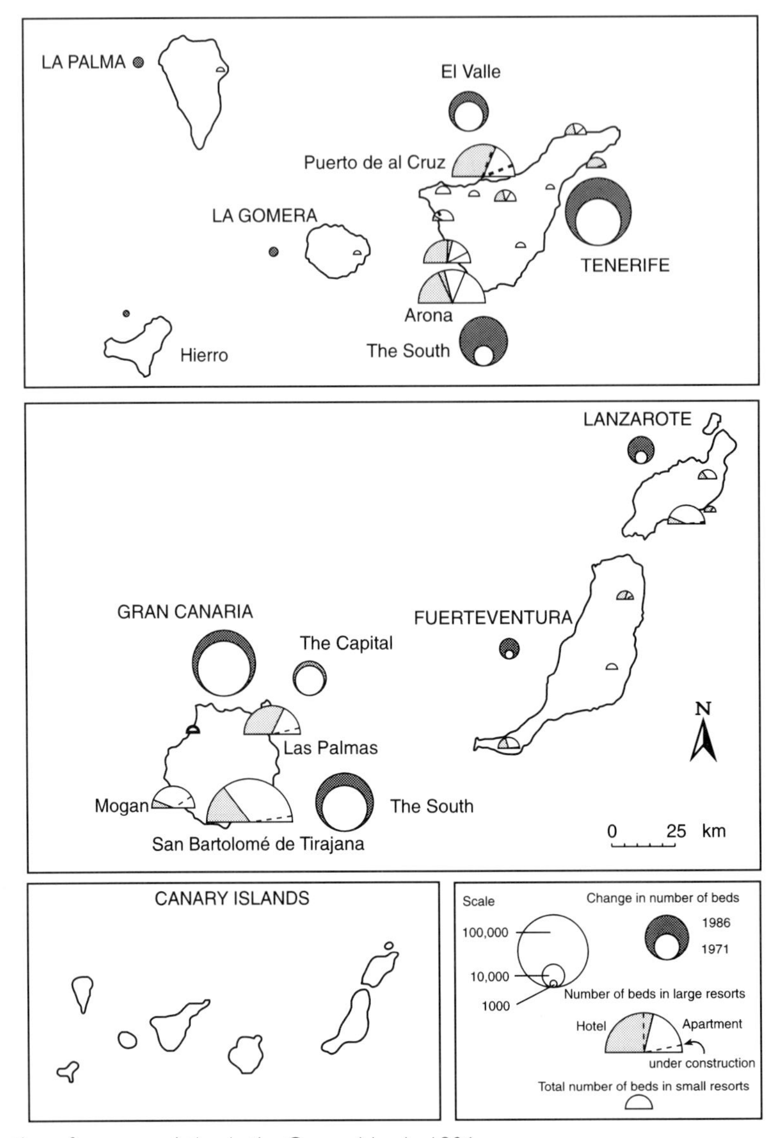

Figure 8.5 Distribution of accommodation in the Canary Islands, 1986.
Source: Adapted from Odouard (1989).

development of Nusa Dua proceeded more slowly than anticipated but by 1991 over 3500 rooms were in place in eight hotels averaging more than 400 rooms each (Plates 5 and 6). In the mean time substantial growth had occurred closer to Denpasar and the international airport at Sanur (3100 rooms) and Kuta (11,100 rooms). While the development of Sanur was foreshadowed in the early plans and a mix of institutional and local development occurred there, Kuta grew in a much more spontaneous fashion as local entrepreneurs responded to the growing demand from low budget travellers, particularly Australian attracted by its wide sandy beach and surfing potential (Fig. 4.3). Over time much of the local *losmen* accommodation has been transformed, with some major international class hotels being built there in the early 1990s, but half the accommodation at that time remained in unclassified hotels. These different processes have not only given rise to the pronounced concentration of demand shown in Fig. 8.7 but also produced in a small area three distinctive types of resorts whose visitor profiles vary quite significantly.

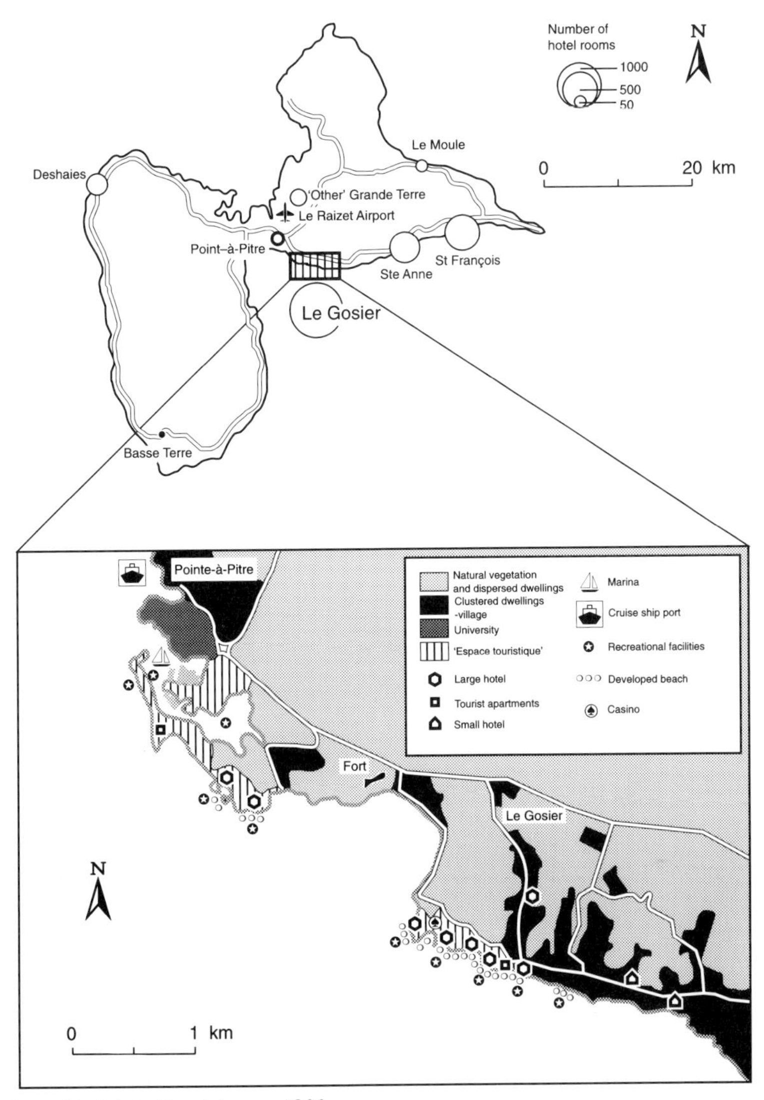

Figure 8.6 Distribution of hotels in Guadeloupe, 1988.
Source: Adapted from Larroque-Chounet (1989).

Insularity

Islands usually generate a positive image for the tourist. In contrast to many mainland regions they possess a distinct and readily recognized and marketable identity. Garcia (1976: 88), for example, writes:

> Being islands, the Balearics provided a suitable place for political and military exiles and for exotic trips in keeping with the romantic ideal prevailing at the time [the nineteenth century]. Famous travellers . . . were the first to provide an image of the island at the international level, an image which was seized upon by the elite tourism of the period.

And later, during the period of isolation which followed the Spanish Civil War:

> Mallorca, which was the best prepared for tourists, managed to attract to itself what little domestic tourism there was in the period 1940–50. Once again, its island character proved to be a decisive attraction since, for the domestic tourist, this was the nearest thing to going abroad, which was out of the question.

Similarly, in the case of Guernsey, Girard (1968: 186–7) notes:

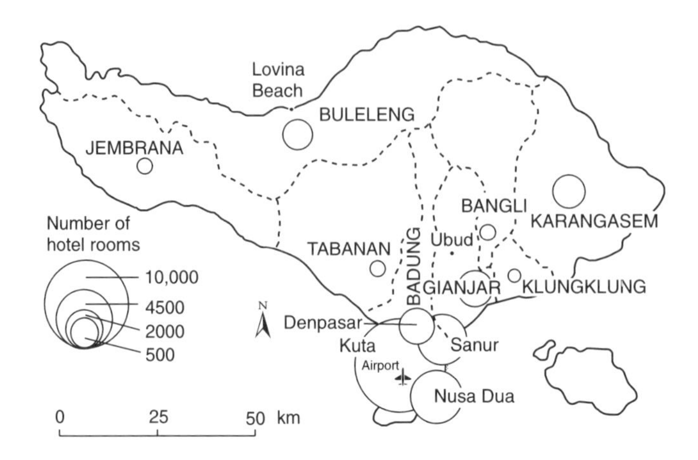

Figure 8.7 Distribution of accommodation in Bali, 1991.
Data source: Dinas Pariwisata Bali.

Plate 5 One of the two gateways in Nusa Dua, Bali: a symbolic and functional entrance into the resort enclave.

Probably Guernsey's most important asset is the fact that it is an island. This alone conjures up a special image in the mind of the average Briton, living in one of the many industrial towns of England, particularly when the island is in a southerly position with suitable sea and air communications with the mainland of Britain.

In particular, the isolation of islands can contribute to the feeling of breaking with routine, most markets being located in continental areas, while the physical attributes associated with tropical and subtropical islands may reinforce the search for relaxation or regression (Chapter 2). These images have often been carefully cultivated by tour operators and tourist organ-

Plate 6 The Bali Hilton, one of the large up-market hotels characteristic of Nusa Dua.

izations (Mirloup 1982; Cazes 1989). Mirloup goes so far as to speak of the 'myth of the island', noting that while islands may be symbolic of escape, once there tourists may be rather captive and experience less flexibility in their holidays and travel plans. Cazes sees the enclaves as islands within islands, citing the extreme cases of the '*îles-hôtels*' or holiday resorts of the Maldives, small islets given over entirely to tourism.

From a structural point of view it is the reliance on sea and especially air communications which distinguish islands from mainland destinations where, depending on the distances involved, overland travel usually plays a significant role. An immediate effect of this dependence is a reduction in the number of gateways, commonly to a single international airport. A sole point of dispersion constrains the amount of internal movement (see Chapter 4). Reduction of internal travel costs, in time and effort as well as money, becomes especially important in mid-ocean locations where airline schedules often impose late night and early morning arrivals and departures. For the tour operators, reducing internal travel also enables the cost of the package they can offer to be kept as low as possible. For small islands in particular, low traffic volumes will normally mean higher fares over equivalent distances to larger countries so that reductions in travel cost become very important.

Reliance on air travel also means that the type of tourism which develops on islands is generally more structured and less diversified than that found in many mainland regions. To arrive at their destination, typical island visitors have at least to enter the formal structured tourist industry by buying their ticket and taking their seat on the plane or boat. From there it is a short step to have the airline or travel agency also arrange the accommodation, for the airline in turn to ensure that it can sell its seats by engaging in hotel operations, and for tour operators to emerge to cut the relatively high costs of this travel by arranging special packages or charter flights. Moreover, sunlust tourism, where small islands are especially competitive, lends itself to this form of packaging, with many tourists being content to stay put at their hotel and beach. This is not to deny the interest of foreign- or metropolitan-based companies in fostering this form of tourism as emphasized by Hills and Lundgren (1977) and S. G. Britton (1980a, 1982) but to suggest their ability to do so arises out of the conditions of insularity and smallness as well as because of economic, political and technological dominance.

Where overland travel is possible, a more diffuse, varied and informal type of tourism is likely to develop, featuring small hotels, camping grounds and holiday homes, both in big countries such as France and Spain, and smaller ones like Luxembourg. Luxembourg, Andorra, Monaco and other small mainland territories also attract a passing wanderlust traffic more readily than many insular destinations where island hopping becomes expensive. Large countries may also develop a highly structured form of tourism as in the case of chartered tours to Spain. There, however, mainland regions such as the Costa del Sol, in contrast to the Balearic Islands, experience a less formal type of tourism based on domestic tourists and overland travellers as well as the package tours to resorts such as Torremolinos

(Fig. 7.7). Where, however, overland travel is difficult, the major markets are long-haul ones and the levels of economic development are relatively low, tourism in continental regions, for example in many parts of Africa, may share many of the features characteristic of tourism on islands.

In terms of spatial organization then, the general effect of insularity is to reinforce the tendency towards concentration noted earlier, and in particular to encourage the location of hotels close to the international airport. This in turn is often, but not always (e.g. Tonga, Western Samoa), close to the major urban centre. Such a pattern is clearly seen in Mallorca, Puerto Rico, Guadeloupe, Bali, Tahiti and Fiji. In the case of the Dominican Republic, construction of an international airport in the north of the country (La Union) was essential to the development of the Puerto Plata area (Symanski and Burley 1973; Banco Central de la Republica Dominicana 1977). The dispersion of the tourist traffic to the Neighbour Islands in Hawaii, including some direct mainland flights, was facilitated by the network of military air bases, some of which were later converted to civilian use. Corsica, with its multiple gateways, is again an exception to the general pattern. The island is situated at a convenient overnight ferry trip from metropolitan France, its major market, and half of all visitors travel by ferry and car, staying in camping grounds, holiday homes and apartments as well as in hotels (Richez and Richez-Battesti 1982).

Land Tenure, Land Use and Land Values

The importance of land tenure systems, patterns of land use and land values, which can be decisive factors on a local scale in mainland areas, especially mountainous ones (Pearce 1989b), is often accentuated on islands by the limited availability of land and pressure on resources. S. G. Britton (1980b) stresses the distribution of freehold land, 'a legacy of nineteenth century European enterprise', in the selection of specific sites for tourism plant in Fiji. In some cases a freehold title encouraged an owner to move into the tourist industry; in others, foreign investors sought scarce freehold land to enhance the security of their investment. 'Ownership or favourable purchase of freehold land' emerged as the top-ranked factor in Britton's survey of hotel location preferences. Ease of acquisition of land has also been a specific locating factor in French Polynesia where multiple ownership and fragmentation of titles are common. One of the major hotels on Tahiti, for instance, was built on land leased from a single titleholder while on Bora Bora public land associated with the airport has been leased for hotel construction.

In the Balearic Islands, Bisson (1986) cautions against explaining the dominance of Mallorca solely in terms of the larger size and resource base of the island and the earlier development there of international air connections. Rather he points to other less obvious factors such as the underlying agricultural and tenurial structures. In the early 1960s when charter tourism to Mallorca 'took off', much of the coast in the commune of Calvia had already fallen into the hands of financial and property development companies as a result of the indebtedness which had arisen out of the break-up of previously large estates among a number of heirs. Such a situation lent itself readily to the urbanization of the coast. In contrast, tourist development on Menorca was delayed by a more conservative landholding system with succession there being to a sole heir and share cropping prevailing on the large estates. Industry was also relatively more important on Menorca, further delaying diversification into tourism.

At Waikiki, changing land values fuelled the rapid growth of tourism and exacerbated the concentration of hotels and condominiums as Farrell (1982: 35–6) notes:

> In the 1940s, before the real impact of air travel, Waikiki's beautiful golden beach provided the stage for several low, elegant and charming hotels, green lawns and well-cared-for parks. . . .
>
> Tourism's development was dramatic. It grew from an enthusiastic roar by the 1950s to a resounding crescendo in the 1960s when concrete monoliths were constructed in every available space. Still, from a room count of 10,700 in 1967, the number grew to 25,000 in a little more than a decade. The rate of increase was unbelievably fast until 1971 when growth flattened out after most prime areas, zoned appropriately, had been taken up. By this time two new regulations were enforced; apparently foreknowledge of these encouraged the earlier frantic activity.
>
> Land values and taxes rose so high that spindly highrises grew from handkerchief-sized lots, and older hotels were dismantled to allow for new structures.

Farrell also relates how the location and development of the large masterplanned resorts on the Neighbour Islands (Fig. 8.2) was directly related to pre-existing land use and tenure. Their scale and unity have been made possible by the existence of large tracts of former plantation or ranch land held under sole ownership. The developers have usually been 'venerable old' Hawaiian companies and landowners: Amfac (Kaanapali), Alexander and Baldwin (Wailea) and the Kamehameha Development Corporation (Keahou).

In the Dominican Republic, landownership and land use have given rise to an enclave within an enclave at La

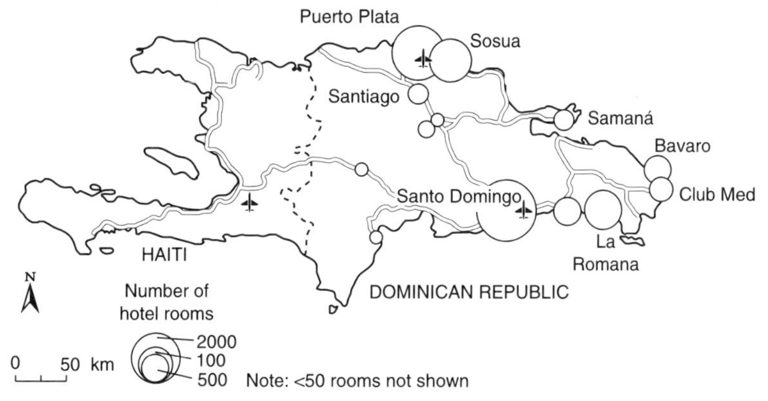

Figure 8.8 Distribution of hotel capacity in the Dominican Republic, 1987.
Data source: Secretaria de Estado de Turismo.

Table 8.2 Hawaii: distribution of tourist rooms; 1955–89

	1955		1960		1970		1980		1989	
	Rooms	(%)	Rooms	(%)	Rooms	(%)	Rooms	(%)	Rooms	(%)
Oahu	2,628	76.1	5,716	83.8	21,217	70.6	34,173	61.3	36,467	53.6
Hawaii	451	13.1	581	8.5	3,486	11.6	6,299	11.3	8,161	12.0
Kauai	152	4.4	237	3.5	2,609	8.7	4,707	8.5	7,398	10.9
Maui	222	6.4	291	4.2	2,720	9.1	10,521	18.9	15,439	22.7
Total	3,453	100	6,825	100	30,032	100	55,700	100	68,034	100

Note: Rooms comprise hotels, condominiums and apartments.
Sources: Farrell (1982); Pearce (1992b).

Romana (Fig. 8.8). By absorbing the South Puerto Rico Sugar Company in 1967, Gulf and Western, the American giant, acquired 1118 square kilometres of land, chiefly canefields, in the east of the country. This represented 2 per cent of all Dominican territory and, depending on the definition used, 5–10 per cent of the country's arable land. In a process of diversification, and one which has parallels in the conversion of plantation 'great houses' into hotels on Antigua (Weaver 1988), the sugar factory management's club was transformed into a hotel (Hotel Romana) and a large luxurious country club complex, La Casa de Campo, was built. Diversifying beyond their enclave, Gulf and Western also acquired and built two hotels in Santo Domingo and another in Santiago (Girault 1980). Hotel construction at Samaná, on the other hand, appears to have resulted from the personal interest of a former president of the republic (Symanski and Burley 1973). Despite his support and an attractive site, Samaná has been handicapped by difficulties of access and the project has been unsuccessful. Puerto Plata, a planned complex on the north coast, however, has experienced considerable growth.

In other instances coastal lands or particular stretches of the littoral may have had comparatively little agricultural value or competition from other land uses prior to the advent of tourism. As noted earlier, the relative infertility of the Nusa Dua area contributed to its selection as the main resort site on Bali. With regard to Kuta, Hussey (1989: 314) observes: 'Balinese held beachfront property, the shoreline, and the sea in low esteem. Beach land not only was agriculturally unproductive, but also was considered spiritually impure'. Likewise, in his study of the Thai islands of Phuket and Ko Samui, Cohen (1982: 200–1) noted: 'beachland was traditionally of little, if any value. Village landowners would be happy to sell it, but in the past there were few takers'. In the case of Phuket, far-sighted Chinese urban elite acquired the land which they subsequently developed for tourism whereas on Ko Samui, as at Kuta, more local entrepreneurship was evident. In the Canaries, the development of large complexes on the south of Gran Canaria and Tenerife took place on land with little agricultural value as the landholders took advantage of the growth in tourism which began in and around the cities several years earlier (Odouard 1989).

Plate 7 Waikiki, Oahu: one of the densest concentrations of tourist accommodation on any island in the world.

Evolving Patterns

The spatial concentration of accommodation continues to be a dominant feature of tourism on small islands but some evolution is also evident, both within island groups and on individual islands. In Hawaii, Oahu's share of accommodation capacity dropped from over 80 per cent in 1960 to just over half in 1989 due to developments on the Neighbour Islands (Table 8.2). The most dynamic growth in the 1970s was on Maui and although more rooms were added there in the 1980s the most rapid rates of growth in that decade were on Kauai. Factors accounting for this dispersion include improved accessibility, reaction to the concentration on Waikiki, the attractiveness of the Neighbour Islands, development initiatives and changing market demand. Marked differences have emerged between the islands in terms of the type, quality and cost of accommodation, market composition and the types of holidays experienced (Plates 7 and 8). Oahu is now a highly urbanized destination with a greater range of night-life and shopping but maturity and urbanization have brought difficulties, ageing plant has led to the need for major refurbishment programmes, congestion and overcrowding are problems. In contrast, the more modern facilities of the Neighbour Islands are complemented by good stretches of less crowded beaches and high-quality golf courses, an air of relative tranquility replaces the hectic activity of Oahu and the spectacular natural scenery has been joined by the fantasy resorts of the Kauai Westin and the Hyatt Regency Waikaloa.

In French Polynesia Tahiti retains just under a half of the hotel accommodation, with significant growth occurring on Moorea in the 1970s (Table 8.3). Official emphasis has gone from luxury hotels on Tahiti to bungalow type accommodation on Moorea to encouragement for small family units on the more distant islands, particularly in the Tuamotu archipelago (Blanchet 1981). The latter policy was designed to maintain the population in these islands, but the small units developed through these state subsidies have proved difficult to market.

A similar dispersion is also evident, to greater or lesser degrees, in other island groups. The Riviera Sud maintains its dominance in Guadeloupe but some growth has occurred in other smaller islands in the département, particularly on Saint Martin during the 1980s (Larroque-Chounet 1989). While large new hotels have been built on Saint Martin, other islands have been characterized by more modest forms of accommodation. In the Canaries, the southwards expansion of tourism on Gran Canaria and Tenerife was followed by tour operators opening up new complexes on the eastern islands of Lanzarote and Fuerteventura,

Plate 8 Looking across the golf course at Kaanapali, a masterplanned resort on former plantation land in Maui.

Table 8.3 French Polynesia: distribution of hotel rooms, 1960–91

	1960		1970		1980		1991	
	Rooms	(%)	Rooms	(%)	Rooms	(%)	Rooms	(%)
Tahita	48	79	894	73	1133	54	1318	48
Moorea	13	21	243	20	593	28	965	35
Iles Sous le Vent	—	—		7	313	15	387	14
Other islands	—	—		—	75	3	94	3
Total	61	100	1230	100	2114	100	2764	100

Sources: Blanchet (1981); Service du Tourisme.

about 10 per cent of all accommodation in 1989, but the three smaller western islands in the group remain largely undeveloped, except for some day trips. Attempts have also been made in the Seychelles to spread tourism beyond the island of Mahé, particularly to Praslin, but these have been hampered by the lack of direct air connections and infrastructure (Doumenge 1989). In other island groups, such as Fiji and Vanuatu, governments anxious to limit certain social and cultural impacts of tourism have developed rather more cautious policies towards the spread of tourism and have continued to favour international hotel development on the main islands, even if some potential for alternative forms of tourism on the outer islands exists (National Planning Office 1983; McLisky 1992).

The spatial structure of tourist accommodation on individual islands has also evolved. The movement away from the urban areas of the Canaries to resort complexes (noted earlier), is also apparent for example on Phuket. Phuket Town had about two-thirds of the hotel rooms in Phuket in the late 1970s (Cohen 1982) compared with one-fifth in 1991 as the resort traffic rapidly surpassed business and other travellers. San Juan's dominance in Puerto Rico has declined in both relative and absolute terms, due to the loss of tourist and commercial hotels in the city and the encouragement of alternative accommodation on the island, notably small *paradores* (Table 8.4). Bali has also seen the recent emergence of small-scale homestay accommodation away from the coast, notably at Ubud, as well as an increase in modest budget accommodation along the northern coast, for example at Lovina Beach, but these scarcely if at all challenge the resort complexes of Sanur, Kuta or Nusa Dua.

Table 8.4 Puerto Rico: distribution of rooms by type of accommodation, 1968–92

			San Juan	Island	Puerto Rico
Tourist hotels	1968	no.	5347	1461	6808
		%	78.5	21.5	100
	1980	no.	5680	1606	7286
		%	78	22	100
	1992	no.	4865	1806	6671
		%	72.9	27.1	100
Commercial hotels	1968	no.	933	313	1246
		%	74.9	25.1	100
	1980	no.	830	305	1135
		%	73.1	26.9	100
	1992	no.	282	235	517
		%	54.5	45.5	100
Guest houses	1968	no.	308	57	365
		%	84.4	15.6	100
	1980	no.	268	54	322
		%	83.2	16.8	100
	1992	no.	245	89	334
		%	73.4	26.6	100
Tourist villas on the island	1968	no.	—	—	—
		%			
	1980	no.	—	392	392
		%	—	100	100
	1992	no.	—	25	25
		%		100	100
Paradores Puertorriqueños	1968	no.	—	—	—
		%			
	1980	no.	—	195	100
		%			
	1992	no.	—	549	549
		%	—	100	100
Condo Hotel	1992	no.	319	—	319
		%	100	—	100
All accommodation	1968	no.	6588	1831	8419
		%	78.3	21.7	100
	1980	no.	6788	2552	9330
		%	72.6	27.4	100
	1992	no.	5711	2704	8415
		%	67.9	32.1	100

Data source: Tourism Company of Puerto Rico.

In other instances, dispersion is initiated not by the development of these alternative forms of tourism but by the construction of new, up-market hotels in a process similar to that described earlier by Thurot (1973) for the islands of the Caribbean (Chapter 1). Segui Llinas (1991) identifies three stages in this process on Mallorca:

- First, new hotels and apartments are built on virgin sites far from saturated areas. Higher prices can be charged for quality products on magnificent sites, drawing part of the quality market from the saturated established resorts and a large share of new arrivals.
- Second, the hotels in the established zones seek to compensate their losses by lowering prices to attract less-well-off tourists, thereby setting in train a process of progressive degradation in the quality of service offered.
- Third, the increased purchasing power in the new zone does not offset that lost in the established zones so that there is no overall improvement in tourism in the Balearics.

General processes which recur from island to island may in some places be completely overshadowed by specific factors or events. Cyprus provides a graphic illustration of this (Gillmor 1989; Lockhart 1993). The de facto division of the island which followed Turkish military intervention in 1974 has resulted in a major redistribution of hotels on Cyprus. New development, including purpose-built resorts at Ayia Napa and Paralimni, has occurred on selected coastal areas in the south, replacing and extending those along the better beaches of the Turkish occupied north.

Conclusions

Many interrelated factors influence the amount, nature and distribution of tourism on small islands, but the combination of smallness and insularity produces spatial structures which are more evident there than in most mainland countries and destinations. In particular, tourist accommodation on islands tends to be concentrated in a small number of coastal localities in or adjacent to the major urban centre and close to the international airport. The predominance of sun-sand-sea tourism, especially on tropical and subtropical islands, is a direct consequence of an insular situation and the limited range of other possible tourist resources. The absence of relatively low-cost land access reduces the range of possible markets and the types of development which may occur. This pattern is reinforced by the limited local demand which stems from the size of the islands' own markets and the small distances involved. The emphasis on air travel and structured holidays for foreign tourists not only leads to the concentration of hotels close to the airport and major urban centre (the two usually go together) but also increases the likelihood that the island tourist industry may be controlled by external developers.

The effect of these factors on small islands may also be reinforced by relatively low levels of economic development which in turn result in varying degrees from the conditions of smallness and insularity. Elsewhere, notably in many parts of Africa and Asia, low levels of economic development may produce patterns similar to these. Very low standards of living in a number of contiguous continental countries will give rise not only to very small domestic markets but also to a limited demand for short- and medium-haul international travel. The tourism which has developed under these circumstances has often involved structured air travel from more distant, developed markets and the development of coastal enclaves for package tourists. Nevertheless, in many such cases other national and cultural resources may also be exploited and an alternative form of lower budget overland travel creating different demands and patterns of development may also emerge, as in parts of South East Asia and South America. More systematic work is needed in these areas, however, on past and present processes of tourist development and the resultant spatial structure before these issues can be clarified and the factors and patterns compared to those on islands or in more developed countries.

Land-related issues may also be more accentuated on small islands due to the pressure on resources. The specific situation in any one case – former plantation land, communal ownership, land values – may reinforce some of the general patterns outlined above or result in variations from them.

Some island governments have recently encouraged alternative forms of tourism, especially on outer islands. While such projects may correspond to local development needs, they have often met with limited success as it has not been easy to overcome some of the constraints and conditions which insularity has imposed. The general structure of tourism on many islands thus appears unlikely to change in the near future.

CHAPTER 9

Coastal Resorts and Urban Areas

Earlier chapters have shown that the tourist industry at the national and regional level is comprised of a network of different resorts, urban centres and rural areas. This chapter focuses on the spatial structure of tourism at the local level, and in particular on two major and contrasting types of tourist centres: coastal resorts and urban areas. The former, based on sunlust tourism, have generally developed a distinctive form and function and have attracted a certain amount of attention from geographers and others since the mid-1940s. Urban tourism is more diverse in character, its significance has been recognized rather more recently and a less coherent body of literature has been devoted to it. Finally, some comparisons between the structure of coastal resorts and of tourism in urban centres are made. While some of the effects of such developments on surrounding rural areas are discussed here, no attempt is made to review the structure of tourism in the countryside, in national parks or in alpine areas. Nevertheless, the general approach adopted in this chapter might be applied, with relevant changes in emphasis, to such areas.

Coastal Resorts

Although the Romans frequented towns such as Frejus primarily for recreational purposes, the modern coastal resort has its origins in the seaside towns which began to develop in England, France and other parts of Europe in the late eighteenth and early nineteenth centuries. Later in the nineteenth century New World resorts started to emerge on the Atlantic coast of the United States. Domestic resorts have proliferated along the coasts of developed countries throughout this century while international tourism, often on a large scale, has become widespread in tropical and subtropical regions, as illustrated by Chapter 8.

Gormsen (1981) suggests that variations in the extent and length of development of tourism will give rise to different types of resorts (Fig. 1.11). Variations are seen to occur not only in the range of accommodation offered but also in the degree of involvement of local and external developers and in the extent to which the resorts cater for different social classes. These differences are seen to be part of an evolutionary process with an implicit progression along the same path from different starting-points. Other writers have commented more explicitly on the nature of this evolution. Stansfield (1978), in the case of Atlantic City, refers to a 'resort cycle', a concept developed subsequently by Butler (1980) (Fig. 1.10).

Of particular importance for the spatial structure of the resorts, however, will be the date or period at which development begins as much as at what stage the process has reached, whether in Gormsen's or Butler's terms. Technological changes, whether in transport or the building industry, might be expected to have important morphological implications. Attitudes and expectations have also changed over the past century. Demand has multiplied rapidly since the 1950s as a holiday at the coast, both within the home country and increasingly abroad, has come within the reach of a larger proportion of households in developed countries. Mass tourism has brought with it the development of large-scale resorts, virtually overnight.

Discussion of resort structure is further complicated by definitional problems, with a variety of coastal urban centres having some tourism function. In most cases, tourism has been grafted on to some existing centre, whether a small fishing port, as at St Tropez, or a much larger and growing city such as San Juan. Tourism may have subsumed these earlier functions or remain dominated by them. Other resorts have been planned as such. Some of the most striking examples of these are the large new resorts created *ex-nihilo* in the 1960s and early 1970s – La Grande Motte, Cancun, La Manga del Mar Menor. It should be recalled, however, that a number of the nineteenth-century resorts were planned and developed specifically for tourism, for example Deauville in France and Atlantic City in New Jersey. Over the years such centres have gradually acquired other functions, particularly residential zones. Freeport in the Bahamas was developed as a multifunctional centre from the outset (Bounds 1978). It is debatable too whether clusters of hotels, as for example at Le Gosier in Guadeloupe (Fig. 8.6), constitute resorts in themselves for they depend heavily on the services and other functions of nearby cities, in this case Point-à-

Pitre. However, there is also a growing tendency for individual hotels to be referred to as 'resorts' (Ayala 1991). In effect, a spectrum of coastal resorts exists, ranging from those with a wholly tourist function, notably the new planned resorts, to those where a significant amount of tourist activity occurs alongside a variety of other urban functions.

Examples from different parts of this spectrum are found in the literature. The most frequently cited studies deal with what might be called the traditional seaside town, particularly those of England (Gilbert 1939, 1949; Barrett 1958; Wall 1971; H. Robinson 1976; de Haan and Ashworth 1985) and the United States (Stansfield 1969, 1978; Stansfield and Rickert 1970; Funnell 1975). Detailed descriptive accounts of established resorts in France have come from Burnet (1963) and Clary (1977), while Pearce (1978c) has considered newer developments there. Aspects of Spanish resorts have been examined by Garcia (1976), Mignon and Heran (1979), Dumas (1982) and Vera Rebollo (1987). B. Young (1983) has presented a generalized model of the 'touristization' of Maltese villages while Dewailly (1990) provides a comparative overview of resorts in northern Europe. The more recent expansion of tourism in other parts of the world has resulted in studies of new coastal resorts, both planned and unplanned, in South East Asia (Franz 1985; Wong 1990; R. A. Smith 1991, 1992), Mexico (Gormsen 1982; Pearce 1983b; Meyer-Arendt 1990), Australia (Pigram 1977; Baker 1983) and Hawaii (Farrell 1982). Some writers, for example Burnet, Pigram and Smith, have adopted a fairly comprehensive approach while others have focused on particular features, for instance commercial functions (Stansfield and Rickert 1970; Dumas 1982).

Whatever the resort, analysis of its morphology needs to take account of three main sets of factors: site characteristics, tourist elements and other urban functions. Analysis of the site characteristics is important for establishing the context in which the resort develops and the effect of physical conditions and cultural features on its spatial structure (Clary 1993). The tourist elements to be examined include the types of attractions, the range of accommodation, the means of circulation, tourist-related shops and services, and accommodation for those providing these services and facilities. The role and range of other urban functions will depend on the size of the centre and how much it is oriented towards tourism. Factors to be considered here are more general residential and commercial functions, industrial activities, the transport network and other forms of land use.

Much of the distinctiveness of coastal resorts arises from their location along the beach or seashore for, as Stansfield (1969) pointed out, growth outwards from a coastal core is limited to approximately 180° as opposed to a full 360° in most urban centres. On top of this general pattern may be superimposed variations arising from different physical characteristics, such as the presence or absence of beaches, cliffs and lagoons or the form of the coastal environment. Dewailly (1990), for example, distinguishes between the cross-sectional profiles of English, Belgian and Dutch resorts on the basis respectively of the impact of cliffs, developed dunes and protected dunes. He also underlines the impact of existing urban nuclei and lists three ways that tourism may be grafted on to these: concentric growth, lateral extension and development in depth. Wong (1990) examined the geomorphological basis of beach resort sites along the east coast of peninsular Malaysia and identified several characteristic sites: zetaform bays with a protected upcoast curve in the lee of a headland; non-zetaform bays which are usually small; barrier beaches which are straight and exposed; low linear coasts; sandy spits; and estuaries. While Wong's emphasis is on aspects of the physical implications of site selection such as coastal erosion rather than resort form, his study underlies the potential which this approach has to complement the more traditional resort morphologies which have concentrated on built forms.

Given the factors outlined earlier, the morphology of coastal resorts might be expected to vary widely from place to place and period to period. The literature, however, suggests that while details do differ, the basic structural features are frequently repeated from resort to resort. Imitation, particularly within countries, may account for some of these similarities but the form of these resorts generally reflects their specialized function. The basic structure to emerge from these studies is the seafront or *front de mer*. Linear in form, this typically consists of a parallel association of the beach and maybe port, a promenade, road or railway and a first line of buildings comprised of the densest and most expensive forms of accommodation and a core of tourist-related shops, bars and restaurants. A gradation of accommodation in terms of height, density and price occurs behind this seafront as high-priced hotels and high-rise apartments give way to smaller and less expensive hotels, pensions, villas, camping grounds and merge with residential accommodation and other urban functions.

A classic example of this linear structure is given in Plate 9, an aerial view of Nieuport on the Belgian coast (see Fig. 7.6). Immediately adjacent to the beaches is the *digue* or *dijk*, a pedestrianized protective sea-wall, the Belgian equivalent of the promenade. Given the vicissitudes of the Belgian summer, the *digue* has assumed a major role in the life of Nieuport and other Belgian resorts (Plate 10). It is a zone of activity where holidaymakers walk, meet, shop, eat their *frites* and

Plate 9 An aerial view of Nieuport on the Belgian coast showing a gradation in the density of development from top to bottom right, with a zone of villas separating the high-rise apartments overlooking the *digue* from the polder land below. A camping ground is evident in the lower left.
Source: Institut Géographique National

where children play a variety of games. The *digue* has even spawned its own transport, a range of pedal-driven *cuistax*. For much of the day, however, the adjoining high-rise apartment buildings cast their shadows across the *digue* and out on to the beach. The first two blocks back from the beach are densely built-up, development in the third is intermittent, then the grid-like street pattern gives way to a less uniform zone of villas among the dune land. In parts, the more extensive dunes break up development completely, although several centres of social tourism have encroached on the open dunes to the south. The transition from the tourist-occupied dunes to the farmland of the polders is generally abrupt, but along an inland road a large camping ground has developed. The coastal tramway runs along the main street, one block back from the *digue*. This pattern, with minor variations, is repeated the length of the coast.

Plate 10 The seafront at Knokke: the promenade or *digue*, a major zone of activity, is backed by a zone of high-rise hotels and apartments.

The linear form of the coastal resort or seaside town reflects its orientation towards the main centre of attraction, the beach. The gradation in land use and density away from the beach is in large part a response to economic forces. In general, land nearest the attraction commands the highest rent and thus generates a more intensive form of land use. Further away, prices decline, density decreases and forms of accommodation yielding a lower return become viable. Planning regulations, such as those in force along the Belgian coast, may reinforce these economic factors. The price of land will be higher in zones where high-rise buildings are permitted, and the higher the building allowed, the higher the price of land will be (personal communication, N. Vanhove 1985). Variations on the general pattern outlined here will occur from resort to resort as the result of the interplay of a variety of factors including the range of attractions, site characteristics, other land uses, dunes, the period at which development began and most growth took place, the extent of planning and the size of the resort.

The seafront

Much of the variation occurs in the sequence and form of the various elements which comprise the seafront or *front de mer*. In the earliest American resorts, the promenade took the form of a 'boardwalk', the first of which was constructed at Atlantic City in 1870. This pedestrianization of the seafront contrasts with many of the English and French resorts where a road separates the promenade from the first line of buildings. Stansfield (1978: 243) attributes this difference to the early dependence of Atlantic City on rail transport:

> the railroad was not eager to encourage stage or wagon traffic to the new resort. Early vacationers thus couldn't drive their own carriages to the resort, a factor which may have encouraged a seafront pedestrian promenade rather than a seafront vehicular drive for the traditional display of fancy coaches and equippages which characterized older resorts such as Cape May, or even such resorts as Brighton and Nice in Europe.

The railway was also to shape the morphology of English resorts as it was at the period that railways developed that 'many resorts acquired a characteristic "T-shape", the stem of the T being the main street leading from the railway station to the promenade' (H. Robinson 1976: 162). In Belgium, for similar reasons, Knokke and Blankenberg exhibit this same 'T-shape', though at Koksijde-Bad the stem is along the main thoroughfare to the inland town of Koksijde.

Later as the motor car replaced the railway as the dominant means of transport, the coastal road increased in importance, both as a means of access to the resort and for circulation within it. In established resorts the car often contributed to a lengthening of the *front de mer* while in many new resorts, such as St Jean de Monts on the Atlantic coast of France, ribbon development along the coastal road was inevitable. As Clary (1977) points out, the coastal drive initially continued the *m'as-tu-vu* aspect of the seafront. However, as traffic volumes increased, major congestion and circulation problems arose. At worst, the motor car has seriously jeopardized the integrating function of the promenade, as Garcia (1976) notes in the case of Palma de Majorca:

> The seaside promenade not only links the organization of tourist residential space with the beach and the sea, but becomes in its turn a place of consumption, of both objects and places. To some extent the seaside, beach and hotel are not strictly separate, and all the space is structured as a single entity.
>
> However, planning errors and the proliferation of the motor car have all but destroyed this scheme of things, and have converted the seaside promenade into a veritable wall tending to isolate tourist consumption of natural assets (beach and sea) from the other consumption zones (residential and recreational).

Not all resorts, however, have experienced these problems. The effects of the decision by the developers of Deauville in the 1860s to create a 'calm and aristocratic resort' (Burnet 1963) by directing the main access route outside the new town continues to be felt today. However, many resort planners in the 1960s and 1970s reacted specifically to the problems caused by the motor car and have attempted in various ways to limit its impact (Pearce 1978c). In some of the smaller resorts, especially marinas, virtually all vehicle circulation within the resort has been prohibited, with the cars being consigned to parks at the entrance to or on the outskirts of the resort, as at Port Grimaud and the Marines de Cogolin. While some of the larger resorts, e.g. La Grande Motte, certainly have been conceived at the scale of the motor car, the location of the major access route some distance inland has overcome traffic problems along the seafront itself. In other new resorts such as Cancun and La Manga del Mar Menor which have been built on a sandy bar separating a lagoon from the sea, the access road bisects the accommodation zone (Plate 11). Such sites, which essentially have two seafronts, one facing the sea, the other overlooking the lagoon, lend themselves to intensive tourist and residential development, as has been shown at Miami Beach. The trend to high-rise development is often reinforced in resorts where visitors arrive by air and have no on-site transport of their own.

The traditional promenade is absent from many of the new resorts, with the hotels and apartments giving directly on to the beach, public or private, a reflection of the hedonistic obsession with the three Ss which characterizes many of today's tourists. It should be recalled here that the renowned promenades of the Riviera developed at a time when the winter season was the most important. Significantly, where one or more of the three Ss is less attractive or reliable, the promenade continues to play a vital role in the life of the resort, as noted earlier in the case of the Belgian *digue*. The rejection of the interior and the complete orientation to the sea in modern resorts is graphically described by Baptistide (1979: 160) with reference to Le Gosier in Guadeloupe (Fig. 8.6):

> the buildings are built as close as possible to the sea (*pieds dans l'eau*), as if the tourist were disabled and incapable of walking a few metres. The hotel itself, less than 50 metres from the sea, is oriented exclusively towards the water. The tourist's domain is that which separates the room from the beach. The buildings, standardised and soul-less, could have been built anywhere; they all take little account of the relief and none at all of the tropical nature of the environment.
>
> The inland facing side is unattractive, an area of car parks, a no man's land of irregularly mown wild grasses.

However, in many of the masterplanned beach resorts such as Kaanapali in Hawaii or Nusa Dua in Bali, increasing attention has been given to resort landscaping, often designed to enhance the tropical nature of the resort or evoke local flavour, though contrasting and exotic elements may also be incorporated (Ayala 1991). Beachfront hotels may thus be set within spacious grounds and gardens with an abundance of palms and tropical vegetation and which may be maintained 'almost to the point of being manicured' in the case of Nusa Dua (R. A. Smith 1992). With one exception, planners at Sanur have effectively managed to implement the much-vaunted policy of keeping hotels below the height of the palm trees (Plates 12 and 13). This policy, coupled with the use of gardens, an important feature of Balinese culture, and the construction of hotels perpendicular to the beach rather than parallel to it has resulted in a very discreet beachfront which contrasts markedly to the intensive high rises which characterize many other coastal resorts (compare Plates 7, 10 and 12). Elsewhere, for example with several of the new large resort hotels in Hawaii, prolific landscaping, the abundant use of statues and artworks

Plate 11 La Manga del Mar Menor, Costa Blanca: the access road bisects the accommodation zone with apartments and villas to the left overlooking the Mediterranean and those to the right facing La Mar Menor.

and the construction of artificial water complexes has led to resorts being turned in on themselves and away from what is often a modest beach (Plate 14).

Other aspects

Discussion of the seafront, the most distinctive element of the coastal resort, has dominated research on resort morphology. More work is required on other morphological features, particularly in the larger resorts and in those centres where tourism is but one of several urban functions. Burnet's (1963) descriptions of French resorts contain many relevant details but he makes no attempt to generalize from his many examples. His description of Cannes, for example, sheds light on how the resort has developed in size and depth behind the seafront. There the railway lies some 500 metres inland, the third of three axes running east–west parallel to the sea, the two others being the rue d'Antibes, on the traditional route from France to Italy, and the boulevard de la Croisette, which follows the promenade. His map of hotel location in the early 1960s shows that the better class hotels overlook the Croisette, 2- and 3-star hotels border the route d'Antibes, many of the 1- and 2-star hotels lie between it and the railway, and others are found inland of the railway. The city has grown up a valley whose sides were colonized at an early stage by villas. Burnet also noted that Cannes, unlike its larger neighbour Nice, had neither industrial nor workers' quarters: 'The city exists not to produce but to consume, not to manufacture but to sell.'

The seminal study of the effect of consumption patterns on resort morophology is that by Stansfield and Rickert (1970) who distinguish between general commercial functions found in the Central Business District (CBD) and the more specific ones grouped together in the Recreational Business District (RBD). The latter is defined (1970: 215) as 'the seasonally oriented linear aggregation of restaurants, various specialty food stands, candy stores and a varied array of novelty and souvenir shops which cater to visitors' leisurely shopping needs'. In their examples of two New Jersey seaside resorts, the RBD is found along the boardwalk with the CBD and a major shopping thoroughfare located several blocks further back. Stansfield and Rickert (1970: 219) also point out that 'the RBD is a social phenomenon as well as an economic one'. The social aspects of the Golden Mile, an English variation on the RBD, are graphically described by Hugill (1975). Using trend surface analysis, de Haan and

Plate 12 With the exception of the Bali Beach hotel in the distance, the beachfront at Sanur is notable for the absence of high-rise hotels – the palm trees to the left conceal some of the resort's 3000 rooms.

Ashworth (1985: 46) confirmed the existence and relationships of the RBD and CBD in the British resort of Great Yarmouth, noting

> two asymmetric regions dominate the resort, the central area in the inner part of the historic core on the one side and on the other the tourist functions on the waterfront, with a relatively sharp delimitation on the adjacent borders.

Dumas (1982) has examined in detail the commercial structure of Benidorm and identified several distinct districts. The old centre on the promontory separating the two bays has virtually been taken over by tourism. Small tourist-oriented shops comprise three-quarters of all businesses here, with an emphasis on bars, nightclubs, restaurants and souvenir shops. The modern tourist quarter lies to the east, overlooking the Levant beach. This is the zone of high-rise hotels, banks, travel agencies, jewellers' shops, fashion boutiques, shops selling leatherware and other luxury goods. On the eastern and western fringes of the seafront are found less hectic tourist districts, where the ground floors of apartment blocks house souvenir shops, restaurants, bars and the occasional supermarket. Further out still, the built-up area gives way to olive groves and camping grounds. The residential and shopping district now stretches inland from the old centre. A similar structure, particularly the displacement inland of the residential quarters, is also found on the Costa del Sol (Mignon and Heran 1979).

In many Asian beach resorts the RBD often takes the form of shops and stalls, restaurants, rental car agencies and other tourist-oriented businesses living both sides of the street one block back from the beach behind the hotels. This may be complemented by hawker stalls selling souvenirs flanking one or more passage, ways leading from this street onto the beach and by hawkers on the beach itself as at Sanur and Kuta. The 'strip' at Patong in Phuket is something of an exception to this pattern, as it is located along the street immediately behind the beach and in front of most of the hotels (Fig. 8.3). The cluster of bars and associated night-life located in Patong highlights the significance in some of these resorts of the fourth 'S', sex. The location of nocturnal activities, which are an important element for many visitors, is not dependent on proximity to the beach but rather the clustering of bars, restaurants, discotheques and nightclubs in a particular area.

Many of the masterplanned resorts may incorporate shopping complexes but the business which visitors to these resorts may generate may attract other enterprises to set up immediately adjacent to the enclave, as with

Plate 13 Sanur's unobtrusive beachfront is due to hotels such as the Shanti Village which is designed in a distinctive Balinese style with ample gardens, is only two storeys high and built perpendicular to the beach.

Plate 14 The Westin Kauai is characteristic of Hawaii's recent fantasy hotels which incorporate large-scale water features.

Bulau village at Nusa Dua (R. A. Smith 1992). The development of Kaanapali has led to a massive growth in tourist-oriented retail outlets in the nearby former whaling port of Lahaina.

Changing land values resulting from the new or expanded tourist demand have been a major factor in the relocation of pre-existing urban functions and the separation of tourist and residential quarters. Elsewhere this pattern may be reinforced by planning goals and established at the very outset. In Cancun, planners have designated a triangular zone on the mainland for commercial, administrative and residential purposes, with the tourist accommodation zone beginning 3 kilometres away and stretching along the coastal bar (Gormsen 1982). This separation, presumably designed to maximize the tourist-carrying capacity of the seafront, has added little to the life of the resort, particularly in the evenings. The planners of the new resorts in Languedoc-Roussillon also sought to maximize the amount of on-site tourist accommodation and have not encouraged the development of a large resident population. But in Languedoc-Roussillon, unlike Cancun which was developed in a sparsely populated area, the new resorts are able to draw on commuting labour from nearby urban centres. These resorts, nevertheless, tend to lack a sense of community. Elsewhere in Mexico, existing communities have been relocated to enable resort development to proceed, as with the population of Santa Cruz Bay in the Las Bahias de Huatulco (Long 1991). Mitigation measures to reduce the impacts on the affected residents have not been entirely successful however.

Kermath and Thomas (1992) have examined how informal sector vendors have been displaced from beach sites in Sosua, the Dominican Republic, as the resort has grown, with residential relocation also being slated. Other similar examples are cited in Puerto Plata and Samaná (Fig. 8.8). Kermath and Thomas (1992: 187) see the maintenance of a healthy tourist image as the driving force behind this process, noting

> Where 'generic' attractions exist, such as scenic, white, sandy beaches lined with palm trees and bathed with warm crystal waters, image is crucial. Governments and formal sector operators view street vendors serving food under unsanitary conditions from unsightly stands as problems with one solution – eliminate them. It may be desirable to expel or relocate low-income housing as well. In short, remove or improve as much poverty as possible to create conditions for favourable tourist impressions.

Ports, particularly small fishing ports, are perhaps the most typical of the pre-existing coastal communities which have been transformed by tourism and seen traditional functions replaced by new ones. Such resorts have often assumed a crescent-shaped form rather than the linear morphology of the beach-oriented resort. In many cases, the attraction initially lay in the charm of the site. Some, such as St Tropez, achieved early prominence through being 'discovered' by artists and writers and owed their growth as resorts to subsequent visits by filmstars and other celebrities which conferred on them a certain fame or notoriety (Christ 1971). As pleasure boating expanded in the 1960s, yachts and cabin cruisers began to share the port with fishing vessels, then to replace them or to occupy extensions to the harbour. At St Tropez, in particular, the quaysides have been invaded by the easels of artists who 'capitalize on the good name of this former painter's corner and on the gullibility of tourists' (Christaller 1964: 103). However, no good beaches are found at St Tropez itself to cater for today's sun worshippers, so they flock to the immense sandy beach at Pampelonne 6 kilometres away. This separation of beach and port has contributed to the *mitage* of the surrounding hillsides and countryside by second homes.

At Albufeira on the Algarve, Portuguese fishermen still draw their boats up on to the central beach around which the town initially developed but they are now flanked by the bronzed or burnt bodies of scantily clad British holidaymakers. Much of the initial nucleus of shops and fishermen's houses still remains here, but hotels, apartments and villas have progressively colonized the surrounding hillsides with adjoining areas now being subdivided at lower densities for second homes (Plate 15). Form and process here are similar to those described by B. Young (1983) with reference to the transformation of Maltese fishing villages.

As the growth of boating has exceeded the capacity of existing ports, new resorts have been developed expressly for this new demand, with very functional forms (Pearce 1978c). Whether designed around basins or islets, resorts such as Port Camargue, Port Grimaud and the Marines de Cogolin bring together nautical holidaymakers, their boats and holiday homes. Elsewhere, the boat harbour or marina has been used as a second focus of activity to the beach, either to spread the prime real estate sites in new resorts (e.g. La Grande Motte) or to generate new demand in existing centres (e.g. Carnon, Deauville and Brighton).

In other cases, golf courses have been used to develop a new resort in depth. Fairways are constructed behind the seafront hotels, creating both an additional recreational attraction and a pleasant open environment. The developer benefits from this strategy in several ways: by enhancing the attractiveness of the hotels and broadening their clientele, by reducing problems of seasonality, by direct returns from the golf course and, most significantly, by greatly enlarged real estate

Plate 15 Albufeira: Portuguese fishermen still draw their boats up to the central beach but hotels, apartments and villas have progressively colonized the surrounding hillsides.

opportunities, notably for the construction of condominiums and villas. This strategy has been used successfully in Hawaii (Farrell 1982) and is being applied in some of the new *urbanizaciones* on the Costa del Sol and elsewhere (Vera Rebollo 1991). Valenzuela Rubio (1981) suggests that in the case of the Costa del Sol the golf courses, like the new marinas, are directed at the top of the market, which has been less affected by the economic recession, while the substantial investment required for such projects perpetuates conditions of external dependency.

The evolution of coastal resorts

Many of the studies reviewed so far have recognized historical features of resorts and discussed their evolution in general terms. However, comparatively few writers have documented in detail the morphological changes which have occurred from one period to another (Lacroix, Roux and Zoido Naranjo 1979; Franz 1985; Priestley 1986; Meyer-Arendt 1990; R. A. Smith 1991, 1992). More emphasis still needs to be given to changes in morphology so that the processes involved can be better understood and the evolution of future forms more readily anticipated.

One of the early studies in this field was by Pigram (1977) who examined the changing morphology of two pairs of twin towns on the Gold Coast of Australia: Coolongatta-Tweed Heads and Southport-Surfers Paradise. Pigram presents land-use maps for each pair of towns for 1958 and 1975 based on published maps and local body records for the earlier period and fieldwork for the latter. Fig. 9.1 shows that important changes in both form and function have occurred in Southport-Surfers Paradise over this period. Pigram (1977: 539) suggests that 'their urban morphology has become effectively fused, with shared attractions and facilities and little duplication of services'. Southport's more traditional resort function has declined (the pier is now gone) with the closing of the railway and improved access to the beaches, but its administrative, commercial and residential roles have expanded. Surfers Paradise, whose coastal bar site is reminiscent of others discussed earlier, has emerged as the dominant tourist centre. Pigram (1977: 538–9) describes the major structural elements and changes in Surfers Paradise thus:

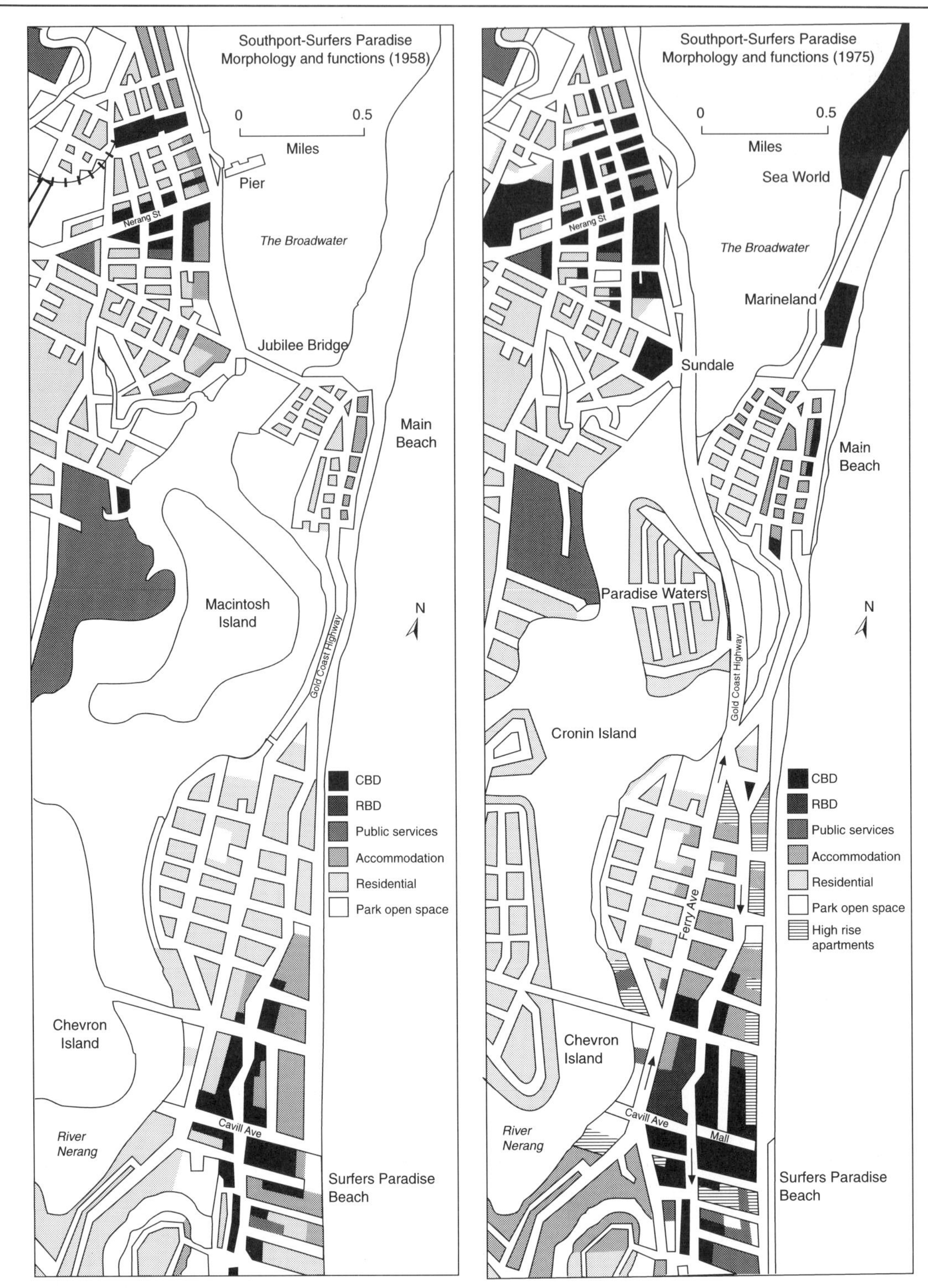

Figure 9.1 Morphological changes in Southport-Surfers Paradise, 1958–75.
Source: Pigram (1977).

> Morphologically, land-use zones are, once again, aligned north–south parallel to the beachfront. The highest density tourist accommodation is found adjacent to the front, where high-rise apartment blocks are progressively replacing lower-density units. A second zone of high-rise development is beginning to emerge further inland overlooking the Nerang River estuary reflecting both the attraction of water-based site and the growing shortage of beachfront land. . . .
>
> In general, the zone of permanent residence in Surfers Paradise has migrated from its original location near the beachfront and is now confined to the area west of the highway and well removed from the resort core. The relatively highly valued land flanking the RBD and the front strip has been taken up by medium and lower-density tourist accommodation to form a useful buffer separating two functional zones. . . . At the same time, Surfers Paradise remains highly regarded as a residential environment for second homes and retirement . . . Chevron Island and Paradise Waters are only part of an extensive canal estate development program designed to provide high status and expensive home sites.

Some of these changes, however, have not been without environmental problems. Figure 9.1 also shows changes in the RBD, the development of a one-way street system and the emergence of new attractions to the north in the form of Marineland and Sea World.

Calella, a Spanish resort located 40 km north-east of Barcelona, provides a second example of resort development. Figure 9.2, which is based on vertical aerial photographs taken in August 1966 and January 1980, shows the morphological changes which have occurred over this period, particularly those which have taken place on the margins of the resort, a zone frequently neglected in resort studies. Fig. 9.2 shows that in 1966 Calella consisted essentially of a contiguous built-up area, aligned with the coast and lying between the railway and a steeply rising hill to the north. Some more scattered hotels and apartments are found to the west and in the adjoining commune of Pineda to the east where farmhouses can also be seen. Much of the development after 1966 has consisted of infilling on the margins and some inland expansion up the valley in the west. The new apartments and hotels are generally several storeys taller than the buildings in the core. Both the infilling and expansion on the margins has involved the encroachment of the urban area on to the surrounding market gardens. Plate 16 provides a good example of the rural/tourism fringe to the west of Calella in 1982 and shows the market gardens separated from apartment buildings by several private houses. If demand for tourist accommodation continues, further encroachment can be expected here and on the horticultural land lying along the Calella–Pineda boundary as market gardening can no longer support the increased land rent.

Earlier research by Ferras (1975) enables these developments to be put into the context of the wider region, the Maresme. Ferras's analysis of building permits for 1969, for example, distinguishes Calella from the other communes in terms of the virtual absence of individual dwellings and the height and size of the new buildings. Much of the Maresme is frequented by second home owners and holidaymakers from Barcelona, but Calella stands out as a resort favoured by foreigners, particularly Germans.

The utility of aerial photograph interpretation involving time-series photography constitutes a valuable means of analysing morphological changes as has been demonstrated elsewhere in Spain by Menanteau and Martin Vincente (1979), Fourneau (1983) and Priestley (1986). Using this technique complemented by statistical data on population, dwellings, accommodation and employment, Priestley, for example, documented the changes which had occurred in the built environment of Lloret de Mar on Spain's Costa Brava from 1956 to 1981. During this period the urban centre trebled in size, in large part due to hotel construction, and *urbanizaciones* came to cover a quarter of the commune's total surface area.

Other longitudinal studies have been undertaken along the Gulf Coast of both the United States and Mexico (Meyer-Arendt 1985, 1990) and in South East Asia and Australia (R. A. Smith 1991, 1992). Common features in the models proposed by these authors include lateral extension and intensification of development of the beachfront from an initial site favoured for its accessibility. Meyer-Arendt (1985: 462–3) describes the process on Grand Isle, Louisiana, thus:

> From an initial settlement nestled among the higher beach ridges and focused toward the back bay and away from the environmentally hazardous beachfront, tourism development extended out to the shore and into the exposed vicinage zone. With the construction of a beach highway, settlement spread laterally and gradually intensified.

Based on the unplanned South East Asian examples he examined, R. A. Smith (1991) emphasizes the importance of second homes in the earliest phases, with the street pattern which formed the basis of subsequent hotel development being defined in terms of these. Drawing on a study of Nusa Dua, R. A. Smith (1992: 216) later extended this work to propose a Model of Integrated Resort Development in which a planned core is 'surrounded by unplanned resort functions,

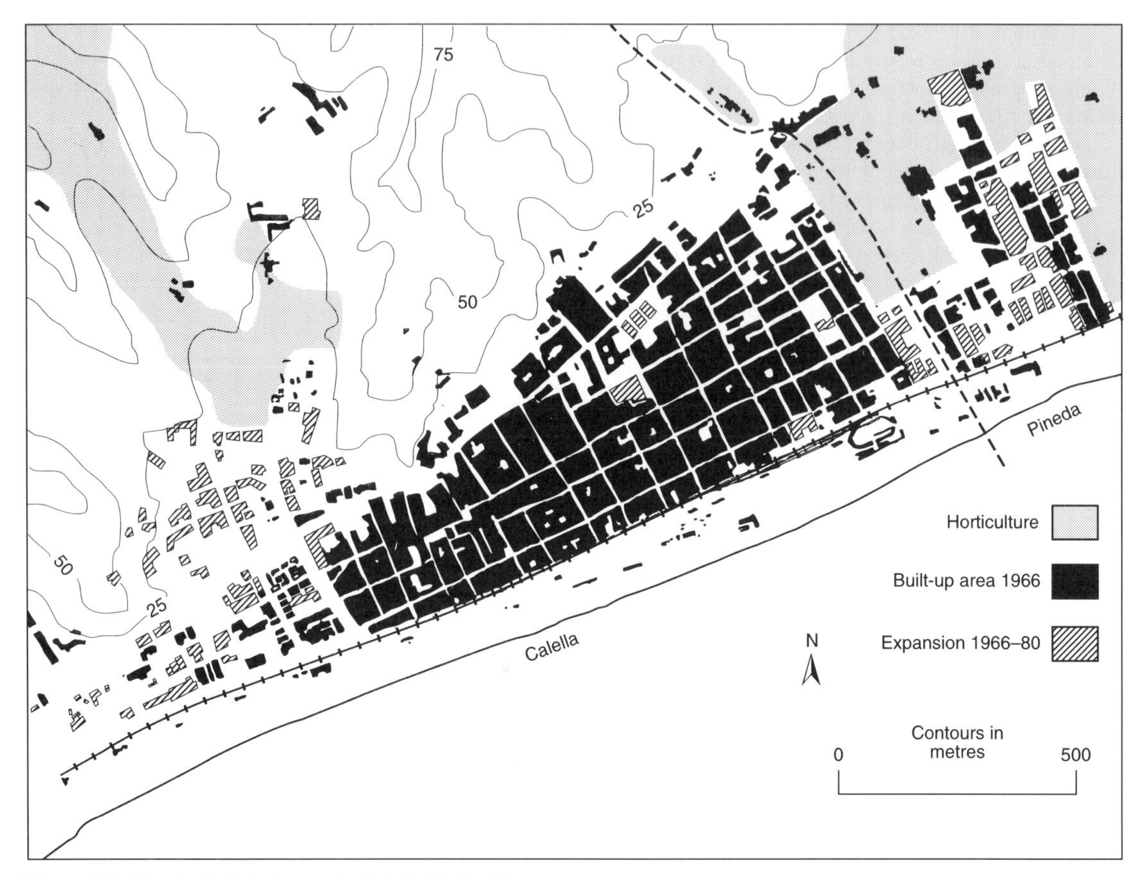

Figure 9.2 Morphological changes in Calella 1966–80.

hotels, guest houses, restaurants, business and residential areas'.

Alternative representations

Jeans (1990) complements studies of the physical aspects of coastal resorts with an innovative interpretative semiotic model (Fig. 9.3a). Drawing heavily on an early physical model of the English seaside resort (Wall 1971) together with more recent Australian examples, Jeans makes a fundamental distinction between the resort, representing Culture, and the sea, representing Nature. The former, he suggests, is 'controlled and structured by class and function' while the latter provides 'tantalising dangers' and 'compensating sensuous pleasures'. Culture and Nature, according to Jeans are separated by a symbolic zone of transition transversed by signposted paths and a zone of ambiguity, the beach, where 'people behave and undress "naturally", while culture attempts to maintain control'. Jeans further asserts (1990: 281) that 'a second axis of meaning runs parallel to the coast, where a centre is surrounded by peripheries at each end of the beach', the latter occupied by the young and the non-conformist (surfers, nude and semi-nude bathers). The relative importance of each of these zones might be gauged by the time-budget studies discussed in Chapter 4, where it was shown, for example, that the beach accounted for between one-fifth and one-quarter of visitors' non-sleeping time (Gavíra *et al.* 1975; Pearce 1988b; Hottola 1992).

In a similar fashion, but with their emphasis more on physical rather than social features, Pearce and Kirk (1986) distinguished different forms of carrying capacity associated with specific resort zones (Fig. 9.3b). The implications of this were that the assessment of the carrying capacity of any resort is more complex than might at first be supposed given the different environmental, perceptual and physical thresholds which might exist as one moves from the coastal waters, across the beach and dunes to the immediate hinterland and the corresponding range of assessment procedures required. Assessing the resilience of dune systems or water quality necessitates a different expertise than that

Plate 16 The rural/tourism fringe of Calella: apartment buildings encroach on former horticultural land.

needed to establish acceptable social levels of beach use or the physical possibilities of hotel construction but all may be used to establish appropriate planning norms.

Urban Areas

The structure of tourism in urban areas remains a comparatively neglected area of research, both by tourism researchers and those interested by broader urban processes (Ashworth 1989; Mullins 1991; Law 1992). In general there has been little attempt to build systematically upon some of the early theoretical and empirical studies, for example (Burnet and Valeix 1967; Yokeno 1968; Eversley 1977), the comprehensive account of the tourist-historic city by Ashworth and Tunbridge (1990) being a notable exception. Much of what has been written on urban tourism derives from a continuing interest in the links between tourism and urban redevelopment or conservation (Ford 1979; Hoyle, Pinder and Husain 1988; Page 1989; Law 1992) or results from work on related special interest topics, for example restaurants (Huetz de Lemps and Pitte 1990).

Tentative theoretical contributions to the problem were made by Yokeno (1968) but his work, published in a French series, was not picked up and developed in the English language literature. More comparative and systematic studies appear a few years later. Chenery (1979) considered planning for tourism in six European cities and, in a significant departure for an urban geography text, Burtenshaw, Bateman and Ashworth (1981) devoted a chapter of their book on the West European city to 'The Tourist City' and a second to 'Urban Recreation Planning'. Pearce (1981a) adopted a more conceptual approach and illustrated theoretical patterns with the example of Christchurch (Fig. 1.7). Ashworth and de Haan (1985) and Ashworth and Tunbridge (1990) have elaborated a 'tourist-historic city' model which incorporates the notion of a partial migration of the CBD and the emergence of a 'tourist city' in a zone of overlap between the historic core and the contemporary CBD. Weaver (1993) presents a more specific model of urban tourism space for small Caribbean islands, one which reflects the impact of their cruise-ship port function. This concentric zone model is characterized by the emergence of specialized tourist retail functions close to the cruise-ship dock and the development of accommodation and attractions in the CBD and more distant neigbourhoods.

The comparative neglect of tourism in urban areas can be attributed in part to differences in the role of such areas in national tourist industries and to the relative importance of tourism within cities. Earlier

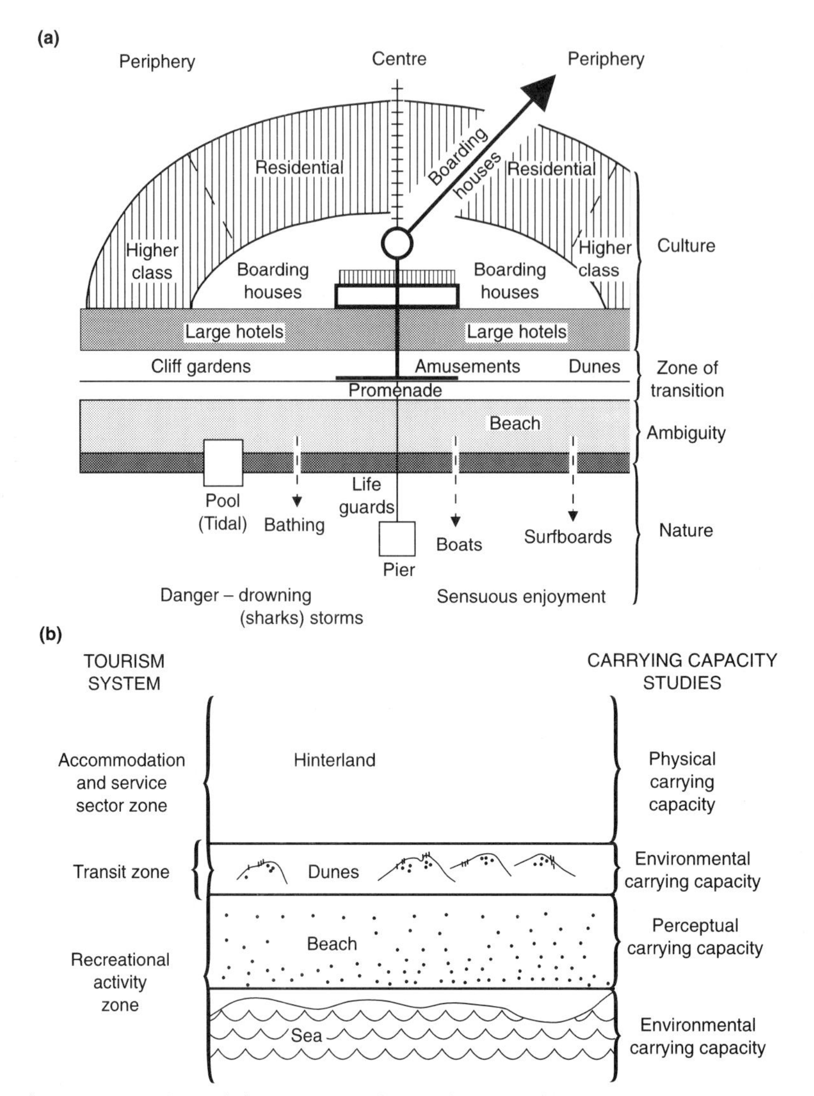

Figure 9.3 Alternative representations of the structure of coastal resorts: (a) a semiotic model of the beach resort; (b) schematic representation of the spatial relationship between elements of the tourism system, the coastal environment and carrying capacity studies.
Sources: Redrawn from Jeans (1990) and Pearce and Kirk (1986).

chapters have highlighted the importance of major urban areas within countries, both as gateways and as destinations in their own right (see, for example, Figs 1.6 and 1.7, Chapters 4 and 7). However, even in cities such as Paris, London and New York with their large tourist industries (Paris has more than double the number of hotel rooms of Waikiki) tourism is far from being the dominant activity, being surpassed by commercial, administrative, industrial and residential functions. An immediate consequence of this is that tourism in cities is less visible than in coastal, alpine or thermal resorts which have a more explicit, distinctive and dominant tourist function. This lower profile has undoubtedly contributed to urban areas being overlooked in a field of research which is still comparatively young. The more cynical might also stress the greater appeal to urban-based academics of fieldwork on the coast, in the mountains or in the countryside.

At the same time, the multifunctional nature of cities and the wide range of visitors they attract makes any analysis of tourism there more complicated. Some vacationers will come for the entertainment and nightlife, others for the historical and cultural features, or to shop in the large department stores or boutiques. Many may be attracted to cities by what Matley (1976: 30) refers to as 'the individual character and atmosphere which transcend the mere sum of their buildings and other physical attractions. An obvious example is Paris'. With the exception of the accommodation sector, few of the facilities and services used by tourists are purpose-built or provided especially for them. Rather they share in varying degrees with local residents the transport services, shops, restaurants, cathedrals, museums, theatres, and so on. In addition to a variety of vacationers, many cities also receive a considerable volume of traffic generated by their other functions. Administration, commerce and industry attract large numbers of business travellers while a sizeable resident population will generate a significant VFR traffic. Conference and special events will draw other visitors. Together, these groups may well exceed the pleasure-oriented tourists, the mix in any particular case depending on the attractions and functions of the city. For example, those with an international vocation, such as Brussels and Geneva, will have a much greater proportion of business travellers. Multipurpose trips are also likely to be important. Deriving accurate purpose of visit figures in urban areas is, unfortunately, not an easy task. Furthermore, for each of these groups the city may play a variety of roles as was discussed earlier with reference to Fig. 1.7, notably those of a gateway, staging post, stopover or destination.

Given the complexities of tourism in urban areas, it is not surprising to find that many of the early studies which discuss the structure of tourism in cities have focused on the distribution of hotels, the most visible and pure manifestation of tourism in the city. The following section will review and extend work on the structure of accommodation, exemplified by a case study of the hotel industry in Paris. Subsequent sections will discuss attractions and other sectors of the industry, then developments beyond the city.

Distribution of accommodation

Yokeno (1968) proposes a concentric model of urban land use, based on land rent curves, in which he locates a hotel zone in the central city, between the innermost administrative centre and the commercial zone. Empirical studies of inland urban areas generally confirm the concentration of hotels in the central city, on the margins of the historic core. These central hotels are not of such density, however, that they constitute a uniform single-use hotel zone similar to those found in the seafront of coastal resorts or to the linear concentrations of financial institutions or specialized shopping streets which may characterize other parts of the city. Rather, one or more clusters are usually found where hotels are interspersed with other commercial, administrative and sometimes residential functions. The number and location of these clusters will depend on several interrelated factors including the size of the city, its functions, pre-existing form, means of access and whether it is mononuclear or polynuclear in character.

A central city hotel responds to the needs of a variety of urban visitors whose stay is generally relatively short, typically in the order of two to four days. Many business travellers value proximity to the CBD so as to transact their affairs promptly and efficiently. Here too the shopping-oriented visitor can descend rapidly on a range of high-order shops. It is also in the centre that the sightseeing tourist will find many of the city's historic buildings, monuments and other cultural attractions. The overlapping of the focus of activity of each of these different markets reinforces the concentration of hotels in the central city. Moreover, the clustering of hotels enables the visitor to 'easily choose one according to his taste or budget' (Yokeno 1968: 16).

Tourism on a large scale, however, is a relatively recent phenomenon and few hotels, especially large modern ones, occupy sites in the very core of the city. When the major period of expansion in hotel construction occurred in Europe's cities in the 1960s and 1970s, there was a scarcity of available inner-city sites sufficiently large enough to accommodate hotels on a scale which would make them viable. Moreover, building restrictions and planning regulations designed to preserve the character of the historic core also effectively limited or discouraged hotel construction in many cities (Chenery 1979; Burtenshaw, Bateman and Ashworth 1981). However, as Chenery points out, in relatively small and compact cities such as Copenhagen and Amsterdam, this is not such a disadvantage since 'nowhere is far from the centre'.

Recency, rent and restrictions thus combine to lead to the central concentration of hotels around rather than within the core itself. Eversley's (1977) map of the location of hotels of 200 rooms or more in London shows none within the City, but major concentrations in Westminster, Kensington, Chelsea and, to a lesser extent, Camden. The scale and rate of hotel construction in London during the 1960s and early 1970s was so great as to spill over into residential areas such as Mayfair, causing much public and political concern over the loss of housing stock (Hall 1970; G. Young 1973). Chenery's (1979) comparative study shows this 'creeping conversion' of older housing into hotels was not a feature of the other cities he studied, except Amsterdam. In London it led to a tightening up of permission to build hotels in the most affected areas, particularly Westminster and Kensington. Such regulations coupled with financial complications posed major problems for growth in London's hotels. A study in 1978 (BTA 1979a) found that the most viable hotels would be 5- and 4-star hotels in central locations and 3-star and budget hotels in peripheral locations but noted (p. 7 of Appendix):

- application of planning controls, particularly in central locations, creates a serious impediment to the development of new hotels
- the building of quality hotels in central locations is unlikely to solve the projected shortfall problems because the bulk of the projected additional demand is for hotels of lower standard and price
- it is doubtful whether developers will be attracted to build lower standard hotels in peripheral locations which would be the ones most vulnerable to changes in overall demand.

As the central location of hotels reflects demand generated by other central functions, it follows that the changes in the distribution of hotels will result from an expansion of the CBD or the appearance of new centres as well as from the shortage of central sites. This is evident not only in the medium-sized tourist-historic city, such as Norwich, on which Ashworth and de Haan (1985) developed their model, but also in larger cities where the tourist industry has expanded rapidly at the same time that major morphological changes have occurred. Gutiérrez Ronco (1977, 1980) has traced in detail the evolution of hotels in Madrid and shows how there has been a progressive shift from the Puerta del Sol to the north and north-east as administrative and commercial activities have developed in that direction, particularly in the 1960s and 1970s. This has given rise to the current pattern where the smaller, older hotels are found clustered around the Puerta del Sol and Avenida Jose Antonio while the large, modern quality hotels are located along the axis formed by the Paseo del Prado-Paseo de le Castellana-Avendida del Generalisimo, where many office blocks have recently been built and many government ministries relocated. A similar pattern occurs in Lisbon, where the Avenida da Libertade links the older smaller hotels of the centre with the more recent, larger and better hotels in the streets surrounding the Parque Eduardo VII, also an area of recent commercial and administrative expansion.

Comparable patterns and processes are to be found in Asian cities such as Hong Kong. Many of the early hotels were located on Hong Kong Island in Central District, Wanchai and Causeway. Much hotel expansion occurred on Kowloon in Tsim Sha Tsui which developed as a thriving retail centre, specializing in duty-free shopping, perhaps the prime attraction for visitors to Hong Kong. By 1982 the district accounted for almost two-thirds of the colony's registered hotel rooms. From 1975 to 1982 more than 4000 high-tariff rooms were built in Tsim Sha Tsui while no other hotels had been constructed anywhere else in Hong Kong except for one at the airport. Half of these new rooms were in the four hotels built in Tsim Sha Tsui East, on government land zoned commercial in the mid-1970s following the relocation of the railway station. Two further hotels were built on reclaimed land in the New World complex.

Changes in maritime technology that have induced the growth of separate maritime industrial development areas have also led to functional changes in city waterfront sites through major redevelopment projects that often include hotels and other tourist functions (Hoyle 1988). Hotels were successfully attracted into the Baltimore downtown area through the inner harbour redevelopment project (Law 1985) and ten new hotels were proposed for London's Docklands project (Page 1989) though few of these have since materialized.

Changes in the pattern of hotel location have resulted not only from modifications in land use but also from developments in the transport sector. Sometimes these changes have occurred together. Earlier in the twentieth century when travel by rail was more important, clusters of hotels sprung up around city railway stations, located on the margins of the city centre. These were usually rather small and modest although some major terminus hotels were built. In his study of hotels in Brussels, de Ganseman (1982) notes that in the period 1930–58, the city's three railway stations along with the city centre (Place de Broukère) constituted the major poles of hotel location. However, little new construction occurred there in the period of hotel expansion which followed in the 1960s and 1970s as Brussels developed its international role (EEC, Nato). Rather there was a *delocalization* towards the *haut de la ville* due to the growth in night-time entertainment there and the attraction of a chic commercial quarter. Burtenshaw, Bateman and Ashworth (1981) observe that in pre-war Berlin there was a distinct clustering of hotels around the city's three main railway stations. By 1976 these had been abandoned in favour of a westwards shift along the Tauentzienstrasse and Kurfurstendam, a process 'strongly reinforced by the political division of the city'. Subsequent reunification is likely to bring further changes.

Since the 1960s other hotels have been located more specifically to meet the needs of the motorists, especially the business person travelling by car. These are commonly located on the outskirts, on or immediately adjacent to the highways leading into or out of the city. In France, such a pattern is to be found not only in Paris where such hotels are located at the major gateways, e.g. the Porte de Sevres or the Porte Maillot (Cadart 1975), but also in smaller cities such as Metz

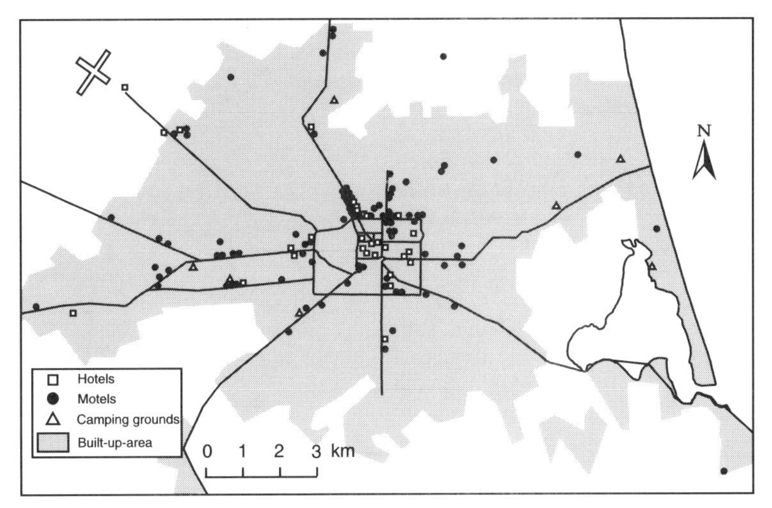

Figure 9.4 Distribution of accommodation in Christchurch, 1993.

(Spack 1975). Certain French hotel chains even tend to specialize in providing such non-central accommodation (Cadart 1975).

Elsewhere, notably in North America and Australasia where rates of car ownership are particularly high, motels (the motor-hotel) have been developed for the motoring public. Many of these simply offer non-serviced accommodation but in New Zealand most provide fully contained self-catering facilities, that is, they are equipped with a small kitchen and a dining-cum-living room as well as bedrooms. Motels in New Zealand tend to attract a range of guests, from travelling business people to families on holiday, for whom they offer a cheaper alternative to hotels.

Figure 9.4 shows that while hotels in Christchurch, New Zealand, exhibit the characteristic centralized clustering seen elsewhere, many of the motels, particularly the more recent ones, are distributed in a linear fashion along the major highway axes, a pattern that is repeated in other New Zealand cities. Access, however, is not the only factor here, for the lower densities of motels mean they are unable to support central city rents. Moreover, they may be excluded from the centre itself by land use controls. In Christchurch, and elsewhere, camping grounds, whose densities and rent-paying ability are lower than motels, are found in more peripheral locations, while still being handy to the major access roads.

A similar pattern to that shown in Fig. 9.4 was also observed by Liu (1983) in metropolitan Victoria, British Columbia. Three-quarters of the hotels there were located in the Downtown Market area, while the majority of motels occurred in a 'corridor-type' development along Highway 1A leading out of the centre. Other North American studies have also stressed the highway orientation of motels (W. R. White 1969; Ohler 1971). In Toronto, Wall, Dudycha and Hutchinson (1985) noted some decline in the proportion of motels in Toronto over the period 1965 to 1979 (down from 42 per cent of all Toronto establishments to 37 per cent), a decline which also resulted in less overall clustering (p. 615):

> Motels now constitute a small proportion of the total accommodation supply and their spatial pattern has been influenced as much by closures as by new construction. Since much of their business is casual, they have strict locational requirements and they concentrate in areas with a high traffic density. They benefit from the proximity of similar establishments for this creates greater combined visibility. ... There are still benefits to be gained from agglomeration but the large hotel has greater locational flexibility than the motel and this helps to explain the recent trend towards dispersal. However, while not necessarily in the downtown area or airport strip, large hotels are still likely to be in areas of high accessibility, such as close to major highway interchanges.

The airport hotel results from the large-scale development of air travel since the 1960s. Some are located within the airport complex itself (as, for example, at Los Angeles), but more often they form a distinct cluster on the margins of the airport, frequently on the road leading into the city (Fig. 9.4; see also Cadart 1975; Eversley 1977; Chenery 1979; de Ganseman 1982). Burtenshaw, Bateman and Ashworth (1981: 169) observe that at the city-region scale in south-east England, 'successive rings can be identified including a hotel-rich central city, a hotel-poor suburban ring, and a hotel-rich outer zone including distinct hotel clusters around Heathrow and Gatwick airports'. In Toronto, the development of the airport strip on the perimeter of the city has contributed to some reorientation of the distribution of hotels to the northwest (Wall, Dudycha and Hutchinson 1985). The number of airport hotels will depend on the volume and nature of the air traffic with, for example, demand for accommodation close to the airport increasing with the proportion of transit passengers and the importance of the gateway and staging post roles of the city.

Other factors have contributed to the growth of sub-centres of hotel accommodation within the city, particularly in larger ones. As de Ganseman (1982) noted with Brussels, evening entertainment may be as significant a location factor for some visitors as day-time interests. Kings Cross in Sydney, for example, famed (or ill-famed) for its restaurants and night-life, has developed a cluster of hotels second only to the CBD (Council of City of Sydney 1980).

Site factors may also break up the central concentration of hotels. This is especially true of cities on the coast, a point which has been largely overlooked due to the concentration in the literature on inland European capitals. The distinction between a coastal resort and a city is not always easy to make although purpose of visit rather than location is a useful starting-point. Few primarily sunlust visitors are likely to be attracted to harbour cities such as Hong Kong or even Sydney, but this group no doubt forms a major part of the traffic to San Juan whose beach-oriented hotels form part of Puerto Rico's primate city (Fig. 8.4). Intermediate cases require more detailed analysis. Rio de Janiero can hardly be considered a coastal resort in a traditional sense, yet two major linear concentrations of hotels overlook the magnificent beaches of Copacabana and Ipanema. A third major cluster is found in the Flamengo district which is on the margins of the downtown area, in close proximity to Santos Dumont airport, and overlooks the Flamengo beach and Guanabana Bay.

Many of these different locational patterns and factors are combined and summarized in Ashworth and de Haan's (1985) typology of urban hotel locations (Fig. 9.5). Although derived in the context of the tourist-historic city, it would appear to have more general application. Ashworth and de Haan make a distinction between those sites that are 'transport' oriented (sites B, C and F) and those that are 'location' or 'market' oriented (sites A, D and E).

While the models of Yokeno (1968) and Ashworth and de Haan (1985) provide some conceptual basis for explaining the spatial structure of urban tourism, much interpretation of observed patterns of accommodation has, as noted, been based largely on general factors such as access, land rent and zoning regulations with

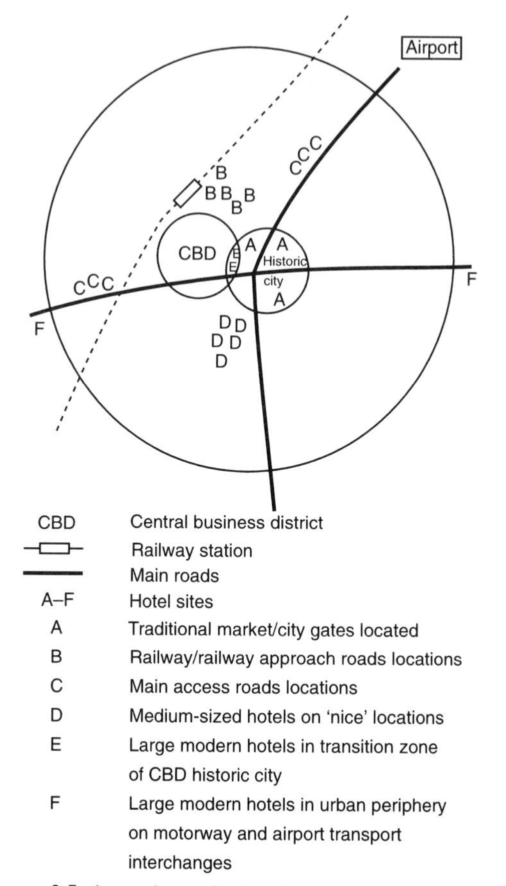

Figure 9.5 A typology of urban hotel locations.
Source: Redrawn from Ashworth and de Haan (1985).

assumptions being made about which markets are served by different clusters of hotels or motels. Few direct attempts have been made to look at intra-city variations in demand or the locational choice behaviour of visitors. Mayo (1974) presents an interesting general motel-choice model and the results of a survey of 748 American motorists to determine which factors influenced their choice of motel. Mayo's respondents ranked 'being not far from main route' and 'convenient parking' as the second and third most important factors behind moderate prices and ahead of sixteen other variables including a quiet or attractive setting and proximity to tourist attractions. However, his results were aspatial in nature and not applied to any particular city. Later, Pearce (1987d) examined differences in the profiles of motel guests in Christchurch and their choice of a motel in one of three locations in the city – the central city, suburbs and along major access routes. Central city motels tended to attract a larger share of vacationers, many of whom were from overseas and on their first visit to Christchurch. Centrality was a determining factor in their choice of motel. Suburban motel guests come to Christchurch for a wide variety of purposes, are more likely to be domestic in origin, coming from the South Island in particular, and to be on a repeat visit to the city. Proximity to a particular venue or event is a key factor in the selection of a motel in the suburbs. Differences between the major route respondents and those in the other two groups were more muted. Further behavioural studies such as these are needed to complement the purely morphological analyses and to fully understand the processes at work within the city.

With these general patterns outlined, the case study of hotel accommodation in Paris will now be examined to show in more detail variations in the structure of its hotel industry, the interrelationships of the different factors which contribute to the patterns observed and how these reflect both general tendencies elsewhere and factors more specific to the city itself.

Paris

Since the 1970s Paris has experienced a significant amount of hotel construction and by 1991 had a capacity of approximately 70,000 rooms in 1400 classified hotels (different official sources give somewhat varying figures: Paris Promotion 1991). A boom in hotel construction occurred in the early 1970s, pulling the city's hotel industry out of a dormant post-war period. From 1947 to 1970, 133 hotels representing 7000 rooms had disappeared while only about 10 new hotels had been built. Over 80 per cent of Paris's 2- and 3-star hotels in 1970 had been constructed before the First World War (Anon 1971). The opening of the Hilton in 1966 and the Sofitel Bourbon in 1970 were the first additions to the upper end of the market since the Georges V in 1935. These hotels marked a resurgence in the hotel industry as Paris sought to meet the growing demand for quality accommodation, particularly from the business sector. Many of the new hotels were built by chains, both national and international (Cadart 1975). From 1970 to 1982 the city's hotel capacity expanded by some 7000 rooms, with the proportion of deluxe and 4-star hotels increasing from 17 to 25 per cent while the share of 1-star hotels dropped from 33 to 24 per cent. In a process opposite to the 'creeping conversion' experienced in London, a number of 1-star hotels were turned into flats (Chenery 1979). A further 7000 rooms were added in the period 1982 to 1991. By that date almost three-quarters of the city's hotels were classified as 2-star (38 per cent) or 3-star (36 per cent) with a decrease occurring at both the top and bottom ends of the scale (4-star and 4-star deluxe 14 per cent; 1-star, 12 per cent). While real growth has occurred in the middle of the range, analysis of the evolution of the hotel industry is complicated by the introduction of a new classification system in 1986 and changes in grading associated with variations in differential VAT rates (Paris Promotion 1991). In particular, this has affected the upper end of the hotel industry and resulted in an apparent demise of the deluxe sector, although the hotels themselves remain.

The resurgence in hotel construction affected the spatial structure of the industry as new locational factors came into play but, given the age of many of the hotels, the imprint of the past is still very apparent (Fig. 9.6). The distribution of hotels in Paris also reflects both the centralization found elsewhere and the broader east–west structure of the city. While the west, particularly the north-west, is dominated by commercial activity and higher quality residences, the east, especially the north-east, is more industrial and characterized by sprawling lower quality residential areas.

Figure 9.6 shows that the hotel industry is still largely concentrated in central Paris, although not to the same extent as at the time of Burnet and Valeix's (1967) study. In 1991, the 1st, 2nd, 8th, 9th and 10th arrondissements accounted for 39 per cent of the city's capacity, compared with 43 per cent in 1982 and 49 per cent in 1967. Burnet and Valeix attributed this concentration to 'the co-existence of two functions, work and leisure'. This is not only the commercial heart of the city, drawing many business travellers, but also the prime tourist destination attracting sightseers, shoppers and those seeking entertainment. The Champs-Elysées, for example, bisects the 8th arrondissement while the Tuileries, Louvre and Comédie Française are found in the 1st. The 9th has not only the Opera but also the Folies Bergère. In the 10th, clusters of hotels are found around the Gare du Nord and Gare de l'Est. Older terminus hotels around the Gare de Lyon also account for much of the capacity in the 12th arrondissement. Demand along the Left Bank has been generated by the government ministries in the 6th and the university quarter in the 5th as well as the general atmosphere of the Latin Quarter. Some demand in the outer ring results from an extension of the inner-city activities, notably in the 17th where the scale of analysis conceals the concentration of hotels just off the Place Charles de Gaulle in the sector bounded by the Avenues de la Grande Armée and de Wagram. Hotel development is still limited in the north-east (the 18th, 19th and 20th) but to the south and west in this outer ring are found the only three arrondissements to experience any significant increase in their share of the city's accommodation capacity over the period 1967–91: the 14th (3.5–7.1 per cent); the 15th (4.5–9.3 per cent) and the 17th (7–9.4 per cent).

Significant spatial variations also occur in the quality

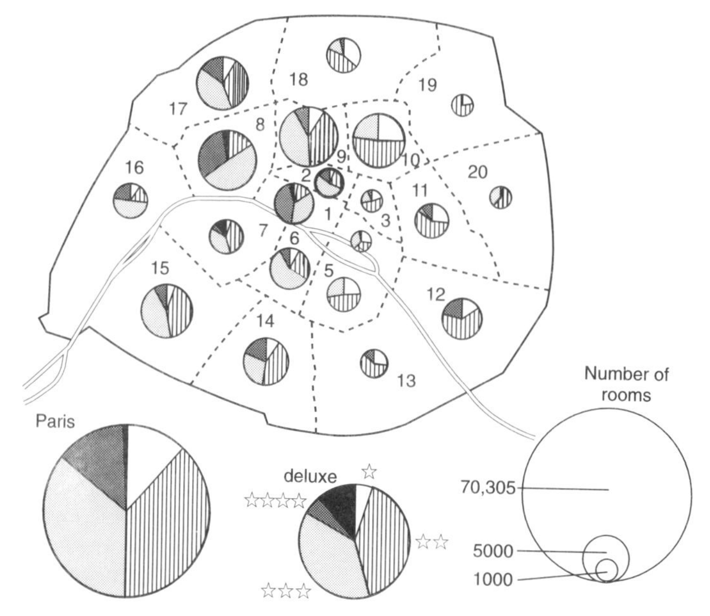

Figure 9.6 Distribution of hotels in Paris, 1991.
Data source: Paris Promotion (1991).

of accommodation. In 1991 the small number of deluxe rooms were found in just three arrondissements, the top five arrondissements in each category accounted for three-quarters of the 4-star rooms, 60 per cent of the 3-stars, half of the capacity in 2-star hotels, and 54 per cent of that in 1-star hotels. The majority of the 4-star hotel capacity is located in the north-west quadrant of the city, in the 1st, 8th, 9th and 17th arrondissements with other clusters to be found in the 14th and 16th. Almost half of the 1st arrondissement's rooms are now classified in the 4-star or deluxe category and one-third of those in the 8th. On the Left Bank, the 5th is more down market than the 6th, which attracts a greater share of official business travellers; 1- and 2-star accommodation predominates in the northern and eastern districts.

The growth of large, quality hotels in the 14th, 15th, 16th and 17th arrondissements represents the major locational change in the distribution of hotels in Paris during the 1970s. Several interrelated factors have contributed to this process, with land rents being the most fundamental (Cadart 1975). Changes and intensification in the central city throughout the 1960s meant that new hotels were rarely able to support the new land rents which often would have been equivalent to one-third of the cost of a hotel. Not only were larger, quality hotels likely to be more viable in these conditons but it was this sector which the authorities were encouraging in order to re-establish the image of Paris and which an expanding market was starting to demand. The city was also seeking to extend its role as a convention centre, notably by the construction of the Palais de Congrès at Porte Maillot. At the same time, Paris was undergoing major urban changes involving the expansion of the office sector and large renewal or redevelopment projects. The planning authorities sought some diversity in these projects and encouraged hotel construction among the office complexes through offering land for hotels at half the rate as for office buildings. Sites in these projects offered not only the advantage of more attractive rents and planning and infrastructural assistance but also the prospect of some new custom generated by the office complexes themselves. As congestion in the inner city grew, more peripheral sites close to the ring-roads or motorways began to appeal to many motoring business travellers who could drive there readily, park and use public transport if necessary to commute to the centre. The lower rents on such sites also meant that developers could increase the attractiveness of their hotels by providing swimming pools and other facilities not available in the centre: only one hotel in the 8th and none in the 1st has a pool.

The location of the large, quality hotels reflects these different factors and has had a marked effect on the arrondissements concerned. The Meridien Montparnasse Paris (953 rooms) and the Pullman Saint-Jacques account for more than one-third of the 14th arrondissement's rooms. The Meridien (formerly the Sheraton) forms part of the Ilot Vandamme renewal project, standing adjacent to the Tour Maine-Montparnasse. The

Pullman St Jacques is located on a site formerly occupied by a warehouse. The Hilton (456 rooms), which is on the northern margins of the 15th arrondissement close to the Eiffel Tower, began the decentralization of luxury hotels into this sector. It was later followed by the Nikko (768 rooms) in the Fronts de Seine development project and the Sofitel Paris (635 rooms) and the Mercure Paris Vaugirard (formerly the Holiday Inn, 91 rooms) at the Porte de Sevres. Both these latter two hotels are very accessible to the motorist, being located respectively on the Boulevard Periphérique and the Boulevard Victor; they are also close to the Palais des Sports. Together, these four hotels account for 30 per cent of the arrondissement's hotel capacity. Paris's two largest hotels – the Meridien Paris Etoile (1025 rooms) and the Concorde La Fayette (990 rooms) – have been built at the Porte Maillot (17th arrondissement) in association with the Palais de Congrès and the new air terminal with its shuttle service to Charles de Gaulle airport. These two hotels also sit astride the western axis which links the established commercial centre with the new complex at La Defense, all of whose projected hotels did not eventuate. The relative size of these new hotels, and thus the importance of these site factors, can be more readily appreciated when compared to the 50 rooms of the average Paris hotel, or the 292-room Georges V, which is the largest hotel in the 8th arrondissement.

This trend did not continue into the 1980s when the tendency within Paris was towards the construction of more modestly sized 2- and 3-star hotels, particularly in the west. The large new hotels have been built outside Paris at Marne-la-Vallée where four 1000-room theme hotels form part of the new Euro Disney® Resort complex. The Euro Disney® hotels add a new attractions dimension to the hotel structure of the Ile de France, one previously dominated by the impact of transport, notably the highway-oriented hotels on the outskirts of the city and the airport hotels at Orly and Charles de Gaulle. As with the Hilton leading the deconcentration of quality hotels within the city, the new impetus provided by Euro Disney® again has a strong international component.

Attractions and other aspects

Attractions and other facilities used by the tourist are more diverse and less well defined in the city than in coastal resorts, with most being shared with local residents. Surveys and other techniques have provided useful insights into where visitors go within particular cities but these rarely provide a complete picture (Chapter 4). Mapping the distribution of urban attractions also has its problems (Chapter 6). Until more detailed information becomes available on the activity patterns of city visitors, perhaps the most useful approach is to consider the different types of attractions and facilities and their distribution, without attempting to assign any relative weighting to them.

The spatial structure of urban tourist attractions and other facilities might be considered in terms of a network of nodes, or clusters of nodes, and routes linking them. While comparatively little is known about the intra-urban movements of tourists in general, it is possible to recognize the more formalized routes and paths. City guides, such as the Guide Michelin and Baedeker, have for long suggested itineraries which tourists are advised to follow to make the most of their stay. Visitors' maps issued by city tourist offices often include a recommended circuit, with brief notes on the major stops and how to get from one to the other. In some cities, such as Boston, these are reinforced by permanent markers to reassure visitors that they are indeed on the right spot and have not deviated from the set path. Many circuits feature a round trip, beginning and ending in the city square. The circuit becomes even more formalized when it is regularly followed by the bus tour of the city. Examination of current itineraries suggests there has been little change in the formal circuit in Paris since it was mapped by Burnet and Valeix (1967). A more flexible variation on the city tour is provided by the Sydney Explorer bus system whereby buses follow a set 17 km route around central Sydney at regular intervals, enabling passengers who have purchased a day ticket to be picked up or set down at any of twenty stops during the day. In cities built along rivers and canals, or around harbours, such as London, Paris, Amsterdam, Bruges, Venice, Bangkok, Sydney and Hong Kong, boat cruises complement these bus tours. Both have their 'city by night' variation, often featuring spectacular illuminations and concluding with a visit to a nightclub. Over half of all visitors to Hong Kong are estimated to take at least one organized tour (Hong Kong Tourist Association 1993b). In other cases where tourists find their own way around the city, the mode of travel, e.g. London's underground, the gondolas of Venice, and the cable-cars of San Francisco, may be as much an attraction as a means of moving on to the next sight.

A critical node in any of these systems is the look-out point which allows an overall view of the city or network. Some look-outs such as those at Hong Kong's Victoria Peak or the Sugar Loaf and Corcovada in Rio de Janiero take advantage of natural viewpoints. In other cases the look-out is artificial, forming part of a tower or building generally built for some other purpose, either historically, such as the towers of Gothic cathedrals and the Leaning Tower of Pisa or, more recently, for example, the Eiffel Tower, Empire State Building and the Centrepoint Tower in Sydney.

These, of course, may be attractions in their own right or access to them may offer some novelty (e.g. the cable-car up Rio's Sugar Loaf). But the look-out may also be, quite literally, the highpoint of the city visit, offering tourists magnificent views and possibly their only opportunity to appreciate the scale and layout of the city as a whole. It is here that they may have their first and perhaps only glimpse of well-known sights. The visit to the look-out confirms that a tourist has indeed 'seen' the city, albeit in many cases very briefly and from afar, and for this reason the look-out is almost invariably included in the city tour.

Most cities have a single recognized viewpoint, although in Paris the observation deck on the Tour Montparnasse now complements or competes with the Eiffel Tower and the towers of Notre-Dame Cathedral. Other types of attractions and facilities are usually more numerous and are frequently clustered together in distinct and often overlapping parts of the city. The clustering of like nodes has many advantages for the visitor who may not have a very detailed image of the city and only a short time to spend there. Clustering may occur for a variety of interrelated reasons as Burtenshaw, Bateman and Ashworth (1981: 201) note with reference to the entertainment sector:

> different sorts of entertainment such as theatres, restaurants, nightclubs and bars will tend to cluster near each other, so that together they form the entertainment quarter, or nightlife district, that the customer seeks. Once established this concentration in a particular district will be reinforced both by popular sentiment that advertises the area to those in search of entertainment in the city, and also by planning decisions. Many such facilities are 'bad neighbours' and will tend to repel other functions. Planners will frequently endeavour to contain the nuisance by opposing its expansion into other areas and encouraging all facilities to locate within the existing entertainment quarter.

The authors distinguish between 'west-end' entertainment, which occupies relatively expensive sites in high-prestige districts, and less reputable red-light districts which may be nearby or spatially quite distinct from them. They identify a relatively localized distribution of nightclubs in Amsterdam and three distinct nightclub areas in Paris – off the Champs Elysées, in Montmartre and in the Latin Quarter. Ashworth, White and Winchester (1988) derived a locational model of urban prostitution areas based on a detailed analysis of red-light districts in several West European cities. They distinguish between areas catering primarily for local clients and visitors, the latter being monofunctional districts in which prostitution is linked with other entertainment, display is neon, transactions occur close to display and controls are weak. Pigalle in Paris is typical of such areas (p. 204):

> The Pigalle district differs sharply from the Rue St Denis. Both offer commercial sex but in Pigalle it is part of a complete package of sex-related activities. The tree-lined boulevards between the Place Clichy and the Place Pigalle are animated at the western end with restaurants and late-night supermarkets, and at the eastern end with sex shops, clubs, cinemas and peep shows. The customers are more likely to be tourists who dominate at the Place Pigalle, with coachloads being dispatched into the sanitised pleasures of the Moulin Rouge. In this area almost all the commercial premises are involved in some aspect of the night-life business. Girls, however, are rarely visible, trade being solicited by middlemen. This is clearly formalised, commercialised sex for the tourist market, tolerated as part of Paris's tourist attractions.

Relph (1976: 83) has termed the cityscape of such districts 'pornscape', an example of 'other directed architecture' – 'that is, architecture which is deliberately directed towards outsiders, spectators, passers-by, and above all, consumers'.

Passers-by are potential consumers for restaurants which tend to be located in areas with large volumes of pedestrian traffic or attractions such as theatres, cinemas and other forms of entertainment. Bonnain-Moerdyk's (1975) seminal study on the gastronomic geography of Paris shows that restaurant districts there have relocated as changes in commercial activities and the centres of entertainment have occurred. Restaurants are now dispersed throughout the city although distinct clusters can still be observed, e.g. in the Latin Quarter and off the Champs-Elysées. Bonnain-Moerdyk also notes an over-representation of restaurants from the high-status western districts in the various restaurant guides published, a pattern which persists in later studies (Chemla 1990). Studies in Canadian cities, notably Quebec (Gazillo 1981) and Kitchener-Waterloo (S. L. J. Smith 1983b), show a marked linear concentration of restaurants and other eating houses in streets on or leading to the inner-city area. Much restaurant trade is, of course, generated locally. In Vieux-Quebec, where the majority of restaurants are concentrated on two different streets, one (rue Saint-Jean) reflecting a local popular culture and the other (rue Saint-Louis) an elite culture, Gazillo (1981) suggests that 'the presence of tourists serves to mitigate the differences between the two worlds'. Other detailed studies from Europe, North America and elsewhere are presented in Huetz de Lemps and Pitte (1990).

Linearity is also a feature of major commercial thoroughfares or specialized shopping streets. These may

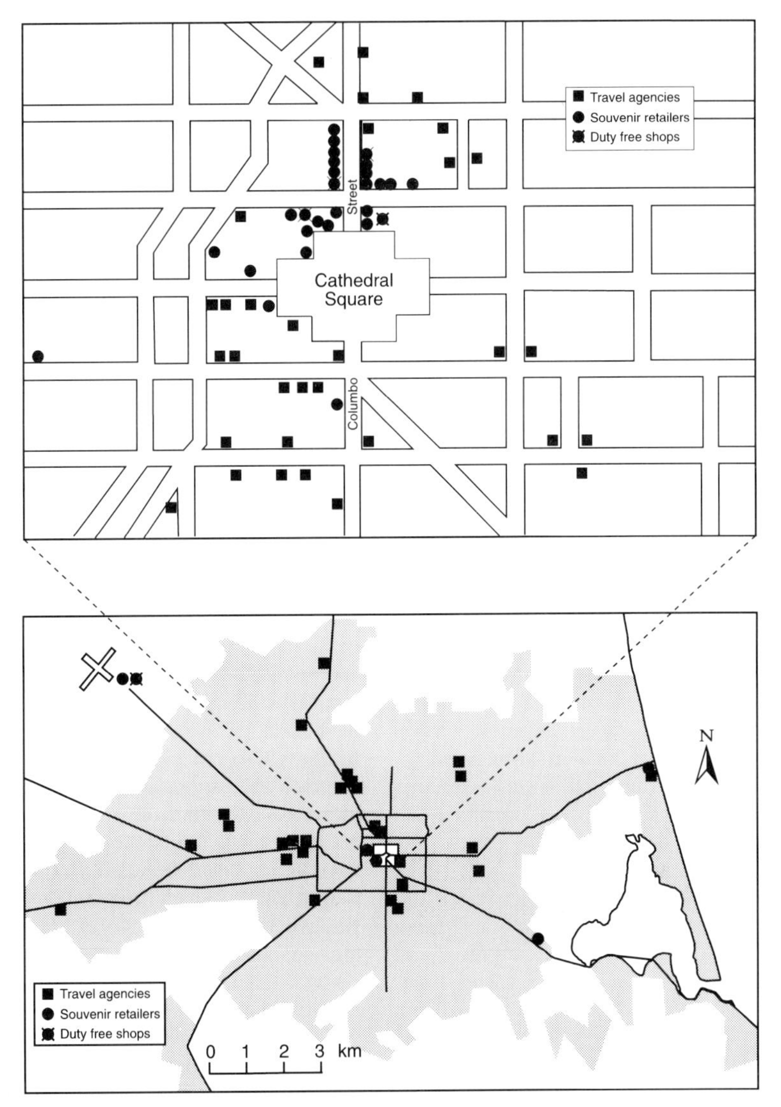

Figure 9.7 Distribution of souvenir retailers and travel agencies in Christchurch, 1994.

not only have an important linking role in the tourist's activity space but also constitute significant attractions in themselves. Presumably because of the greater difficulty of separating tourist business from that originating locally, retail activities in urban areas have not attracted the same attention from a tourism perspective as coastal resorts (Stansfield and Rickert 1970; Dumas 1982) although the recreational function of urban shopping has recently attracted attention (Bak 1992). According to Bak (p. 121) 'comfort and compactness are the most relevant components of funshopping', components which come together dramatically in the case of the Edmonton mall. In some instances it appears to be not so much the desire to purchase goods but the chance to observe the local colour and participate in the general activity that draws visitors to distinctive shopping or trading centres. These range from the flea-markets such as El Rastro in Madrid to the bazaars of Middle Eastern cities.

Souvenir shops, however, are directed specifically at the tourist, although local residents may also buy gifts to send or take abroad. Figure 9.7 shows a pronounced concentration of souvenir and duty-free shops in central Christchurch to the north of Cathedral Square along both sides of Colombo Street, a process which intensified in the late 1980s and early 1990s. The square is the central focus of visits to Christchurch, with the souvenir shops being located between it and several of the large hotels to the north. The concentration of these shops in such a small area, also facilitates comparative shopping. The proliferation of signs in Japanese in these shops, together with their extended evening opening hours, further testifies to the emergence of a specialized tourist shopping strip. Similar concentrations are evident elsewhere, for example in Brussels where tourist-oriented shops line the streets from the Grand Place to the statue of Manneken Pis.

Figure 9.7 also shows that many travel agencies in Christchurch seek a very central location. These businesses serve two sets of travellers, visitors to the city making onward travel or local sightseeing arrangements and local residents travelling away from the city. Travel agencies provide specialized services and like other high-order functions seek a central location so as to be accessible to the largest possible market. Again, the concentration of travel agencies in an industry which has become increasingly competitive allows the potential traveller to shop around and compare fares and prices. However, during the late 1980s and early 1990s the pattern became more dispersed as the number of suburban travel agencies expanded significantly. This would appear to reflect both the continuation of a broader process of the suburbanization of retail outlets and a growth in outbound travel. Suburban travel agencies are most numerous in the more affluent north-western quadrant of Christchurch. In larger cities, airline offices tend to be located together in prestige retail streets, for example along parts of the Avenue de l'Opéra and the Champs Elysées in Paris.

Whereas the central concentration of many retail and entertainment functions reflect contemporary functions, the clustering of many historic buildings and monuments in the inner core is a legacy of earlier times and former functions when the city was confined to a much smaller site, often enclosed by defensive walls. Burnet and Valeix (1967), for example note the concentration of historical buildings and monuments along the Left and Right Banks, with the monumental centre of gravity lying close to the Louvre.

The Römerberg, with its Gothic gables, is one of the most heavily promoted images of Frankfurt but this historic square is but a very small element of Germany's financial centre. Munich has a range of cultural and historic attractions but each day at 11 a.m. and midday (also 5 p.m. June–September) visitors to the city are concentrated, both spatially and temporally, in the Marienplatz to see and hear the Glockenspiel of the Neuesrathaus. In other cities castles were built on defensive sites along major rivers; today the castles are major centres of attraction along with river-boat cruises, for example at Heidelberg and Namur.

In other instances it is not so much a cluster of individual features as the overall impression and general atmosphere of a particular district which appeals to the visitor or is promoted as a 'typical' quarter. Two examples of these are the Barrio Gótico in Barcelona and Lisbon's Alfama district, both medieval quarters characterized by narrow, winding cobblestoned streets and the 'picturesque' houses and buildings of an earlier period. Tour buses are excluded from such confined spaces, with package tour visitors being deposited on the margins and left an hour or two to wander through the streets or to purchase souvenirs and handicrafts (many not of local origin) from the shops which have sprung up in the squares or streets leading into the area. Tourists in shorts and sandals may pass the local residents in the streets, with the older women in the Alfama still often dressed in black, but there is little contact between them. Other tourist districts are identified by their distinctive cultural features, such as the Chinatowns of San Francisco and other North American cities and even of Asian destinations such as Singapore (R. A. Smith 1988).

In more and more cities in Europe, North America and elsewhere, planning regulations have been introduced and funds are being made available to preserve and restore not only individual buildings but also entire quarters (Kain 1978–9; Chenery 1979; Ford 1979; Newcomb 1979; R. A. Smith 1988; Ashworth and Tunbridge 1990). In most cases this is not specifically to enhance the city's tourist attractiveness, athough the increased revenue from tourism may help justify such policies which tend to be rather costly. While hotels have often been excluded from such districts, other tourist facilities have frequently been encouraged. The English Tourist Board (ETB 1981b: 24) in its review of tourism and inner-city schemes in England, e.g. St Katherine's Dock in London and Piece Hall in Halifax, noted: 'The primary motive in these cases was conservation rather than the promotion of tourism, tourism related projects being chosen because they are adaptable and well suited to providing uses for old buildings'. Ford (1979) notes that many preservation projects in North American cities have involved the zone of discard on the margins of the CBD. He also points out (p. 233) that many schemes have been criticized for being elitist, as 'the fancy restaurants, boutiques, architects' offices, and night spots serve primarily a tourist and suburban population rather than locals'. Preservation may also bring other social problems, along with economic benefits, as Ford (p. 216) observes in Charleston where 'residents increasingly complain of noise and fumes from parked tour buses, invasion of privacy by visitors seeking a closer look at "historic residences", and the artificiality of a city full of boutiques and quaint restaurants'.

Others have been concerned with the authenticity of what is being preserved and presented to the tourist, and some interesting discussion in the literature has occurred on the transformation which tourism may hasten or bring about in urban and other areas. The sociologist MacCannell (1976) observes that 'often an entire urban structure is operating behind its tourist front'. He later goes on to present some interesting ideas on the social structure of tourist space which, he suggests, can be seen in terms of front and back regions. Front regions are those which are readily open to the visitor and a place where hosts and guests meet; back regions are the preserve of the residents or workers and are essentially non-tourism oriented in their function. Visitors may stroll through the Barrio Gótico but, except for particular places such as Picasso's Museum or the cathedral, they rarely penetrate inside. MacCannell suggests that there are various ways and degrees to which tourists can penetrate the back regions, to get a glimpse of what goes on 'behind the scene'. He distinguishes between a tourist setting, where provision is made for sightseers to observe an activity not directed towards them, e.g. the visitors' balcony in the New York Stock Exchange, and a stage set which is purposefully designed for the visitor. A good example of the latter is the *son et lumière* display and museum which now take up most of the visit on the guided tour of Paris's sewers, surely the epitome of Ford's (1979) 'sanitized, idealized past'.

Whereas MacCannell speaks of 'staged authenticity', Relph (1976) sees tourism as one of the major factors developing what he terms 'an inauthentic attitude to place'. Relph continues (pp. 83–4):

> in tourism individual and authentic judgement about places is nearly always subsumed to expert or socially accepted opinion, or the act and means of tourism become more important than the places visited. Rasmussen (1964: 16) writes of tourists visiting the church of Santa Maria Maggiore in Rome: '. . . they hardly notice the character of the surroundings, they simply check off the starred numbers in their guide books and hasten on to the next one. They do not experience the place.' This is inauthenticity at its most explicit; the guided tour to see those works of art and architecture that someone else has decided are worth seeing.

Such a view might be considered rather elitist or stereotyped but the concept of the tourists' sense of place warrants further research both in terms of how it affects their activity space and how it relates to travel motivations. Is tourism in historic urban centres founded mainly on ego-enhancement as Relph suggests or do tourists experience a new sense of place (artificial though that may be) which satisfies their need to escape their usual mundane environment (see Chapter 2)?

A more comprehensive approach

The majority of the studies reviewed here have focused on a particular aspect of tourism in urban areas, for example accommodation, attractions or restaurants. In general there have been few attempts to integrate these different components, or link them to broader urban processes, except in the more general redevelopment studies. Scope therefore exists for a more comprehensive approach drawing together the different facets of tourism and integrating these more with the city's other functions. In morphological studies attention might also be paid to what is not there and why rather than focusing solely on the distribution of existing and visible features. The spatial structure of any place is a function of the absence of certain elements as well as their presence. Central Christchurch, for example, would be a different place, both for tourists and residents, if a 167 metre tourist tower proposed for Victoria Square in 1986 had gone ahead. The project was abandoned by the developers in April 1988 when the city council withdrew a planning scheme change, the final step in an involved planning process and one which generated considerable public debate and opposition.

The Victoria Square tower was one of six Christchurch case studies examined by Hart (1992) in an innovative exploration of the influence of the planning process on the development of tourism in the city. Six key issues emerged from her research:

1. Site: the physical location of the project and whether it disadvantages any groups.
2. Developers: their identity and image influences the acceptance of a project.
3. Value agreement: the degree to which all the actors share the same attitudes, values and goals.
4. Opposition: the organization, size and make-up of the opposition will affect its effectiveness.
5. Formal planning procedure: is this basic or complex?
6. Finance: who will pay for the project and how will the money be raised?

Table 9.1 summarizes the relative importance and interrelationships between these and their outcome for the six cases examined. Site factors were particularly important in three cases, two of which did not proceed (the tower and an airport hotel in the green belt) and one of which did (the Mount Cavendish gondola), but only after a series of protracted hearings, appeals and finally ministerial intervention. Lack of value agreement and the mounting of well-organized opposition contributed significantly to the failure of the first two projects and to their absence from the city's landscape today. Further studies of this nature, combining in a

Table 9.1 Summary of issues affecting the outcome of selected tourist projects in Christchurch

Case studies	Site	Developers	Value agreement	Opposition	Formal planning procedures	Finance	Outcome
Memorial Avenue hotel (green belt)	Major controversy	Unknown to public	Nil	Major, well-organized	Complex	Not a public issue	Proposal dropped 20 June 1988 following Planning Tribunal decision
Victoria Square tower (central city)	Major controversy	Unknown to public	Nil	Major, well-organized	Complex	Not a public issue	Council vote on 18 April 1988 to withdraw scheme change results in proposal being dropped
Air Force Museum (suburban)	No controversy	Well known to public	Total	None	Basic	Not a public issue	Opened 1 April 1987
Worcester Boulevard (central city)	Little controversy	Well known to public	High	Little, no organized group	Basic	Not a public issue	First stage opens 1991
Mt Cavendish Gondola (periphery)	Major controversy	Well known to public	Moderate	Major, local, uncoordinated	Complex	Not a public issue	Opens 24 October 1992 seven years after proposal first announced
Christchurch Casino (central city)	Little controversy	Mixture of public identification	Moderate	Moderate, local uncoordinated	Moderate	A major issue	Application approved 1993 and construction began

Source: After Hart (1992).

systematic and comparative fashion different components of urban tourism, and those that have succeeded or failed, will provide important insights into understanding the structure of tourism in urban areas and, indeed, elsewhere.

Beyond the city

As Fig. 1.7 shows, the city may act as a base for day and half-day trips into the surrounding region. From Paris, for example, trips are made to Versailles, Fontainebleau, Chartres and to the Loire Valley. Tourists visiting Madrid will also use it as a base to visit the surrounding historic centres of Toledo, Segovia and the Escorial. While some visitors do stay overnight in these centres, much of the accommodation demand that they generate is captured by the metropolitan centre which acts as a convenient hub, removes the need to change hotels and offers a wide range of complementary attractions, particularly evening entertainment. Consequently, peripheral urban areas tend to have a reduced accommodation sector and the main manifestation of tourism there is often the large number of souvenir shops which line the streets leading from the car parks to the main historic or natural attractions.

Airports, which serve both residents and tourists, may have a considerable impact (Bryant 1973), but by far the greatest changes in land use beyond the city usually result from demand generated by its own inhabitants. Such demand can take several forms, such as open space for tramping, picnicking or fishing, but the biggest impacts have come from second homes. Lundgren (1974) has produced a three-stage model based on the Canadian experience, showing changes in the spatial relationships between the urban centre and a second home or cottage region as the urban area expands (Fig. 9.8). In the first phase, demand from a medium-sized centre has generated a small second home region, typically in an area of broken relief or around a body of water. As the urban area grows, so the demand for second homes increases and the second home region expands, mainly away from the city (Phase II). Lundgren suggests that the 'inside expansion is more urban in character, whereas the outside push still retains the features of the typical vacation home development'. In the third and final stage of the sequence,

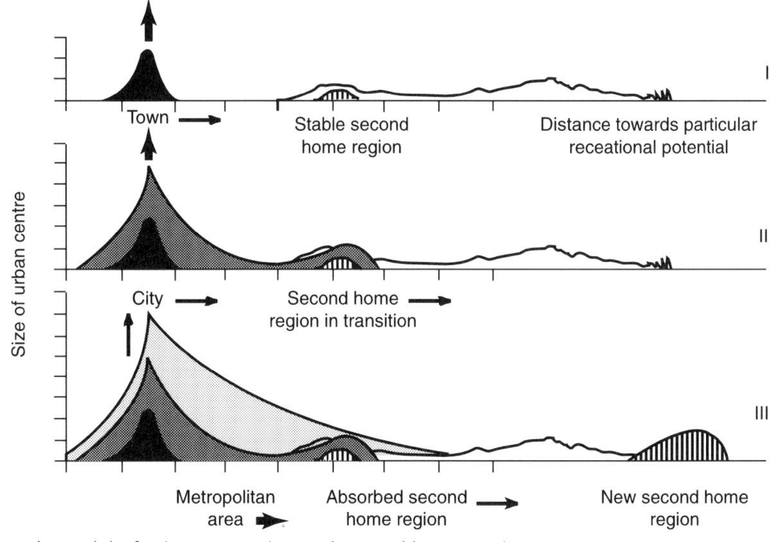

Figure 9.8 Lundgren's model of urban expansion and second home regions.
Source: Redrawn from Lundgren (1974).

the original second home region becomes engulfed by the expanding metropolis and now forms a part of the city itself, with the former second homes being transformed into permanent residences. Meanwhile a new, more distinct second home region has developed, for the demand for weekend or vacation accommodation has not abated but rather increased. This outwards expansion and growth in demand is a function not only of the larger population, but also development of highways, increased car ownership, greater leisure time, and a desire by many local authorities to increase their tax take through intensification of land use as well as the activities of real estate developers promoting speculative subdivisions.

A similar three-stage process has also been identified by Boyer (1980) in the Paris basin. In the first stage, existing rural dwellings are acquired by city residents, mainly for use as second homes. This is followed by a period of new construction on land bought from local inhabitants. At the beginning of this phase, second homes co-exist with primary residences but the latter soon become dominant. In a third phase, individual activity by both city and local residents is replaced by larger subdivisions created either by the local residents, seeking to increase their tax base, or property developers after a profit. Primary residences now become the norm as the peace and isolation sought by the second home owners has now disappeared. The latter group move further outwards and the process is repeated.

Conclusions

Tourism in urban areas has developed largely in response to demand generated directly or indirectly by the city's other functions. As a result, the structure of tourism in the city tends to follow the general form induced by these other functions (as well as by other factors such as site characteristics) rather than shape it. As the CBD has expanded, for example, so the hotel sector has seen its centre of gravity change. Likewise, tourism facilities have formed part of redevelopment schemes but demand for them has rarely initiated those projects which result from more basic urban processes. In these respects the city contrasts with the coastal resort where tourism, characterized by a more homogeneous sunlust market, is a major if not dominant function and where the linear form of the resort reflects this function. Linearity is a feature of the distribution of some tourist facilities in the city, e.g. motels and frequently restaurants, but more often hotels and other facilities are found in distinct and sometimes overlapping clusters concentrated in the central city. Changes in the relative importance of different modes of transport have affected both types of centres, modifying the morphology of coastal resorts and influencing the distribution of hotels and other facilities within the city. Attention in each case has focused on the accommodation sector. More work is now required on other aspects of tourism within the city and on features other than the seafront in the beach resort. Research on urban areas would also benefit from more of the longitudinal studies which have improved our understanding of resort form and, as in the coastal case, extending the geographical range of examples examined. The emphasis in each case on the supply side now needs to be complemented by analysis of tourists' spatial behaviour in these settings, by more interpretative work of the sort undertaken by MacCannell (1976) and Jeans (1990) and by further planning and policy studies investigating the processes that shape the landscapes observed.

CHAPTER 10

Implications, Applications and Conclusions

Previous chapters in this book have examined the fundamental geographical dimensions of tourism. Data sources have been outlined, a range of techniques and methods have been examined, patterns of flows and the spatial structure of tourism at different levels have been presented, and the factors and processes giving rise to them discussed. In this concluding chapter, three major and recurring themes – concentration, spatial interaction and questions of scale – are reviewed and their implications for tourist development, planning, marketing and management explored. Particular emphasis is given to the value of a geographical approach to assessing the impact of tourism. Conclusions are then drawn and suggestions for further research made.

Concentration

The preceding chapters have shown that tourism at all levels is generally characterized by a marked degree of concentration. The extent of this, of course, varies from place to place depending on the kind of tourism involved and the type of measure used. The following observations and generalizations are drawn from the examples cited throughout the book and may therefore not be as applicable to certain forms of tourism, notably that in rural and alpine areas, which have not been given as much emphasis here as tourism in coastal regions and urban areas.

In Chapter 3 it was shown that a small number of developed Western countries account for a large share of all international tourist movements, that most countries draw the majority of their visitors from three or fewer markets and that these are likely to be either neighbouring countries or the United States, Germany, France and Great Britain (Tables 3.2 and 3.3, Figs 3.1 and 3.2). Concentration of demand within Europe, North America and, to a lesser extent, Asia is compounded by significant regional biases within both generating and destination countries. Not only does the general propensity for international travel vary from region to region – departure rates are frequently highest in the main metropolitan regions (Table 2.5, Fig. 4.2) – but different regions also experience a greater or lesser propensity for travel to particular destinations. Certain Caribbean or Pacific countries not only may depend heavily on the United States market but also they may draw the bulk of their visitors respectively from the Atlantic or Pacific coast states (Table 4.1), to the extent that a pronounced 'continental divide' in the American market might be identified (Fig. 4.1). Chapter 8 showed that within small island states tourism plant may be very localized, being concentrated in two or three coastal clusters in and adjacent to the major urban centre (Figs 8.1 to 8.8). Even in larger mainland countries there is a tendency for the top two or three regions to account for the lion's share of the recorded international bednights (Fig. 7.2). The degree of concentration appears to vary between long-haul and short-haul visitors, at least in the European context, but international tourism is generally less dispersed than domestic tourism. Chapter 5 indicated the latter is very localized in nature, with a marked distance decay effect being observed in many studies and a large share of all domestic holidays being spent within the home region itself (Figs 5.3, 5.4 and 5.7).

Further analysis shows that within regions tourism may also be rather concentrated, as evidenced by the localization curves for Scotland (Fig. 7.4) and the distribution of domestic tourism within Florida (Fig. 5.6). In the case of coastal regions, it is often limited to an extremely narrow coastal fringe with relatively little development occurring inland. Ribbon development may give rise to a continuously built-up littoral in the most developed regions but elsewhere development is more nucleated (Figs 7.6 and 7.7). There is also evidence here of more specific concentrations along the coast of different sorts of accommodation, forms of tourism and types of tourists. Hotels may be more important in one locality and camping in another, package tourism may dominate certain resorts while other sectors of the coast are characterized by a less structured form of tourism centred on second homes. Certain resorts or sectors may attract a greater or lesser share of either the international or domestic demand and there is some indication of distinct spatial preferences being exhibited by different groups, both nationals and foreigners, along certain coasts (Figs 7.2 and 7.6). In other regions, tourism may be concentrated in major urban areas as both international visitors and many domestic visitors head for the big city rather than the surrounding hinterland which is the virtual preserve of the city's residents. Further concentration exists at the local level. The densest development in the coastal resorts occurs along a narrow seafront while central city clusters of tourist facilities are found in urban areas (Figs 8.3, 8.4, 8.6, 9.1, 9.2, 9.4 to 9.7).

The progressive concentration at these different levels gives rise in many cases to a pronounced funnelling effect which results in many tourists, particularly international ones, being concentrated within countries along the seafront of a limited number of resorts or clustered together in the central city of a major metropolitan area. Spain, for example, is one of the major tourist-receiving countries in the world, the Balearic Islands constitute the dominant tourist province in Spain; Mallorca is the most important of these islands, most of the accommodation there is found around the Bay of La Palma in a strip adjacent to the seashore. Puerto Rico receives the largest volume of international arrivals in the Caribbean, drawing the vast majority of its tourists from the United States, notably from the Atlantic coast. Three-quarters of the island's hotel capacity is found within metropolitan San Juan, with most of the hotels being concentrated along beachfront sites in the Condado, Ocean Park and at Punta Las Marias. In other words, a significant share of the international sunlust market in Europe ends up along a few kilometres of beach on the outskirts of Palma de Mallorca, and a relatively small number of beachfront hotels in San Juan account for an important portion of the United States traffic to the Caribbean. Such concentration is not limited to sunlust tourism. London dominates international tourism within Britain while Paris is a major destination within France. Tourist activity within both capitals is concentrated in the central city. Both, however, are less important as domestic destinations and domestic tourism in general tends not to be funnelled to quite the same extent as international tourist flows. In other instances, concentration occurs not so much in regions or resorts but along well-defined routes, such as in New Zealand where a distinct tourist circuit can be identified, particularly for overseas visitors.

This concentration results from the interaction of a variety of supply and demand factors which may operate in similar ways at different scales. The resources which attract tourists are numerous and varied but at each level these attractions are generally rather limited in number, distribution, degree of development and the extent to which they are known to the tourist. At the international, national and regional levels, distance plays a major role in influencing which attractions and destinations are accessible to particular markets or tourists, whether in terms of time, money or effort. Accessibility is also determined by other factors such as the nature and extent of the transport network (particularly the number and location of gateways and the modes of transport available), the degree of tour operations, costs within the destination itself and promotion activities. These may modify the broad distance decay effects normally observed and concentrate further or, conversely, disperse the tourist traffic. Accessibility is also a key factor at the local level. Within the resort or city, tourists will seek to be as close as possible to the attractions they desire. This has led to densely developed seafronts within coastal resorts and to the clustering of tourist facilities and activities in the central city. Land rents, planning restrictions, pre-existing activities and site factors may, however, limit the extent to which accommodation, or certain types of accommodation, can locate close to the attractions, whether the beach or historical monuments. Economies of scale in the development of plant and infrastructure are often achieved by concentrating development at a national and regional scale while locally planners may confine certain types of activities to particular zones. A strong regional image can also be promoted in this way while the clustering of particular facilities in certain districts gives them a more readily identifiable image. Success may reinforce the process of concentration as new hotels are built in areas where others have already proved viable or new resort projects are developed within well-established regions to take advantage of existing infrastructure, services and a well-known 'name'.

The number of longitudinal studies undertaken is limited and the quality and quantity of time-series data are variable, but there is evidence of some dispersion in tourist flows over time and signs of a delocalization of tourist plant and facilities although the trends may vary depending on the measure used. Global patterns are evolving, with Asia Pacific emerging as the fastest growing region in the 1980s (Fig. 2.3) and the Japanese and Taiwanese leading a new group of dynamic Asian markets. However, while some reordering in the rankings occurred, the world's leading fifteen spenders and earners saw their share of international travel expenditure and receipts increase in the period 1985–92. Some deconcentration in international demand is apparent in the concentration ratios for 1979 and 1990 although the dependence of each destination on its three major markets generally remains high (Table 3.5). In several of the leading European markets the ratio of international trips to domestic ones has increased and overseas travel has become more dispersed. Most of the growth and redistribution in international travel from Germany, the United Kingdom, Japan and the United States has occurred in the first instance at an intra-regional scale, that is within Europe, Asia and North America respectively, to the benefit of sunlust destinations (Figs 3.6 to 3.9). Growth in longer-haul travel from each of these markets is occurring but much of this reflects overall increases in the outbound traffic rather than pronounced changes in the market shares of more distant destinations though this may change in coming decades.

Tourism within island states is generally very localized but there has been some dispersion as new accommodation has been built away from the gateway cities and movement to the outer islands develops (Tables 8.2 to 8.4). Much of this development is in the form of resort enclaves but smaller scale alternative forms of accommodation are evident in some places. In general, this dispersion has occurred within a period of growth so that although there may have been a relative deconcentration of tourist activity, in absolute terms there has rarely been much if any decline in established centres or resorts. The growth of tourism in major metropolitan centres like London and Paris has been responsible for the delocalization of accommodation, if not of most tourist attractions, as suitable central sites have become increasingly scarce, as new economic factors and planning decisions have come into force or as the importance of different modes of transport have changed. In Christchurch the increasing concentration of souvenir shops oriented to international visitors has been complemented by the suburbanization of travel agencies servicing outbound travellers. The tendency in the design of coastal resorts has generally been to bring accommodation closer and closer to the attractions. This has generally resulted in the intensification of the seafront, the construction of hotels *pieds dans l'eau* and the building of very functional marinas. Development in depth inland has, however, been promoted by some resort developers by building golf courses and encouraged by planners anxious to avoid excessive ribbon development.

The degree of concentration or dispersion at a particular scale is related to the overall stage of development of the country, region, resort or city. The basic pattern might be established at an early stage as the infant tourist traffic quickly identifies the most appealing attractions or the early entrepreneurs acquire and develop the most desirable sites. Or a settling-down period may occur wherein a small number of travellers visit several different regions until, for a variety of reasons, some assume a greater degree of popularity than others. At the same time, developers and entrepreneurs may experiment with a range of different sites and localities, types of attractions and so on, with the success or failure of these early ventures influencing the pattern of location for subsequent development. In either event a process of circular cumulative growth may occur as access to the more popular or successful destinations is improved, as land values rise encouraging further speculative development, as the area becomes more widely known with a feedback system of information developing, as increasing volumes lower travel costs enabling the market to be broadened and so on. During this phase, the pattern may be one of increasing concentration, in both relative and absolute terms, as growth is channelled into a limited number of regions and resorts or cities within those regions and to selected sites and districts within those resorts or cities. At a later stage, other factors may come into play and bring about some deconcentration or dispersion, at least in relative terms. The scarcity of room for further expansion or development is one obvious reason for this, as there are limits to intensification. Moreover, as development occurs, there may be a change in the clientele, with those seeking more novel, exclusive or just less crowded destinations moving on to other areas. New markets may also develop, e.g. that for winter sports, or other areas may seek a share of the growing tourist cake by developing new resources or facilities, upgrading their communications and so on.

Policy-makers may encourage this dispersion as a means of spreading the benefits of tourism or reducing its negative impacts. Regional development strategies incorporating tourism are the most common expression of these policies (Pearce 1989b, 1992c), the new coastal resorts of Languedoc-Roussillon discussed in Chapters 7 and 9 being one direct outcome of them. Increasingly, tourism is also being identified as a means of regenerating inner cities and derelict areas, frequently in regions with no strong traditional tourism base (Sabin 1986; Buckley 1989). Conversely, other policies have aimed at containing the growth of tourism geographically, as noted in some of the Pacific islands and in the case of Nusa Dua in Bali discussed in Chapter 8.

National policies and debate over the merits and disadvantages of concentration and dispersion or indeed any comprehensive geographical perspective on tourism are frequently lacking. Many tourist development policies stress overall growth in tourist arrivals and foreign exchange earnings with relatively little attention being paid to where that growth is going to occur or what effects it will have on the regions or places concerned. A major New Zealand study in the late 1980s, for example, established three scenarios for the future growth of tourism: the Base Scenario centred on recent trends, a Dispersal Scenario which dissipated growth and a Concentration Scenario which saw growth increasingly focused along the main axis (McDermott-Miller Group 1988). Subsequent policies, however, have tended to stress market forces and national goals, with 3 million visitors annually by the year 2000 being signalled as an optimistic target. National tourism policy in the United Kingdom has not been particularly strong but during the 1970s and 1980s some debate did occur over the need for regional policies to counteract the prominence of London, though again effective strategies to do this never fully materialized (C. Cooper 1987; Heeley 1989; Pearce 1992b).

Regional or local reaction to concentration and the dominance of leading centres or regions, on the other hand, is very common (Pearce 1992b). The emergence of regional and local tourist organizations is often a response to those areas seeking to develop and market tourism in their own regions, cities or resorts, believing they can do this more effectively than national policies which they often perceived to favour regions or places other than their own, usually the major ones where demand is concentrated. Thus Scotland struggled for the right to promote itself abroad, the 'other Holland' campaign was promoted in the Netherlands by several more peripheral provinces, and in Hawaii the Neighbour Islands established their own organizations or chapters of the Hawaii Visitors Bureau to counter the prominence of Oahu. Much of the reaction in these cases was prompted by the growing recognition that the regions concerned offered different products and attracted different segments of the market, notably in terms of the geographic origins of their visitors. For similar reasons, dominant metropolitan areas frequently go it alone, opting out of commitment to joint activities with their surrounding regions, for example Munich in Bavaria or Amsterdam in North Holland.

While the merits of these different strategies can be debated – national tourist organizations will often assert the need for a co-ordinated approach which avoids dilution of effort and confusing the market by creating multiple regional and local images – the more general point to be underlined here is that both the source of the issue (regional differentiation of product and market) and the response (creation of sub-national organizations and regionalized marketing strategies, have a clear geographical dimension. Better understanding of these questions can result from a deeper appreciation of this dimension through the application of the various techniques outlined in earlier chapters. The practical application of this point is underscored by Hamilton (1988: 32) with regard to the three scenarios proposed in New Zealand:

> Analysis of this type takes us further than mere documentation of changing trends. The fundamental issue that it raises is that alternative futures constructed so as to highlight contrasting demands on tourism resources imply a *choice* between them. The scenarios illustrated here need not just eventuate, they can be pursued in the light of detailed knowledge of their implications.

If, for example, a decision is made to opt for a more dispersed pattern of tourism growth then selectively targeting those market segments that have been identified as having a greater propensity to stay longer and visit more peripheral places is one way of implementing this policy. The patterns analysed in Chapter 7 have presented indications of which markets these might be. In particular, the role of domestic tourists in this regard should not be overlooked.

The possibilities of manipulating tourist flows and influencing the spatial structure of the tourist industry are not, however, unbounded. In many cases, it seems conditions are such that existing patterns may well be perpetuated, if not intensified. Delalande (1991), for example, cautions against being over-optimistic about the possibilities of restructuring coastal tourism in France while in Scotland attempts to refocus marketing efforts around larger regional entities based on perceptions in the market met opposition from local areas and led to the original five proposed regions becoming eleven (Pearce 1992b). Despite these limitations, attempts at planning and managing the growth of tourism will be enhanced if a broader spatial perspective is adopted and greater attention is paid to patterns of spatial interaction.

Spatial Interaction

This book has systematically explored and analysed the inherent geographic dimensions of tourism, emphasizing the different types and levels of spatial interaction which underlie, indeed constitute, this phenomenon. It is clearly insufficient to ask simply where visitors come from and leave the 'geographical breakdown' at that as most visitor surveys and marketing studies do. We must also know how visitors have come to that destination, their movements within it and their subsequent travel patterns. Likewise, studies of flows generated by a given market should not be restricted just to where they go, but to how they get there and back and where they go once at a destination. Moreover, the role of any given area in a larger tourist system can be appreciated more fully by undertaking both types of studies, that is, examining both its generating and receiving functions. The study area should be placed in a broader geographical framework and considered in terms of how it is linked to other markets and destinations. These patterns of interaction can be analysed at different scales, e.g. country-to-country or city-to-resort flows, as well as between different levels, for instance the intranational travel patterns of international visitors.

Analysis of patterns of spatial interaction has often been limited by the paucity of relevant data and the need to develop or adapt appropriate techniques to investigate what can be rather complex problems.

Nevertheless it has been possible in Chapters 3 to 5 to identify a wide variety of data sources and to provide examples of how these can be analysed with respect to different types of interaction, though in Chapter 6 it was noted there have been few attempts to include transport or any other dynamic element in studies concerned with the geographical differentiation of tourism. Other examples will also be used to show how a spatial perspective can be and has been used for planning, marketing, development and impact assessment.

By establishing how one destination relates to another in terms of particular markets, a geographical perspective can be especially useful in the field of international circuit tourism. Chapter 3 showed much international tourism is characterized by a large proportion of single destination trips or of circuits limited to two or three countries but the significance of circuit tourism varied significantly from destination to destination and often between markets for any one destination (Figs 3.3 to 3.5). Knowledge of whether a place is visited as a single destination or part of a wider circuit, for example, is important for developing appropriate marketing campaigns. Secondary destinations, for instance, may be able to develop and exploit linkages with regional gateways or destinations as parts of Indonesia and Malaysia are attempting to do with Singapore. Such strategies are becoming increasingly important given the trend in international civil aviation to develop major regional hub airports. Small states in the South Pacific that are able to tap into larger circuits may also draw off visitors to their benefit or generate new business through the joint creation of circuits which are more attractive than visits to single islands alone.

Some tension may exist, however, over likely trade-offs between increasing total arrivals through such co-operative strategies and a decline in the length of stay of circuit tourists in any one country on the circuit. Some of the effects of these inter-scale relationships are evident in Fig. 4.4, with the level of inter-island mobility within French Polynesia declining as the proportion of destination travellers decreased. More work is required in this area, particularly in terms of evaluating the effectiveness of joint campaigns and the effects of changing accessibility resulting from airline schedules and itineraries. Fuji, Im and Mak (1992) report that the introduction of direct flights to the Hawaii's Neighbour Islands resulted in a modest gain in the overall number of visitors, relieved congestion at Honolulu International Airport and diverted travellers away from Oahu, thereby redistributing income derived from them as well. Earlier Pearce and Johnston (1986) had noted that multiple island visitors within Tonga had a marked economic impact disproportionate to their numbers, not only generating a disproportionately high share of bednights and total expenditure, but also distributing them both more widely throughout the kingdom.

At the intra-national scale, attention has also been directed at the framework that routes may provide for promotion, planning and development. Theme itineraries promoting visits to a series of like attractions have proved a useful way of generating traffic to sites which may not, individually, warrant a trip. One of the most popular types of such circuits are the wine trails which link up different vineyards. Historical sites, buildings and monuments can also be promoted in this way. Such a strategy has been enthusiastically endorsed in Germany where holiday routes promoted include the German Fairytale Road, the Castle Road and the Romantic Road (Pearce 1992b). Other examples of such an approach include the Meusebanks project (Pearce 1992b) in the Netherlands and attempts to create a tourism marketing region united by Highway 89 in Northeastern California (V. L. Smith, Hetherington and Brumbaugh 1986).

Elsewhere, planning and development has been based on tourist 'corridors'. Such is the case in the Yukon where Wolman Associates (1978: 23) suggest that 'in view of the distances between existing and potential attractions and services, the corridor system is the key to the tourist experience in Yukon, and is the foundation for the proposed development strategy for Yukon's tourism industry'. Various corridors are then identified, along with the segments of the market they are likely to attract, proposals for product development and constraints on the expansion of tourism. Gunn (1972) also emphasizes the role of the 'circulation corridor' – 'the entire visual sweep along all the travel ways' – in the design of regional tourist plans. He notes (p. 92): 'In a broad sense the design is heavily dependent upon two basic functions, sometimes overlapping: (1) a means to an end (to get there and back) and (2) an end in itself (a part of an attraction experience). In both cases the emphasis is upon the people being moved, the means is incidental'.

Analysis of patterns of interaction may also help clarify the role and function of particular places at different scales. Analyses of circuit tourism at the international level, for example, may contribute to our understanding of how places such as Hong Kong and Fiji fit into and function within their broader regional settings (Figs 3.3 to 3.5). Within countries, applications of the Trip Index and other intra-national analytical techniques have proved useful in market segmentation and developing a functional classification of areas for marketing and other purposes (Chapter 4). Particular attention has focused on those places that may act as gateways. Two basic sorts of gateways might be identified, those which act simply as points of entry which visitors pass through en route to a destination elsewhere in the region or country and those, chiefly capital cities, which also constitute major destinations in themselves. In the first instance, local interest may be directed at generating more business by increasing the visitors' length of stay in the area. In the second, attempts may be made to disperse the traffic away from the gateway in order to ease pressures there and to benefit other parts of the region or country. Where a city draws on several different markets, as in the case of Christchurch, a variety of interests and potential strategies might exist.

In the case of gateways which act basically as entry points which are passed through on a larger circuit, the potential for retaining visitors may be limited, as in the cases of Cherbourg (Soumagne 1974; Clary 1978) and the Aosta Valley (Janin 1982) discussed in Chapter 4. In the latter case, Janin suggests that the number of stays in the area might be increased by improving the distribution of information and systems of hotel reservation for passing motorists. He also stresses that local interests should not lose sight of two different types of clientele: the touring visitor who might be attracted by more modern but competitively priced hotels built along the major axes, and the longer-stay visitor who might be tempted by some form of alternative accommodation, such as *gîtes ruraux*.

Where the attraction of a gateway is such that it absorbs much of a region's or country's tourist traffic, efforts might be directed more to enhancing the city's staging post role, encouraging visitors to travel beyond the city to visit other destinations throughout the country or region. Such is the case with Vancouver, British Columbia, discussed in Chapter 4, with Murphy and Brett (1982) suggesting that there is potential for developing touring visits to other parts of the province. Given the different travel patterns of the two major groups they identified – those visiting friends and relations and independent vacationers – it is arguable, however, whether the metropolitan areas are in effect acting as intervening opportunities or whether they are just magnets for a particular sector which is not greatly interested in touring in any case. Murphy and Brett (1982: 160) suggest that other areas could be developed if repeat visits were encouraged, noting: 'The United Kingdom experience indicates London has become less of an interceptor to overseas visitors as their number of visits increases.' Elsewhere, Murphy (1980) notes only limited success in redirecting visitors within Victoria. If congestion in the gateway is a major problem, then efforts to spread the traffic will be welcomed by the industry there. If not, local interests may seek to capture a larger share of the visitors' trips, arguing that their city is 'more than just a gateway' and has a lot to offer in its own right as a destination.

Questions of Scale

Questions of spatial scale have arisen throughout this book. Several of the models introduced in Chapter 1 included a scale dimension and emphasize the hierarchy of tourist flows and the structure of the tourist industry (Figs 1.6 to 1.8). Subsequent analysis of tourist travel patterns and the spatial structure of tourism have also been presented at a range of scales from the international to the local. While this organization reflects in large part a means of ordering the wide array of information and material now available, these earlier chapters have also shown how different research problems emerge at each level, different sorts of data exist and different analytical approaches and techniques may be appropriate. At the same time, linkages between the various scales have also emerged. This is especially so in the preceding two sections where the systematic reviews of concentration and interaction have revealed similar patterns existing and processes operating at different scales, with functional linkages occurring from one level to another. Likewise, the discussion of domestic tourism in Chapter 5 demonstrated the utility of setting the examination of flows at any one scale against those at higher and lower levels. Other authors have interpreted dimensions of tourism on islands in the context of broader political-economic processes (Chapter 8).

While particular problems may be appropriately examined at a specific scale, what emerges from a number of these studies and findings is that many issues, both basic and applied, would benefit from a more multi-scale approach, one that incorporates and integrates different levels of analysis. This is already being recognized in certain areas, as testified by the increasingly refined geographical breakdowns of marketing data used in Chapter 4, but scope exists for this to be undertaken on a more regular and systematic basis. One such approach was developed recently in the preparation of a tourism masterplan for the state of Sarawak (Pearce 1995: 229–44).

The conceptual framework for the masterplan recognized that a state plan involves planning at an intermediate scale, setting Sarawak in the broader national and regional context while at the same time coordinating and integrating activities at the local level. All facets of the plan from marketing, to analysis of transport services and institutional frameworks were thus addressed at these different scales and integrated in the overall strategy. While all these had a geographical dimension, the general process can perhaps best be illustrated in terms of issues raised in earlier chapters by reference to some of the market and flow analyses.

In an effort to identify the particular strengths and opportunities of Sarawak, tourism in the state was first set within the context of the recent expansion of the industry in Malaysia and the broader ASEAN region. The hierarchical market analysis thus began with a review of recent trends in visitor arrivals within the region and Malaysia's position *vis-à-vis* its neighbours. This analysis was largely based on published arrival statistics, but regional growth projections, both of demand and supply, were also reviewed. This procedure not only revealed differential rates of growth and market preferences throughout the ASEAN countries but also indicated that although region-wide opportunities to develop tourism would continue through the 1990s, many areas in the region were seeking to exploit this demand and so competition would correspondingly be keen. Profiles of visitors to Sarawak were then set against those for the country as a whole, again revealing differential rates of growth and market composition. A more comprehensive comparative analysis of competitive destinations at a regional and local level in other parts of Malaysia, Indonesia and Thailand provided further insights into Sarawak's strengths and weaknesses.

Attention then focused more closely on the changing composition of the tourist traffic to Sarawak itself using the state's arrival statistics as well as on the volume and profile of visitors to particular destinations within the state as recorded in national park permits, admission statistics and other local data. Specially commissioned exit surveys generated more specific information on travel patterns to and within the state. Trip Index values enabled the extent of destination and circuit travel to be gauged for the different markets. Key regional gateways and linkages with other destinations were also identified in this way.

The overall outcome of this approach was the presentation of a marketing and development strategy which would channel selected external markets progressively into localized areas within Sarawak, make provision for local demand and plan for the physical development and integration of these areas. While the perspective here has been that of an intermediate level state or regional plan, the underlying multi-scale aproach can be readily adapted to national or local level planning. With the former, there is both a need to look increasingly outwards to set the focal country in its international setting, particularly with the growth in longer-haul traffic and the increasingly complex and competitive situation which results, and to reach further downwards than is usually found in most such plans. Conversely, many local plans would benefit from a more explicit linking to regional, national and indeed international structures and phenomena. Many useful guidelines as to how this might be achieved have been provided by the concepts, techniques, data sets and examples presented in earlier chapters.

The Impact of Tourism

A considerable body of literature has been developed on the impact of tourism. Indeed it is perhaps the most widely studied aspect of tourism, contributions having been made from a wide range of disciplines from anthropology and sociology, through geography and economics to ecology and biology. This diversity reflects the diversity of impacts that the development of tourism is seen to produce: economic, environmental, socio-cultural (Mathieson and Wall 1982; Pearce 1989b; Butler 1993). Costs as well as benefits are now attributed to tourism but much debate still occurs over the nature and extent of these impacts and the ways in which they might be assessed. Geographers have been to the fore in advocating a more balanced approach to impact assessment, drawing in particular on their traditional strengths of synthesis and studies of the interrelationships between societies and their environments. Applications of a spatial perspective to the assessment of tourism's impacts, however, have been less common but such a perspective can contribute significantly to clarifying many of the issues and formulating solutions to the problems identified.

There seems little doubt that many of the negative impacts attributed to tourism have been accentuated by the process of concentration outlined, whether in terms of the degradation of beaches through the discharge of inadequately treated effluent, the congestion of inner-city streets by tour buses, the disruption of local lifestyles or the over-dependence on one or two major markets. At the same time the benefits of tourism, such as increased job opportunities and higher revenues, may be limited to only a few areas. In some cases where the concentration of tourist activities drains investment capital, services and population away from other areas, tourism may lead to a dual economy or heighten regional imbalances. Conversely, where this concentration occurs in areas away from established industrial or agricultural centres, it may help redress regional inequalities. Likewise, concentration may be deemed beneficial if by grouping tourist facilities in selected areas or concentrating on specific regional markets, a more viable and competitive industry can be established and economies of scale achieved so that adequate infrastructure can be provided, thus eliminating or reducing such problems as water pollution. Concentration may also reduce the social impacts of tourism to a few areas, but if not done carefuly the tourists may also overwhelm the local residents in the areas concerned.

Concentration too cannot be divorced from the types of operation and forms of development from which it results. In particular, spatial concentration in the form of tourist enclaves, both in developing and developed countries, is frequently symptomatic of large-scale integrated operations, organized by the state or controlled by a limited number of powerful external operators and developers. Policy-makers in some areas, especially multiple-island states, have attempted to counter this trend by encouraging smaller-scale indigenously owned projects, but these have often been unsuccessful. In Paris, on the other hand, large chain-owned hotels have led the delocalization of the hotel industry in the city.

Geographical analyses of the type discussed in this book can shed light on where the different impacts generated by tourism are occurring and who is being affected. These are important issues and ones which have often been neglected. Economic research, for example, has traditionally focused either on aspatial national studies which have stressed costs and benefits in overall terms or on more detailed local case studies with little or no attempt to bridge the gap between the two.

Likewise, social impact studies rarely examine the geographical context of the processes or impacts they are describing, and environmental research often does not consider the extent of the impacts identified. Analysis of spatial variations in the nature and extent of tourist development can thus provide essential inputs into the assessment process. The analysis of tourist travel patterns can also contribute to an understanding of why particular impacts are experienced in some areas and not others and suggest measures which might be taken to spread the benefits of tourist development or alleviate its negative impacts.

In particular, the techniques evaluated in Chapter 6 constitute a large kitbag of tools which might usefully be applied more frequently to impact assessment. The economic studies cited are already being used in this way and show the potential which exists in this area. The utility of the quantitative composite measures has also been mentioned in this regard. Other more readily available relative measures, whether accommodation or tourist based, might be used much more widely in assessing spatial variations in the intensity of impacts experienced. Even single measures of spatial variations in tourism may provide very useful insights when used in conjunction with the distribution of some other variable. Gosar (1989), for example, identified three basic types of demographic change in former Yugoslavia through comparing maps showing the volume of international tourism and rates of population change. Later chapters provide further examples of applications and studies which elucidate issues of impact assessment. Aerial photograph interpretation has been used effectively to record changes in the built environment and might be applied to measure the areal extent of other economic and environmental impacts. Likewise, local and regional studies in Spain have furnished detailed evidence of the nature and locational dimensions of regional restructuring wrought by the development of mass tourism (Chapter 7) and might be profitably repeated elsewhere.

Impact assessment studies have focused almost exclusively on the destination areas and ignored effects in associated areas, namely the markets and transit areas. Interaction between these areas is important, however, as the structural models reviewed in Chapter 1 and the flow analyses of subsequent chapters have indicated. This interaction forms the basis of two of the most implicit spatial models of tourism impact assessment. Ironically, neither was developed by geographers but by an interdisciplinary team (Federal Department of Forestry 1981) and a French sociologist (Thurot 1980). Both use an origin-linkage-destination framework to identify different causes of impact and types of problems in different but interrelated places.

The Swiss team were concerned with assessing en-

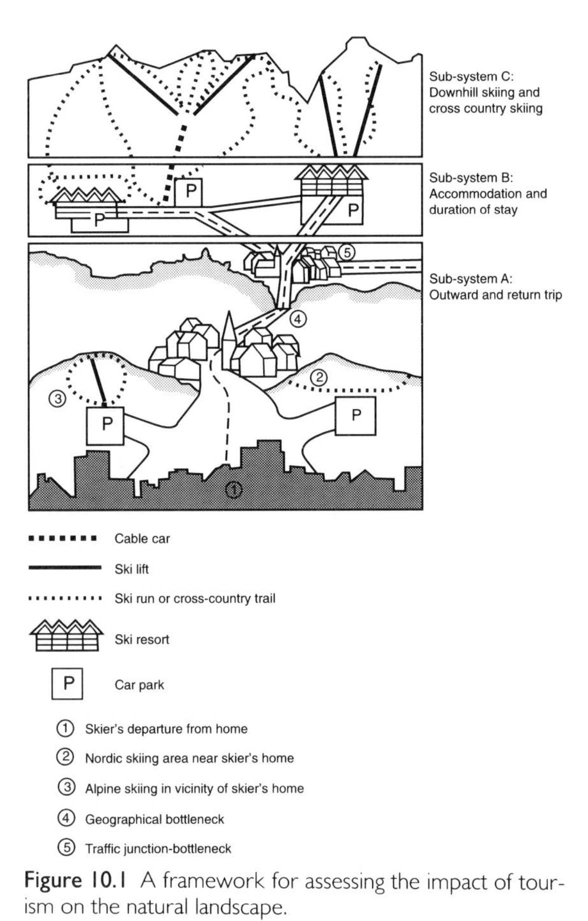

Figure 10.1 A framework for assessing the impact of tourism on the natural landscape.
Source: After Federal Department of Forestry (1981).

vironmental stress arising out of the growth of ski tourism in Switzerland. A central feature of their study was a model of ski tourism and the natural landscape, comprised of three subsystems (Fig. 10.1). The first subsystem involves the outward and return trip; the second, accommodation and the stay in the resort; the third, skiing and associated facilities. Each of these subsystems was then examined in more detail in terms of the specific problems which arose and possible solutions which might be implemented. Problems in subsystem A, for instance, essentially result from the traffic generated by skiers and give rise to such impacts as an overloading of arterial routes, congestion in villages en route to the ski-fields, bottlenecks in passes, increased vehicle emissions and so forth. Solutions to these problems include extending existing roads or building new ones, creating a more efficient public transport system or developing alternative ski-fields and redirecting the traffic to them. Each of these choices may ease the pressure in one area but generate new stress elsewhere. Bypasses, for instance, may relieve congestion caused by transit traffic in the villages but at the same time directly result in the loss of cultivated land and, by improving access, attract additional traffic which may cause greater emissions and increase demand on the other two subsystems.

Activities within the skiing subsystem are seen in terms of a cycle of events: 'the individual's skiing activity begins at a specific point of departure, progresses via intermediary stages and terminates with the unbuckling of the skis after the return.' The processes at work in the development and expansion of the ski-field are then analysed in terms of this 'run-cycle'. Ecological impacts associated with the development of a ski-field may include soil compaction, disappearance of plant species and the transformation of the existing drainage system. Such impacts can generally be reduced by careful planning at the outset while the unnecessary expansion of the ski slopes may be avoided by improving the efficiency and capacity of existing facilities and promoting other types of skiing with different requirements, notably cross-country skiing.

The use of a spatial interaction model such as that shown in Fig. 10.1 can thus enable a broader range of interrelated impacts to be identified and a more comprehensive approach to solving them adopted. That is to say, not only may skiing generate stress outside the resort itself, the traditional focus for impact studies, but also solutions to the pressures in the resort may lie elsewhere, for instance, in the manipulation of traffic along feeder routes.

Thurot (1980) too adopts a more comprehensive approach in his review of carrying capacity, arguing that most studies to date have been too limited as they have focused only on conditions and impacts in the destination region. He distinguishes between internal factors in the destination region associated with the host society and economy and the local environmental context, and external ones whose origins lie elsewhere, either in the generating region or the transit region (Fig. 10.2). In particular, not only will conditions in the generating region affect total demand, but also factors such as institutionalized holidaymaking and traditional habits may give rise to a marked seasonality in demand and to distinct preferences for certain types of holidays and destinations. These conditions can accentuate the impacts felt in the destination region by concentrating demand spatially and at certain times of the year, week or day. How the tourist industry is organized may also reinforce or reduce these patterns, with much domestic tourism being unstructured and more dispersed and international tourism being more organized and directed. The nature, capacity and organization of the transport network in the transit region may accentuate further this spatial and temporal concentration. Chapter 8, for instance, outlined the significance of air transport for the spatial organization of tourism on islands. Conditions in the transit region, notably those which influence the cost of travel, may also significantly affect the volume of traffic generated and thus the pressure which develops at the destination. This region will also be affected by demand emanating from the generating region in the same way that subsystem A was affected in Fig. 10.1. Severe bottlenecks may be experienced at the interfaces of the different regions; that is, the outskirts of cities and the approaches to resorts, at the beginning and end of the main holiday season and at the start and finish to the weekend. Moreover, feedback loops might be built into the system. If the destination region becomes saturated, this will limit the propensity to travel from the generating region, reduce the amount of choice available to potential holidaymakers or lower the quality of their experience such that the original need to escape may not be fulfilled. Overloading of the transport system in the transit region will aggravate these problems.

A spatial perspective then can add greatly to the assessment of the impact of tourism, first, by indicating the nature and extent of tourist activity and likely problem areas, second, by highlighting the sorts of problems which may occur in different places and arise from different sources. Adopting a framework such as that shown in Figs 10.1 and 10.2 may at first sight seem only to complicate the issue by extending the traditional focus of activity but, ultimately, the broader perspective it gives will enhance the likelihood of finding and implementing more effective solutions. Unfortunately in the mid-1990s there is still little evidence of impact assessors rising to the challenge which these models present.

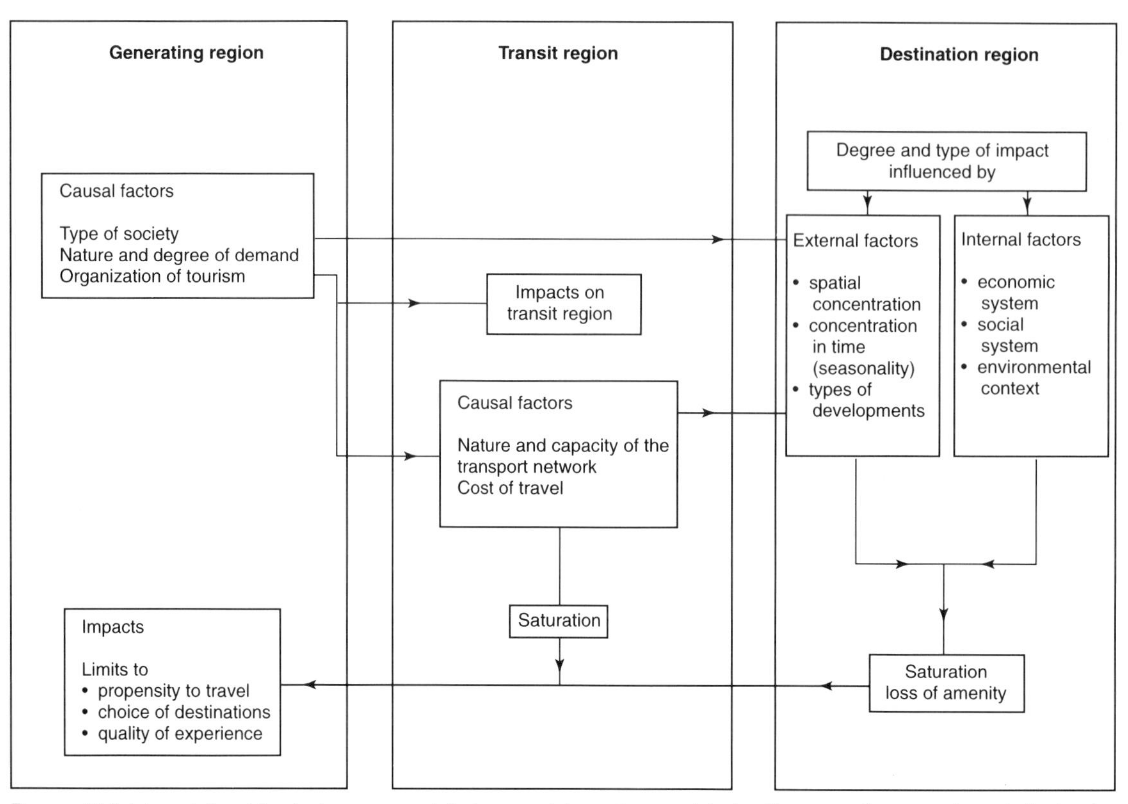

Figure 10.2 Interrelationships between causal factors and impacts associated with generating, transit and destination regions.
Source: Adapted from Thurot (1980).

Conclusions

A substantial if disparate body of literature now exists dealing with the geographical dimensions of tourism. An attempt has been made in this book to structure this material and to present the results of new research in this field. Various models of tourist flows, spatial interaction and development were reviewed in Chapter 1. These generally provide a useful point of entry into particular areas of analysis and interpretation, but the geography of tourism is not yet underpinned by a strong theoretical base. Critical debate is now developing and some replication is occurring but so far these have generally been limited to only a couple of models. More widespread discussion, testing and application are needed. Chapter 2 examined the questions of demand and motivation which underlie and give rise to the phenomenon under study. Much interesting work has been done in this field but many challenges remain in understanding destination choice processes and factors influencing travel from and to particular places. Chapters 3 to 5 dealt with the analysis of tourist flows at different scales. While many data limitations remain, significant progress has been made in this area and it has been possible to start to identify more general patterns, the processes underlying them and some of their implications. A variety of techniques for analysing spatial variations in tourism were reviewed in Chapter 6. This diversity suggested a general lack of direction in the literature, both in terms of questions addressed and approaches adopted, but again there is evidence of a more critical appraisal of these emerging. In Chapters 7 to 9 selected techniques were used to identify general patterns in the spatial structure of tourism at a national and regional level, on islands, in coastal resorts and urban areas. Finally, earlier sections of this chapter have been concerned with the implications of the major recurring themes of concentration, spatial interaction and scale and with the application of a geographical perspective and geographical techniques in planning, development, marketing and impact assessment.

Significant contributions have been made by geographers in a wide range of areas but much scope exists for further research. In the first instance, much more work needs to be done in each of the areas outlined, whether this be intra-national flows or tourism in urban areas.

Several different avenues might be followed here. First, the geographical coverage of topics needs to be greatly enlarged if we are to understand fully the universality or particularity of patterns and processes observed. Reflecting the enormous growth in tourism within the region, an increasing number of Asian references has been added in recent years to the literature once dominated by European and North American studies but much scope exists for this process to be extended further, into Africa, Latin America and other parts of the globe. The emergence of domestic tourism in developing countries over coming decades presents a particular challenge but few of the topics reviewed here can yet be considered to have been widely dealt with and might also be addressed in many other places. Second, related to this is a need for more systematic and comparative research and for the replication of studies from place to place and time to time so that the general might be distinguished from the specific and models and theories can be developed and tested. This would involve the identification of appropriate themes, the adoption of common approaches, the standardization of terminology and the collection of comparable data. Clearly there are dangers if a bland uniformity in tourism research were to develop and initiative and originality were to be lost, but little progress in our understanding of the geography of tourism will be achieved if the fragmentation and diversity such as that outlined in Chapter 6 continues. Third, more longitudinal studies are required, showing how patterns and processes have evolved or developed through time. The value of this approach has been demonstrated by the insights derived from the trends and evolutionary studies discussed in earlier chapters. Fortunately longitudinal approaches are becoming increasingly feasible as more and more time-series data have become available and as earlier studies have provided an ever-increasing base on which today's researchers and those of the future can build.

At the same time, the links between the various topics discussed in the preceding chapters need to be developed further. First, the links between the various geographical levels of research need strengthening. How does the resort fit into the region, what is the region's role in the nation and the nation's place in the world? Or, reversing the focus, how do international trends affect a given country and particular places within that country? Second, more attention needs to be given to how one type of research relates to another and how different research topics are interconnected. Those developing models, for instance, have not widely tested their creations by empirical research and most case studies have not been based on any particular theory. Studies stressing methodology are frequently weak on interpretation while others offering plenty of comment are often thin on technique. Work on motivation needs to be further related to studies of tourist flows so that patterns cannot only be identified but explained more soundly. Likewise research on distribution needs to take fuller account of development processes. Most topics would benefit from a greater integration of supply side and demand perspectives and further matching of dynamic and static elements. Meeting these challenges will become increasingly important and rewarding as tourism continues to expand and evolve.

References

Aldskogius, H. (1977), A conceptual framework and a Swedish case study of recreational behaviour and environmental cognition, *Economic Geography*, **53**(2), 163–83.

Anderson, J. (1971), Space–time budgets and activity studies in urban geography and planning, *Environment and Planning*, **3**(4), 353–68.

Andrieux, D. and Soulier, A. (1980), Physionomie du tourisme en Languedoc-Roussillon: une enquête sur les résidences secondaires, *Etudes et Statistiques Languedoc-Roussillon*, Repères No. 2, Août, INSEE, Montpellier.

An Foras Forbatha (1973), *Brittas Bay: a planning and conservation study*, An Foras Forbatha, Dublin.

Annuaire Statistique de la France, *INSEE*, Paris.

Anon (1971), La situation du tourisme parisien, *Paris Projet*, **6**, 66–75.

Archer, B. and Shea, S. (1973), *Gravity Models and Tourist Research*, Tourist Research Paper TUR2, Economics Research Unit, Bangor.

Arizona Hospitality Research and Resource Center (1992), *The 1990–91 Arizona Visitor Profile*, Arizona Hospitality Research and Resource Center, Flagstaff.

Ashworth, G. J. (1976), *Distribution of Foreign Tourists in the Netherlands*, Occasional Papers no. 2, Dept of Geography, Portsmouth Polytechnic, Portsmouth.

Ashworth, G. J. (1989), Urban tourism: an imbalance in attention, *Progress in Tourism, Recreation and Hospitality Management*, **1**, 33–54.

Ashworth, G. J. and de Haan, T. Z. (1985), *The Tourist Historic City: a model and initial application in Norwich, U.K.*, University of Groningen, Groningen.

Ashworth, G. J. and de Haan, T. Z. (1987), *Regionalising the Resort System: tourist regions on the Languedoc coast*, University of Groningen, Groningen.

Ashworth, G. J. and Tunbridge, J. E. (1990), *The Tourist–Historic City*, Belhaven, London.

Ashworth, G. J., White, P. E. and Winchester, H. P. M. (1988), The red light district in the West European city: a neglected aspect of the urban landscape, *Geoforum*, **19**(2), 201–12.

Ayala, H. (1991), Resort hotel landscape as an international megatrend, *Annals of Tourism Research*, **18**(4), 568–87.

Backman, S. J., Uysal, M. and Backman, K. (1991), Regional analysis of tourism resources, *Annals of Tourism Research*, **18**(2), 323–27.

Bailey, M. (1986), A study of the Japanese travelling overseas, *Travel and Tourism Analyst*, April, 13–23.

Bak, L. (1992), The recreational function of urban shopping, pp. 119–25 in C. A. M. Fleischer-van Rooijen (ed.) *Spatial Implications of Tourism*, Geo Pers, Groningen.

Baker, R. J. (1983), *An Analysis of Urban Morphology and Precincts Within Selected Coastal Resorts of the Port Stephens–Great Lakes Area, New South Wales*, MA thesis (unpublished), University of New England, Armidale.

Banco Central de la Republica Dominicana (1977), La experiencia del Fondo para el Desarrollo de la Infraestructura Turistica (Infratur), pp. 169–82 in *El Turismo y su Financiamiento en Espana, Caribe y Centro America*, Instituto de Credito Oficial, Madrid.

Baptistide, J.-C. (1979), *Tourisme et Développement de la Guadeloupe*, Thèse de Troisième Cycle (unpublished), Institut de Géographie, Faculté des Lettres et Sciences Humaines de Rouen, Rouen.

Barbaza, Y. (1970), Trois types d'intervention du tourisme dans l'organisation de l'espace littoral, *Annales de Geographie*, **434**, 446–69.

Barbier, B. (ed.) (1975), *Atlas de Provence Côte d'Azur, ACTES*, Le Paradou.

Barbier, B. (1978), Ski et stations de sports d'hiver dans le monde, *Weiner Geographische Schriften*, **51/52**, 130–46.

Barbier, B. and Pearce, D. G. (1984), The geography of tourism in France: definition, scope and themes, *GeoJournal*, **9**(1), 47–53.

Baretje, R. and Defert, P. (1972), *Aspects Economiques du Tourisme*, Berger–Levrault, Paris.

Barke, M. (1991), The growth and changing pattern of second homes in Spain in the 1970s, *Scottish Geographical Magazine*, **107**(1), 12–21.

Barrett, J. A. (1958), *The Seaside Resort Towns of England and Wales*, PhD thesis (unpublished), University of London, London.

Bell, M. (1977), The spatial distribution of second homes: a modified gravity model, *Journal of Leisure Research*, **9**(3), 225–32.

Benthien, B. (1984), Recreational geography in the

German Democratic Republic, *GeoJournal*, **9**(1), 59–63.

Berriane, M. (1978), Un type d'espace touristique marocain: le littoral méditerranéen, *Revue de Géographie du Maroc*, **29**(2), 5–28.

Berriane, M. (1990), Fremdenverkehr in Maghreb: Tunesien und Marokko in Vergleich, *Geographische Rundschau*, **42**(2), 94–9.

Berriane, M. (1992), Le role de la mer dans le développement du tourisme au Maroc, *Revue Maroc Europe*, 2, 131–54.

Berriane, M. (1993), Le tourisme des nationaux au Maroc (une nouvelle approche du tourisme dans les pays en développement), *Annales de Géographie*, 570, 131–61.

Bevilacqua, E. (1984), Le tourisme dans les Alpes italiennes, pp. 209–17 in *Les Alpes*, Comité International d'Organisation du 25e Congrès de Géographie, Caen.

Bielckus, C. L. (1977), Second homes in Scandinavia, pp. 35–46 in J. T. Coppock (ed.) *Second Homes: curse or blessing?*, Pergamon, Oxford.

Bisson, J. (1986), A l'origine du tourisme aux îles Baleares: vocation touristique ou receptivité du milieu d'accueil?, Paper presented at the meeting of the IGU Commission of the Geography of Tourism and Leisure, Palma de Mallorca (mimeo).

Blanchet, G. (1981), *Les Petites et moyennes entreprises polynésiennes: le cas de la petite hôtellerie*, Travaux et Documents de l'ORSTOM, 136, ORSTOM, Paris.

Blanchet, G. (1989), Le tourisme en Polynésie Francaise ou le rêve à l'épreuve des faits, pp. 223–36 in *Iles et Tourisme en Milieux Tropical et Subtropical*, CRET, University of Bordeaux; GEGET–CRNS, Talence.

Boerjan, P. (1984), *Les Vacances des Belges en 1982*, Cahiers du Tourisme B31, Centre des Hautes Etudes Touristiques, Aix-en-Provence.

Boerjan, P. and Vanhove, N. (1984), The tourism demand reconsidered in the context of the economic crisis, *Tourist Review*, **38**(2), 2–11.

Bonnain–Moerdyk, R. (1975), L'espace gastronomique, *L'Espace Géographique*, **4**(2), 113–26.

Bonnieux, F. and Rainelli, P. (1979), Une typologie des cantons littoraux de Bretagne, *Etudes et Statistiques Bretagne*, Octant no. 4, November.

Bounds, J. H. (1978), The Bahamas tourism industry: past, present and future, *Revista Geografica*, **88**, 167–219.

Boyer, J.-C. (1980), Residences secondaires et 'rurbanisation' en région parisienne, *Tijdschrift voor Economische en Sociale Geografie*, **71**(2), 78–87.

Boyer, M. (1972), *Le Tourisme*, Editions du Seuil, Paris.

Breuer, T. (1987), Villages de vacances sur l'île de Tenerife – essai de bilan vingt ans après leur création, pp. 85–104 in *Le Développement du tourisme dans les espaces voisins des grandes zones de fréquentation touristique*, Office National du Tourisme Tunisien, Sousse.

British Tourist Authority (1976), *Touring patterns of overseas motorists in Great Britain, summer, 1975*, British Tourist Authority, London.

British Tourist Authority (1979a), *Tourism Growth and London Accommodation*, British Tourist Authority, London.

British Tourist Authority (1979b), *International Tourism and Strategic Planning*, British Tourist Authority, London.

British Tourist Authority (1981a), *Digest of Tourist Statistics No. 9*, British Tourist Authority, London.

British Tourist Authority (1981b), *Regional Spread and Accommodation and Transport Usage of Overseas Visitors to the UK 1978*, British Tourist Authority, London.

British Tourist Authority (1981c), *Digest of Tourist Statistics No. 9*, British Tourist Authority, London.

British Tourist Authority (1989), *Digest of Tourist Statistics No. 12*, British Tourist Authority, London.

British Tourist Authority (1992), *Digest of Tourist Statistics No. 16*, British Tourist Authority, London.

British Tourist Authority/English Tourist Board (1992), *Overseas Visitor Survey 1992*, British Tourist Authority and English Tourist Board, London.

Britton, J. N. H. (1971), Methodology in flow analysis, *East Lakes Geographer*, **7**, 22–36.

Britton, S. G. (1980a), A conceptual model of tourism in a peripheral economy, pp. 1–12 in D. G. Pearce (ed.) *Tourism in the South Pacific: the contribution of research to development and planning*, NZ MAB Report no. 6, NZ National Commission for Unesco/ Dept of Geography, University of Canterbury, Christchurch.

Britton, S. G. (1980b), The spatial organisation of tourism in a neo-colonial economy: a Fiji case study, *Pacific Viewpoint*, **21**(2), 144–65.

Britton, S. G. (1982), The political economy of tourism in the Third World, *Annals of Tourism Research*, **9**(3), 331–58.

Britton, S. G. (1991), Tourism, capital and place: towards a critical geography of tourism, *Environment and Planning D: Society and Space*, **9**(4), 451–78.

Brunet, R. (1973), Structure et dynamisme de l'espace française, schèma d'un système, *L'Espace Géographique*, **2**, 249–54.

Bryant, C. R. (1973), L'agriculture face à la croissance métropolitaine: le cas des exploitations de grande culture expropriées par l'emprise de l'aéroport Paris Nord, *Economie Rurale*, **95**, 3–35.

BTR (1990), *International Visitor Survey 1990*, Bureau of Tourism Research, Canberra.

BTR (1991), *Australian Tourism Trends* 1991, Bureau of Tourism Research, Canberra.

Buckley, P. J. (1989), Tourism in difficult areas, pp. 499–501 in S. F. Witt and L. Moutinho (eds) *Tourism Marketing and Management Handbook*, Prentice Hall, Hemel Hempstead.

Buckley, P. J. and Klemm, M. (1993), The decline of tourism in Northern Ireland: the causes, *Tourism Management*, **14**(3), 184–94.

Burac, M. (1989), Tourisme et utilisation du littoral dans les petites Antilles, pp. 87–93 in *Iles et Tourisme en Milieux Tropical et Subtropical*, CRET, University of Bordeaux; CEGET–CRNS, Talence.

Burkart, A. J. and Medlik, S. (1974), *Tourism: past, present and future*, Heinemann, London.

Burnet, L. (1963), *Villégiature et Tourisme sur les Côtes de France*, Hachette, Paris.

Burnet, L. and Valeix, M.-A. (1967), Equipement hôtelier et tourisme, pp. 833–43 in *Atlas de Paris et de la Région Parisienne*, Berger-Levrault, Paris.

Burtenshaw, D., Bateman, M. and Ashworth, G. J. (1981), The City in West Europe, Wiley, Chichester.

Butler, R. W. (1980), The concept of a tourist area cycle of evolution: implications for management of resources, *Canadian Geographer*, **24**(1), 5–12.

Butler, R. W. (1993), Pre- and post-impact assessment of tourism development, pp. 135–55 in D. G. Pearce and R. W. Butler (eds) *Tourism Research: critiques and challenges*, Routledge, London.

Cadart, C. (1975), *Les Nouvelles implantations hôtelières à Paris et dans la région parisienne*, Mémoire de Maitrise (unpublished), Centre d'Etudes Supérieures de Tourisme, Paris.

Cals, J., Esteban, J. and Teixidor, C. (1977), Les processus d'urbanisation touristique sur la Costa Brava, *Revue Géographique des Pyrénées et du Sud-Ouest*, **48**, 199–208.

Campbell, C. K. (1967), An approach to research in recreational geography, pp. 85–90 in *B.C. Occasional Papers no.* 7, Dept of Geography, University of British Columbia, Vancouver.

Carlson, A. S. (1938), Recreation industry of New Hampshire, *Economic Geography*, **14**, 255–70.

Carlson, A. W. (1978), The spatial behaviour involved in honeymoons: the case of two areas in Wisconsin and North Dakota, 1971–76, *Journal of Popular Culture*, **11**, 977–88.

Carter, M. R. (1971), A method of analysing patterns of tourist activity in a large rural area: the Highlands and Islands of Scotland, *Regional Studies*, **5**(1), 29–37.

Cazes, G. (1980), Les avances pionnières du tourisme international dans le Tiers-Monde: réflexions sur un système décisionnel multinational en cours de constitution, *Travaux de l'Institut de Geographie de Reims*, **43–44**, 15–26.

Cazes, G. (1989), L'île tropicale, figure emblématique du tourisme international, pp. 37–53 in *Iles et Tourisme en Milieux Tropical et Subtropical*, CRET, University of Bordeaux; CEGET–CNRS, Talence.

Chadefaud, M. (1981), *Lourdes: un pèlerinage, une ville*, Edisud, Aix-en-Provence.

Chemla, G. (1990), L'évolution récente des restaurants gastronomiques parisiens, pp. 39–58, in A. Huetz de Lemps and J.-R. Pitte (eds) *Les Restaurants dans le monde et à travers des ages*, Editions Glénut, Paris.

Chenery, R. (1979), *A Comparative Study of Planning Considerations and Constraints Affecting Tourism Projects in the Principal European Capitals*, British Travel Educational Trust, London.

Choy, D. J. L. (1992), Life cycle models for Pacific Island destinations, *Journal of Travel Research*, **30**(3), 26–31.

Christ, Y. (1971), *Les Metamorphoses de la Côte d'Azur*, Balland, Paris.

Christaller, W. (1964), Some considerations of tourism location in Europe, *Papers, Regional Science Association*, 95–105.

Ciaccio Campagnoli, C. (1975), Développement touristique et groupes de pression en Sicile, *Travaux de l'Institut de Géographie de Reims*, **23–24**, 81–7.

Ciaccio Campagnoli, C. (1979), The organisation of tourism in Sicily, *Weiner Geographische Schriften*, **53/54**, 132–42.

Clary, D. (1977), *La Façade littorale de Paris: le tourisme sur la côte normande, étude géographique*, Editions Ophrys, Paris.

Clary, D. (1978), La frontière maritime de la Normandie et l'impact régional du tourisme internationale, *Etudes Normandes*, **2**, 39–54.

Clary, D. (1993), *Le Tourisme dans l'Espace Français*, Masson, Paris.

Cockerell, N. (1989), West Germany outbound, *Travel and Tourism Analyst*, **4**, 38–59.

Coggins, C. (ed.) (1990), *Florida visitor Study 1990: executive summary*, Office of Tourism Research, Tallahassee.

Cohen, E. (1972), Towards a sociology of international tourism, *Social Research*, **39**, 164–82.

Cohen, E. (1982), Marginal paradises: bungalow tourism on the islands of Southern Thailand, *Annals of Tourism Research*, **9**(2), 189–228.

Collin-Delavaud, A. (1989), Tourisme et environnement aux îles Galapagos, pp. 243–52 in *Iles et Tourisme en Milieux Tropical et Subtropical*, CRET, University of Bordeaux; CEGET–CNRS, Talence.

Commission of the European Community (1986), *Europeans and their Holidays*, Commission of the European Community, Brussels.

Cook, S. D. (1989), US leisure patterns and outbound travel, *Travel and Tourism Analyst*, **3**, 33–50.

Cooper, C. (1981), Spatial and temporal pattems of tourist behaviour, *Regional Studies*, **15**, 359–71.

Cooper, C. (1987), The changing administration of tourism in Britain, *Area*, **19**(3), 249–53.

Cooper, C. and Jackson, S. (1989), Destination life cycle: the Isle of Man case study, *Annals of Tourism Research*, **16**(3), 377–98.

Cooper, M. (1980), The regional importance of tourism in Australia, *Australian Geographical Studies*, 1980, **18**(2), 146–54.

Coppock, J. T. (ed.) (1977), *Second Homes: curse or blessing?*, Pergamon, Oxford.

Corsi, T. M. and Harvey, M. E. (1979), Changes in vacation travel in response to motor fuel shortages and higher prices, *Journal of Travel Research*, **17**(4), 6–11.

Council of City of Sydney (1980), *City of Sydney Strategic Plan*, Council of City of Sydney, Sydney.

Cribier, F. (1969), *La Grande Migration d'eté des citadins en France*, CNRS, Paris.

Cribier, F. and Kych, A. (1977), *L'Hébergement touristique dans les 318 stations du littoral français au coeur de l'été*, Laboratoire de Géographie Humaine, Paris.

Crompton, J. (1992), Structure of vacation destination choice sets, *Annals of Tourism Research*, **19**(3), 420–34.

Crompton, J. L. (1979), Motivations for pleasure vacation, *Annals of Tourism Research*, **6**(4), 408–24.

Cross, M. (1979), *Urbanization and Urban Growth in the Caribbean*, Cambridge University Press, Cambridge.

Crouch, G. I. and Shaw, R. N. (1992), International tourism demand: a meta-analytical integration of research findings, pp. 175–207 in P. Johnson and B. Thomas (eds) *Choice and Demand in Tourism*, Mansell, London.

Cullinan, T. *et al.* (1977), *Central American Panama Circuit Tourism Study*, SRI International, Menlo Park.

Dann, G. M. S. (1977), Anomie, ego-enhancement and tourism, *Annals of Tourism Research*, **4**(4), 184–94.

Dann, G. M. S. (1981), Tourist motivation: an appraisal, *Annals of Tourism Research*, **8**(2), 187–219.

Dann, G. M. S. (1993), Limitations in the use of 'nationality' and 'country of residence' variables, pp. 88–112 in D. G. Pearce and R. W. Butler (eds) *Tourism Research: critiques and challenges*, Routledge, London.

Dann, G., Nash, D. and Pearce P. (1988), Methodology in tourism research, *Annals of Tourism Research*, **15**(1), 1–28.

Deasy, G. F. and Griess, P. R. (1966), Impact of a tourist facility on its hinterland, *Annals Assn. American Geographers*, **55**(2), 290–306.

Debbage, K. G. (1990), Oligopoly and the resort cycle in the Bahamas, *Annals of Tourism Research*, **17**(4), 513–27.

Debbage, K. G. (1991), Spatial behaviour in a Bahamian resort, *Annals of Tourism Research*, **18**(2), 251–68.

Defert, P. (1960), Introduction à une géographie touristique et thermale de l'Europe, *Acta Geographica*, **36**, 4–11.

Defert, P. (1966), *La Localisation Touristique: problèmes théoriques et pratiques*. Editions Gurten, Berne.

Defert, P. (1967), *Le Taux de Fonction Touristique: mise au point et critique*, Cahiers du Tourisme, C-13, CHET, Aix-en-Provence.

de Ganseman, P. (1982), *Hôtels de Bruxelles*, Projet de Licence (unpublished), Université Libre de Bruxelles, Brussels.

de Haan, T. Z. and Ashworth, G. J. (1985), *Modelling the Seaside Resort: Great Yarmouth (U.K.)*, University of Groningen, Groningen.

Delalande, M. (1991), Dépasser les pratiques de plage et de soleil ou au delà des 4S, pp. 75–82 in F. Fourneau and M. Marchena (eds) *Ordenación y Desarrollo del Turismo en España y en Francia*, Casa de Velázquez, Madrid.

Department of Tourism, Posts and Telecommunications (1992), *Tourism in Indonesia 1991*, Department of Tourism, Posts and Telecommunications, Jakarta.

Dewailly, J.-M. (1978), Un révélateur de contrastes régionaux: l'indice de confort des logements touristiques, *Travel Research Journal*, **1**, 23–9.

Dewailly, J.-M. (1979), Fascination et pesanteur d'une frontière pour le tourisme et la récréation: l'exemple franco-belge *Frankfurter Wirtschafts-und Sozialgeograpische Schriften*, Heft 31, 309–17.

Dewailly, J.-M. (1990), *Tourisme et Aménagement en Europe du Nord*, Masson, Paris.

Direccao Geral do Turismo (1980), *O Turismo Estrangeiro em Portugal: inquérito 1979*, Direccao Geral do Turismo, Lisbon.

Dirección General de Politica Turística (1991), Las vacaciones de los españoles en 1990, *Estudios Turísticos*, **109**, 65–96.

Doering, T. R. (1976), A re-examination of the relative importance of tourism to state economies, *Journal of Travel Research*, **15**(1), 13–17.

Donehower, E. J. (1969), *The Impact of Dispersed Tourism in French Polynesia*, MA thesis (unpublished), University of Hawaii, Honolulu.

Doumenge, J. P. (1989), Le tourisme dans les îles Seychelles: effets sur l'économie et l'environnement, pp. 255–64 in *Iles et Tourisme en Milieux Tropical et Subtropical*, CRET, University of Bordeaux; CEGET–CRNS, Talence.

DRV (1989), *Wirkschaftsfaktor Tourismus, eine Grundlagenstudie der Reisebranche*, Frankfurt, Deutscher Reisebüro-Verband.

Duffield, B. S. (1984), The study of tourism in Britain – a geographical perspective, *GeoJournal*, **9**(1), 27–35.

Dumas, D. (1975), Un type d'urbanisation touristique littorale: la Manga del Mar Menor (Espagne), *Travaux de l'Institut de Géographie de Reims*, **23–24**, 89–96.

Dumas, D. (1976), L'urbanisation touristique du littoral de la Costa Blanca (Espagne), *Cahiers Nantais*, **13**, 43–50.

Dumas, D. (1982), Le commerce de détail dans une grande station touristique balnéaire espagnole: Benidorm, *Annales de Géographie*, **506**, 480–9.

Dundler, F. (1988), *Urlaubsreissen, 1954–1987. 34 Jahre Erfassung des touristischen Verhaltens der Deutschen durch soziologische Stichprobenuntersuchen*, Starnberg, Studienkreis fur Tourismus.

Eagles, P. F. J. (1992), The travel motivations of Canadian ecotourists, *Journal of Travel Research*, **31**(2), 3–7.

Edwards, A. (1991), The reliability of tourism statistics, *Travel and Tourism Analyst*, **1**, 62–75.

Elliott, J. M. C. (1981), *Tourism in Christchurch*, MA thesis (unpublished), University of Canterbury, Christchurch.

Ellis, J. A. (1976), *Industrial Concentration*, Research Paper no. 20, NZ Institute of Economic Research, Wellington.

Elson, M. J. (1976), Activity spaces and recreational spatial behaviour, *Town Planning Review*, **47**(5), 241–55.

Equipo Investigador del IET y CIS (1980), Comportamiento vacacional y turístico de los Españoles (Enero–Septiembre 1979), *Estudios Turísticos*, **66**, 17–110.

Equipo Investigador del IET (1981), Comportamiento vacacional y turístico de los extranjeros: encuesta a extranjeros que visitaron algunas zonas de Espana, diciembre de 1980 y enero de 1981, *Estudios Turísticos*, **70/71**, 41–107.

ETB (1981a), *England's Tourism*, English Tourist Board, London.

ETB (1981b), *Tourism and the Inner City*, English Tourist Board, London.

Etzel, M. J. and Woodside, A. G. (1982) Segmenting vacation markets: the case of distant and near-home travellers, *Journal of Travel Research*, **20**(4), 1982, 10–14.

Eversley, D. (1977), The ganglion of tourism: an unresolvable problem for London?, *London Journal*, **3**(2), 186–211.

Farrell, B. H. (1982), *Hawaii, the Legend that Sells*, University Press of Hawaii, Honolulu.

Federal Department of Forestry (1981), A case study of the growth of ski tourism and environmental stress in Switzerland, pp. 261–318 in *Case Studies on the Impact of Tourism on the Environment*, OECD, Paris.

Ferras, R. (1975), Tourisme et urbanisation dans le Maresme Catalan, pp. 135–42 in *Tourisme et Vie Régionale dans les Pays Méditerranéens, Actes du Colloque du Taormina*, Centre Géographique d'Etudes et de Recherches Méditerranéens and Scuola di Studi Turistici in Rimini dell Universita di Bologna, Rimini.

Fesenmaier, D. R. and Lieber, S. R. (1987), Outdoor recreation expenditures and the effects of spatial structure, *Leisure Sciences*, **9**(1), 27–40.

Fisher, R. J. and Price, L. L. (1991), International pleasure travel motivations and post–vacation cultural attitude change, *Journal of Leisure Research*, **23**(3), 193–208.

Flament, E. (1973), *Capacité d'Accueil et Fréquentation Touristique du Littoral Belge*, Les Cahiers du Tourisme, Série B, no. 19, Centre d'Etudes du Tourisme, Aix-en-Provence.

Ford, L. R. (1979), Urban preservation and the geography of the city in the USA, *Progress in Human Geography*, **3**(2), 211–38.

Forer, P. C. and Pearce, D. G. (1984), Spatial patterns of package tourism in New Zealand, *New Zealand Geographer*, **40**(1), 34–42.

Fornairon, J. D. (1978), Note sur l'origine des touristes fréquentant le littoral languedocien, *Economie Méridionale*, **25**(100), 67–73.

Foster, D. M. and Murphy, P. (1991), Resort cycle revisited: the retirement connection, *Annals of Tourism Research*, **18**(4), 553–67.

Fourneau, F. (1983), Loisirs de proximité et résidences secondaires autour d'une métropole régionale: le cas de Seville, *Norois*, **120**, 619–24.

Franz, J. C. (1983), Development and growth of seaside resorts in Southeast Asia, Paper presented at the 15th Pacific Science Congress, Dunedin.

Franz, J. C. (1985), Pattaya–Penang–Bali: Asia's leading beach resorts, *Tourism Recreation Research*, **10**(1), 25–9.

Fuji, E., Im, E. and Mak, J. (1992), Airport expansion, direct flights, and consumer choice of travel destinations: the case of Hawaii's Neighbor Islands, *Journal of Travel Research*, **30**(3), 38–43.

Funnell, C. (1975), *By the Beautiful Sea: the rise and high times of that great American resort, Atlantic City*, Knopf, New York.

Garcia, M. V. (1976), Social production and consumption of tourist space: outline of methods applied to the study of the Bay of Palma, Majorca, pp. 83–94 in ECE, *Planning and Development of the Tourist Industry in the ECE Region*, United Nations, New York.

Gardavsky, V. (1977), Second homes in Czechoslovakia, pp. 63–74 in J. T. Coppock (ed.) *Second Homes: curse or blessing?*, Pergamon, Oxford.

Gaviria, M., Iribas, J. M., Monterde, M., Sabbah, F., Sanz, J. R. and Uderia, E. (1974), *España a Go-Go. Turismo charter y neocolonialismo del espacio*, Ediciones Turner, Madrid.

Gaviria, M., Iribas, J. M., Sabbah, F. and Sanz, J. R. (1975), *Turismo de Playa en España*, Ediciones Turner, Madrid.

Gazillo, S. (1981), The evolution of restaurants and bars in Vieux–Quebec since 1900, *Cahiers de Géographie du Quebec*, **25**(64), 101–18.

Generalitat Valenciana (1992), *El Turismo en la Comunidad Valenciana 1991*, Generalitat Valenciana, Valencia.

Getz, D. (1986), Models in tourism planning: towards integration of theory and practice, *Tourism Management*, 7(1), 21–32.

Gibson, L. J. and Reeves, R. W. (1972), The spatial behaviour of camping America: observations from the Arizona Strip, *Rocky Mountain Social Science Journal*, **9**(2), 19–30.

Gilbert, E. W. (1939), The growth of inland and seaside health resorts in England, *Scottish Geographical Magazine*, **55**, 16–35.

Gilbert, E. W. (1949), The growth of Brighton, *Geographical Journal*, **114**, 30–52.

Gilbrich, M. (1991), *Urlaubsreisen 1991: Kurzfassung der Reiseanalyse 1991*, Starnberg, Studienkreis für Tourismus.

Gilg, A. W. (1988), Switzerland: structural change within stability, pp. 123–44 in A. M. Williams and G. Shaw (eds) *Tourism and Economic Development: Western European experiences*, Belhaven, London.

Gillmor, D. A. (1989), Recent tourism development in Cyprus, *Geography*, **74**(3), 262–65.

Ginier, J. (1974), *Géographie Touristique de la France*, SEDES, Paris.

Girard, P. S. T. (1968), Geographical aspects of tourism in Guernsey, *La Société Guernesiaise Reports and Transactions*, **18**(2), 185–205.

Girault, C. (1980), Gulf + Western en République Dominicaine. De l'enclave sucrière au controle d'une partie de l'économie nationale, *L'Espace Géographique*, **9**(3), 223–9.

Goldsmith, O. F. R. and Forrest, J. (1982), Spatial context of holiday travel, pp. 214–18 in *Proc. 11th NZ Geog. Conf., NZ Geog. Soc.*, Wellington.

Goonatilake, S. (1978), *Tourism in Sri Lanka: the mapping of international inequalities and their internal structural effects*, Working Paper no. 19, Centre for Developing-Area Studies, McGill University, Montreal.

Gormsen, E. (1981), The spatio-temporal development of international tourism: attempt at a centre-periphery model, pp. 150–70 in *La Consommation d'Espace par le Tourisme et sa Préservation*, CHET, Aix-en-Provence.

Gormsen, E. (1982), Tourism as a development factor in tropical countries – a case study of Cancun, Mexico, *Applied Geography and Development*, **19**, 46–63.

Gosar, A. (1989), Structural impact of international tourism in Yugoslavia, *GeoJournal*, **19**(3), 277–83.

Graburn, N. H. H. (1983), The anthropology of tourism, *Annals of Tourism Research*, **10**(1), 9–33.

Graf, P. (1984), Le tourisme dans l'espace alpin allemand, pp. 226–231 in *Les Alpes*, Comité International d'Organisation du 25e Congrès de Géographie, Caen.

Gray, H. P. (1970), *International Travel – International Trade*, Heath Lexington, Lexington.

Greer, T. and Wall, G. (1979), Recreational hinterlands: a theoretical and empirical analysis, pp. 227–45 in G. Wall (ed.) *Recreational Land Use in Southern Ontario*, Dept of Geography Publication Series, no. 14, University of Waterloo.

Grimmeau, J. P. (1980), Petite géographie du tourisme étranger en Belgique, *Revue Belge de Geographie*, **104**(5), 99–110.

Grinstein, A. (1955), Vacations: a psycho-analytic study, *International Journal of Psycho-Analysis*, **36**(3), 77–85.

Guitart, C. (1982), UK charter flight package holidays to the Mediterranean, 1970–78: a statistical analysis, *Tourism Management*, **3**(1), 16–39.

Gunn, C. A. (1972), *Vacationscape: designing tourist regions*, University of Texas, Austin.

Gunn, C. A. (1979), *Tourism Planning*, Crane Rusak, New York.

Gutiérrez Ronco, S. (1977), Localización actual de la hostelería madrilena, *Boletin de la Real Sociedad Geografica* 1976, part 2, 347–57.

Gutiérrez Ronco, S. (1980), Evolución en la localización de la hostelería madrileña, pp. 283–8 in *Jornadas de Estudios sobre la Provincia de Madrid*, Diputación Provincial de Madrid, Madrid.

Hall, P. (1970), A horizon of hotels, *New Society*, 12 March 1970, 445.

Hamilton, J. (1988), *Trends in Visitor Demand Patterns in New Zealand: past and future*, New Zealand Tourist and Publicity Department, Wellington.

Hampton, A. (1987), The UK incentive travel market: a user's view, *European Journal of Marketing*, **21**(9), 10–19.

Hart, A. J. (1992), *Planning for Tourism in Christchurch:*

a comparative study, MA thesis (unpublished), University of Canterbury, 1992.

Haywood, K. M. (1986), Can the tourist-area life cycle be made operational? *Tourism Management*, **7**(3), 154–67.

Heeley, J. (1989), Role of national tourist organizations in the United Kingdom, pp. 369–374 in S. F. Witt and L. Moutinho (eds) *Tourism Marketing and Management Handbook*, Prentice Hall, Hemel Hempstead.

Herbin, J. (1982), Le tourisme hivernal dans les Alpes Autro-Allemandes: présentation d'une étude cartographique, pp. 201–10 in *Montagne et Aménagement*, Institut de Géographie Alpine, Grenoble.

Hills, T. L. and Lundgren, J. (1977), The impact of tourism in the Caribbean: a methodological study, *Annals of Tourism Research*, **4**(5), 248–67.

Hinch, T. D. (1990), A spatial analysis of tourist accommodation in Ontario: 1974 to 1988, *Journal of Applied Recreation Research*, **15**(4), 239–64.

Hodge, S. M. (1991), *Pleasure Travel to Texas: geographic patterns of 1989 tourist flow*, Master of Applied Geography thesis (unpublished), Southwest Texas State University, Austin.

Hodgson, P. (1983), Research into the complex nature of the holiday choice process, pp. 17–35 in *Proceedings of the Seminar on the Importance of Research in the Tourist Industry, Helsinki, 8–11 June 1983*, ESOMAR, Amsterdam.

Hong Kong Tourist Association (1993a), *1993 Multi-Destination Analysis*, Hong Kong Tourist Association, Hong Kong.

Hong Kong Tourist Association (1993b), *Visitor Profile Report 1992*, Hong Kong Tourist Association, Hong Kong.

Horwath and Horwath (1981), *L'Hôtellerie Parisienne*, Horwath and Horwath, France, Paris.

Hottola, P. (1992), *Tourist Time–space Budgets in Eilat, Israel*, MSc thesis (unpublished), University of Joensuu, Joensuu.

Hoyle, B. (1988), Development dynamics at the port-city interface, pp. 3–19 in B. S. Hoyle, D. A. Pinder and M. S. Husain, *Revitalising the Waterfront: international dimensions of dockland redevelopment*, Belhaven, London.

Hoyle, B. S. Pinder, D. A. and Husain, M. S. (1988), *Revitalising the Waterfront: international dimensions of dockland redevelopment*, Belhaven, London.

Huetz de Lemps, A. (1976), *L'Espagne*, Masson, Paris.

Huetz de Lemps, A. (1989a), Le tourisme dans les petites îles tropicales et subtropicales, pp. 1–12 in *Iles et Tourisme en Milieux Tropical et Subtropical*, CRET, University of Bordeaux; CEGET–CNRS, Talence.

Huetz de Lemps, A. (1989b), Le tourisme dans l'île de Madère, pp. 127–50 in *Iles et Tourisme en Milieux Tropical et Subtropical*, CRET University of Bordeaux; CEGET–CRNS, Talence.

Huetz de Lemps, A. and Pitte, J.-R. (1990), *Les Restaurants dans le monde et à travers les ages*, Editions Glénat, Grenoble.

Hugill, P. J. (1975), Social conduct on the Golden Mile, *Annals of Assn of American Geographers*, **62**(2), 214–28.

Hussey, A. (1982), Tourist destination areas in Bali, *Contemporary Southeast Asia*, **3**(4), 374–85.

Hussey, A. (1989), Tourism in a Balinese village, *Geographical Review*, **79**(3), 311–25.

Instituto Centrale di Statistica (1977), *Indagine speciale sulle vacanze degli italiani nel 1975*, Note et Relazioni no. 55, Instituto Centrale di Statistica, Rome.

Iribas Sánchez, J. M. (1991), Oferta complementaria al turismo de sol y playa: el caso de Benidorm, pp. 83–89 in F. Fourneau and M. Marchena (eds) *Ordenación y Desarrollo del Turismo en España y en Francia*, Casa de Velázquez, Madrid.

Ishii, H. (1982), Distribution of major recreational regions in Japan, *Frankfurter Wirtschafts-und Sozialgeographische Schriften*, **41**, 187–203.

Iso-Ahola, S. E. (1982), Toward a social psychological theory of tourism motivation: a rejoinder, *Annals of Tourism Research*, **9**(2), 256–61.

IUOTO (International Union of Official Travel Organisations) (1975), *The Impact of International Tourism on the Economic Development of the Developing Countries*, IUOTO/WTO, Geneva.

Jackowski, A. (1980), Methodological problems of functional typology of tourist localities, *Folia Geographica, Series Geographica-Oeconomica*, **13**, 85–91.

Jackson, E. L. and Schinkel, D. R. (1981), Recreational activity preferences of resident and tourist campers in the Yellowknife region, *Canadian Geographer*, **25**(4), 350–63.

Jackson, R. T. (1991), Some geographical aspects of the international tourism boom in Queensland, *Frankfurter Wirtschafts-und Sozialgeographische Schriften*, **59**, 161–76.

Janin, B. (1982), Circulation touristique internationale et tourisme étranger en Val d'Aoste, *Revue de Géographie Alpine*, **70**(4), 415–30.

Japan Travel Bureau (1992), *JTB Report '92: all about Japanese overseas travellers*, Japan Travel Bureau, Tokyo.

Jeans, D. N. (1990), Beach resort morphology in England and Australia: a review and extension, pp. 277–85 in P. Fabbri (ed.) *Recreational Uses of Coastal Areas*, Kluwer, Dordrecht.

Jenkins, J. M. and Walmsley, D. J. (1993), Mental maps of tourists: a study of Coffs Harbour, New South Wales, *GeoJournal*, **29**(3), 233–41.

Johnston, D. C., Pearce, D. G. and Cant, R. G. (1976), Canterbury holiday makers: a preliminary study of internal tourism, pp. 5–19 in R. G. Cant (ed.) *Canterbury at Leisure*, Publication no. 4, Canterbury Branch NZ Geog. Soc., Christchurch.

Jülg, F. (1984), Le tourisme autrichien, pp. 217–26 in *Les Alpes*, Comité International d'Organisation du 25e Congrès de Géographie, Caen.

Jurdao, F. (1990), *España en Venta*, Endymion, Madrid, 2nd edn.

Kain, R. (1978–9), Conservation planning in France: policy and practise in the Marais, Paris, *Urbanism Past and Present*, **7**, 22–343.

Kamp, B. D., Crompton, D. M. and Hensarling, D. M. (1979), The reactions of travellers to gasoline rationing and to increases in gasoline prices, *Journal of Travel Research*, **18**(1), 37–41.

Kearsley, G. W. and Gray, G. (1994), International visitor flows and infrastructure needs: a New Zealand example, *Proc. 17th NZ Geog. Soc. Conf.*, NZ Geog. Soc., Wellington.

Keogh, B. (1984), The measurement of spatial variations in tourist activity, *Annals of Tourism Research*, **11**(2), 267–82.

Kermath, B. M. and Thomas, R. N. (1992), Spatial dynamics of resorts, Sosúa, Dominican Republic, *Annals of Tourism Research*, **19**(2), 173–90.

King, R. (1988), Italy: multi-faceted tourism, pp. 58–79 in A. M. Williams and G. Shaw (eds) *Tourism and Economic Development: Western European experience*, Belhaven, London.

Klaric, Z. (1992), Establishing tourist regions: the situation in Croatia, *Tourism Management*, **13**(3), 305–11.

Knox, A. D. (1967), Some economoic problems of small countries, pp. 35–44 in B. Benedict (ed.) *Problems of Smaller Territories*, Athlone, London.

Kulinat, K. (1986), Fremdenverkehr in Spanien, *Geographisches Rundschau*, **38**(1), 28–35.

Lacroix, J., Roux, B. and Zoido Naranjo, F. (1979), La 'Costa de la Luz' de Cadix: le cas de Chipiona, pp. 117–239 in A. M. Bernal, F. Fourneau and F. Heran, *Tourisme et développement régional en Andalousie*, Editions de Boccard, Paris.

Langenbuch, J. R. (1977), Os municipios turisticos do estado de Sao Paulo: determinàçao e caracterizaçao geral, *Geografia*, **2**(3), 1–49.

Larroque-Chounet, L. (1989), *Les Guadeloupéens et le développement du tourisme*, CENADOM, Talence.

Lavery, P. (1974), The demand for recreation, pp. 22–48 in P. Lavery (ed.) *Recreational Geography*, Davd and Charles, Newton Abbot.

Law, C. M. (1985), *Urban Tourism in the United States*, Dept of Geography, University of Salford

Law, C. M. (1992), Urban tourism and its contribution to economic regeneration, *Urban Studies*, **29**(3/4), 599–618.

Lee, G. P. (1987), Tourism as a factor in development cooperation, *Tourism Management*, **8**(1), 2–19.

Leiper, N. (1984a), Tourism and leisure: the significance of tourism in the leisure spectrum, pp. 249–53 in *Proc. 12th NZ Geog. Conf.*, NZ Geog. Soc., Christchurch.

Leiper, N. (1984b), International travel by Australians, 1946 to 1983: travel propensities and travel frequencies, pp. 67–83 in B. O'Rourke (ed.) *Contemporary Issues in Australian Tourism*, Dept of Geography, University of Sydney, Sydney.

Leiper, N. (1989), Main Destination Ratios: analyses of tourist flows, *Annals of Tourism Research*, **16**(4), 530–42.

Levantis, G. (1981), *Analyse Factorielle du Phenomène Touristique: l'espace touristique grec*, Centre des Hautes Etudes Touristiques, Aix-en-Provence.

Lewis, J. and Williams, A. M. (1988), Portugal: market segmentation and regional specialisation, pp. 101–122 in A. M. Williams and G. Shaw (eds) *Tourism and Economic Development: Western European experiences*, Belhaven, London.

Li, S-M. W. (1992), Outbound travel market of the Republic of China on Taiwan, pp. 2–26 in *ROC (Taiwan) Outbound Market Facts for the Travel Industry*, PATA Asia Division, Singapore.

Lichtenberger, E. R. (1984), Geography of tourism and leisure society in Austria *GeoJournal*, **9**(1), 41–6.

Lihou-Perry, C. R. (1991), *Resort Enclaves and Sustainable Development: a Balinese example*, MA thesis (unpublished), Dept of Geography, University of Waterloo, Waterloo, Canada.

Liu, J. C. (1983), Hotel industry performance and planning at the regional level, pp. 211–33 in P. E. Murphy (ed.) *Tourism in Canada: selected issues and options*, Western Geographical Series vol. 21, University of Victoria, Victoria, Canada.

Lockhart, D. G. (1993), Tourism and politics: the examples of Cyprus, pp. 228–47 in D. G. Lockhart, D. Drakkis-Smith and J. Schembri (eds) *The Development Process in Small Island States*, Routledge, London.

Lockhart, D. G. and Ashton, S. E. (1991), Tourism in Malta, *Scottish Geographical Magazine*, **107**(1), 22–32.

Long, V. H. (1991), Government–industry–community interaction in tourism development in Mexico, pp. 205–22 in M. T. Sinclair and M. J. Stabler (eds) *The Tourism Industry: an international analysis*, CAB International, Wallingford.

Longwoods International (1990), *Colorado's Opportunities in the U.S. Pleasure Travel Market*, Longwoods International, Colorado.

Longwoods International (1991), *New Jersey's Position in the U.S. Touring Vacation Market*, Longwoods International, Colorado.

López Palomeque, F. (1988), Geografía del turismo en España: una aproximación a la distribución espacial de la demanda turística y de la oferta de alojamiento, *Documents d'Anàlisi Geogràfica*, **13**, 35–64.

Lovingood, P. E. and Mitchell, L. E. (1989), A regional analysis of South Carolina tourism, *Annals of Tourism Research*, **163**, 301–17.

Lundgren, J. O. J. (1966), Tourism in Quebec, *Revue de Géographie de Montreal*, **20**(1&2), 59–73.

Lundgren, J. O. J. (1972), The development of tourist travel systems – a metropolitan economic hegemony par excellence, *Jahrbuch für Fremdenverkehr*, 20 Jahrgang, 86–120.

Lundgren, J. O. J. (1974), On access to recreational lands in dynamic metropolitan hinterlands, *Tourist Review*, **29**(4), 124–31.

Lundgren, J. O. J. (1975), Tourist penetration/the tourist product/entrepreneurial response, pp. 60–70 in *Tourism as a Factor in National and Regional Development*, Occasional Paper 4, Department of Geography, Trent University, Peterborough.

Lundgren, J. O. J. (1982), The tourist frontier of Nouveau Quebec: functions and regional linkages, *Tourist Review*, **37**(2), 10–16.

Lundgren, J. O. J. (1984), Geographic concepts and the development of tourism research in Canada, *GeoJournal*, **9**(1), 17–25.

Lundgren, J. O. J. (1987), Tourism development in the northern periphery, *Teoros*, **6**(1), 13–19.

McAllister, D. M. and Klett, F. R. (1976), A modified gravity model or regional recreation activity with an application to ski trips, *Journal of Leisure Research*, **8**(1), 22–34.

MacCannell, D. (1976), *The Tourist: a new theory of the leisure class*, Macmillan, London.

McCool, S. F. (1980), Vacation travel and fuel shortages: a critical comment, *Journal of Travel Research*, **19**(2), 18–19.

McDermott-Miller Group (1988), *The Implications of Tourism Growth in New Zealand*, New Zealand Tourist and Publicity Dept, Wellington.

McDowell Group (1989), *Alaska Visitor Statistics Program II: patterns, opinions and planning, summer 1989*, Alaska Division of Tourism.

McEachern, J. and Towle, E. L. (1974), *Ecological Guidelines for Island Development*, International Union for the Conservation of Nature, Morges.

McElroy, J. L., de Albuquerque, K. and Dioguardi, A. (1993), Applying the tourist destination life-cycle model to small Caribbean and Pacific islands, *World Travel and Tourism Review*, **3**, 236–44.

McLisky, F. P. (1992), *Tourism 'Off the Beaten Track': Alternative tourism as an appropriate form of tourism in Fiji*, MSc thesis (unpublished), University of Canterbury, Christchurch.

Malamud, B. (1973), Gravity model calibration of tourist travel to Las Vegas, *Journal of Leisure Research*, 5(1), 13–33.

Mankour, N. (1980), Localisation et function des équipements touristiques étatiques algériens, *Cahiers Geographiques de l'Ouest*, **4**, 46–57.

Mansfeld, Y. (1992), From motivation to actual travel, *Annals of Tourism Research*, **19**(3), 399–419.

Mariot, P. (1969) Priestorové aspekty cestovnélio rechu a otázky gravitacného zázemia návstevnych miest, *Geografick'y Casopis*, **21**(4), 287–312.

Market Opinion Research (1990), *Las Vegas Visitor Profile – Fiscal Year 1990: composite report*, Las Vegas Convention and Visitors Authority, Las Vegas.

Marsden, B. S. (1969), Holiday homescapes of Queensland, *Australian Geographical Studies*, **7**(1), 57–72.

Mathieson, A. and Wall, G. (1982), *Tourism: economic, physical and social impacts*, Longman, Harlow.

Matley, I. M. (1976), *The Geography of International Tourism*, Resource Paper no. 76–1, Assn of American Geographers, Washington, DC.

Mayo, E. J. (1974), A model of motel-choice, *Cornell Hotel and Restaurant Administration Quarterly*, **15**(3), 55–64.

Menanteau, L. and Martin Vincente, A. (1979), Environnement et tourisme, pp. 241–304 in A. M. Bernal, F. Fourneau and F. Heran, *Tourisme et développement régional en Andalousie*, Editions de Boccard, Paris.

Mercer, D. C. (1970), The geography of leisure: a contemporary growth-point, *Geography*, **55**(3), 261–73.

Mercer, D. C. (1971), Discretionary travel behaviour and the urban mental map, *Australian Geographical Studies*, **9**, 133–43.

Meunier, G. (1985), Étude sur les motivations, les attentes et les habitudes des familles à faible revenu en regard des activités de loisir touristique, *Loisir et Société*, **8**(2).

Meyer-Arendt, K. J. (1985), The Grand Isle, Louisiana resort cycle, *Annals of Tourism Research*, **12**(3), 449–65.

Meyer–Arendt, K. J. (1990), Patterns and impacts of coastal recreation along the Gulf coast of Mexico, pp. 1330–48 in P. Fabbri (ed.) *Recreational Uses of Coastal Areas*, Kluwer, Dordrecht.

Mignon, C. and Heran, F. (1979), La Costa del Sol et son arrière–pays, pp. 53–133 in A. M. Bernal, F.

Fourneau and F. Heran, *Tourisme et développement régional en Andalousie*, Editions de Boccard, Paris.

Mings, R. C. (1982), Classroom use of the National Travel Survey: a recreation travel exercise, *Journal of Geography*, **81**(2), 215–23.

Mings, R. C. and McHugh, K. E. (1992), The spatial configuration of travel to Yellowstone National Park, *Journal of Travel Research*, **30**(4), 38–46.

Miossec, J. M. (1976), *Eléments pour une Théorie de l'Espace Touristique*, Les Cahiers du Tourisme, C-36, CHET, Aix-en-Provence.

Miossec, J. M. (1977), Un modèle de l'espace touristique, *L'Espace Géographique*, **6**(1), 41–8.

Mirloup, J. (1974), Eléments méthodologiques pour une étude de l'équipment hôtelier: l'exemple des départements de la Loire moyenne, *Norois*, **83**, 443–52; **84**, 563–83.

Mirloup, J. (1982), Du tourisme insulaire aux phenomènes d'insularité touristique en milieu continental: éléments de reflexion, pp. 185–94 in C. Ciaccio and L. Pedrini (eds) *Le Tourisme dans les Petites Iles*, Colloqué de la Commission UGI de Géographie du Tourisme et des Loisirs, Lipari.

Mitchell, L. S. (1984), Tourism research in the United States: a geographic perspective, *GeoJournal*, **9**(1), 5–15.

Mitchell, L. S. and Murphy, P. E. (1991), Geography and tourism, *Annals of Tourism Research*, **18**(1), 57–70.

Molnar, E., Mihail, M. and Maier, A. (1976), Types de localités touristiques dans la République Socialiste de Roumanie, *Revue Roumaine de Géologie, Géophysique et Géographie, Série de Géographie*, **20**, 189–95.

Moran, W. (1979), Processes and policies for land use diversification, *Proc. 10th NZ Geog. Conf. and 49th ANZAAS Congress*, NZ Geog. Soc., Auckland, pp. 240–5.

Morgan Research Centre (1982), Domestic Tourism Monitor, 1981–82 (unpublished), Morgan Research Centre, Sydney.

Moutinho, L. (1987), Consumer behaviour in tourism, *European Journal of Marketing*, **21**(10), 5–44.

Mullins, P. (1991), Tourism urbanization, *International Journal of Urban and Regional Research*, **15**(3), 326–42.

Murphy, P. E. (1980), Tourism management using land use planning and landscape design: the Victoria experience, *Canadian Geographer*, **24**(1), 60–71.

Murphy, P. E. (1992), Data gathering for community-oriented tourism planning: case study of Vancouver Island, British Columbia, *Leisure Studies*, **11**(1), 65–79.

Murphy, P. E. and Brett, A. C. (1982), Regional tourism patterns in British Columbia: a discriminatory analysis, pp. 151–61 in T. V. Singh, J. Kaur and D. P. Singh (eds) *Studies in Tourism Wildlife Parks Conservation*, Metropolitan Book Co., New Delhi.

Murphy, P. E. and Keller, C. P. (1990), Destination travel patterns: an examination and modelling of tourist patterns on Vancouver Island, British Columbia, *Leisure Sciences*, **12**(1), 49–65.

Murphy, P. E. and Rosenblood, L. (1974), Tourism: an exercise in spatial search, *Canadian Geographer*, **18**(3), 201–10.

Myrdal, G. (1957), *Economic Theory and Under-Developed Regions*, Duckworth, London.

Newcomb, R. M. (1979), *Planning the Past*, Dawson, Folkestone; Archon, Hamden.

Nickerson, N. P. and Ellis, G. D. (1991), Traveller types and activation theory: a comparison of two models, *Journal of Travel Research*, **29**(3), 26–31.

Noronha, R. (1976), *Review of the Sociological Literature on Tourism*, World Bank, New York.

Noronha, R. (1979), Paradise reviewed: tourism in Bali, pp. 177–204 in E. de Kadt (ed.) *Tourism: passport to development?*, Oxford University Press, Oxford.

NZTD (1991), *New Zealand Domestic Travel Study 1989/90*, New Zealand Tourism Department, Wellington.

NZTP (1989a), *New Zealand International Visitors Survey 1988/89*, New Zealand Tourist and Publicity Dept, Wellington.

NZTP (1989b), *New Zealand Domestic Travel Study 1988/89*, New Zealand Tourist and Publicity Dept, Wellington.

NZTPD (1983), *New Zealand International Visitors Travel Survey 1982, Vol. 1*, NZ Tourist and Publicity Dept, Wellington.

O'Brien, K. (1991), The European business travel market – a market of immense size and value which is difficult to research, pp. 87–98 in *Seminar on Travel and Tourism in Transition: the research challenge*, ESOMAR, Amsterdam.

Observatoire National du Tourisme (1993), *Mémento du Tourisme 1993*, Observatoire National du Tourisme, Paris.

Odouard, A. (1973), Le tourisme et les Iles Canaries, *Les Cahiers d'Outre-Mer*, **102**, 150–71.

Odouard, A. (1989), Le tourisme aux Canaries: organiser l'acquis, pp. 151–62 in *Iles et Tourisme en Milieux Tropical et Subtropical*, CRET, University of Bordeaux; CEGET–CNRS, Talence.

O'Hagan, J. W. (1979), *The dispersal pattern of United States tourists in Europe, 1967–1977*, European Travel Commission, Dublin.

Ohler, K. E. (1971), *The Distribution of Inns, Hotels and Motels in Somerset County, Pennsylvania: a geographical sequential analysis*, M.A. thesis (unpublished), University of Pennsylvania, Indiana.

Opinion Research Corporation (1980), *A Study of Potential U.S. Vacation Visitors to the Pacific Area: executive summary*, PATA, San Francisco.

Oppermann, M. (1992a), *Tourismus in Malaysia*, Sozialwissenschafliche Studien zu internationalen Problemen, Verlag Breitenbach, Saarbrücken.

Oppermann, M. (1992b), Intranational tourist flows in Malaysia, *Annals of Tourism Research*, **19**(3), 482–500.

Oppermann, M. (1993a), Tourism space in developing countries, *Annals of Tourism Research*, **20**(3), 535–60.

Oppermann, M. (1993b), *Intranational Tourist Flows and Changing spatial Demand Patterns in New Zealand – a market segmentation analysis with implications for planning and marketing*, Report submitted to the New Zealand Tourism Board.

Ottersbach, G. (1985), Travel patterns: insights from a regional perspective, pp. 3–13 in *Proceedings 16th TTRA Annual Conference*, University of Utah, Salt Lake City.

Oum, T. H. and Lemire, N. (1991), An analysis of Japanese international travel destination choices, *International Journal of Transport Economics*, **18**(3), 289–307.

Owen, M. L. and Duffield, B. S. (1971), *The Touring Caravan in Scotland*, Scottish Tourist Board, Edinburgh.

Paajanen, M. (1993), *The Economic Impact Analysis of Tourism: a comparative study of the Nordic Model and the Tourist Economic Model*, Helsinki School of Economics and Business Administration, Working Paper W-35, Helsinki.

Page, S. J. (1989), Tourist development in London Docklands in the 1980s and 1990s, *GeoJournal*, **19**(3), 291–95.

P. A. Management Consultants (1975), *The Role of Tourism and Recreation in the Albury-Wodonga Growth Centre*, Australian Department of Tourism and Recreation, Canberra.

Papadopoulus, S. I. (1986), The tourism phenomenon: an examination of important theories and concepts, *Tourist Review*, **40**(3), 2–11.

Paris Promotion (1991), *Le Plan d'aménagement du tourisme Parisien*, Paris Promotion, Paris.

Pearce, D. G. (1978a), Demographic variations in international travel, *Tourist Review*, **33**(1), 4–9.

Pearce, D. G. (1978b), *Tourism in France: regional perspectives*, Canterbury Monographs for Teachers of French no. 3 (Fifth Series), Department of French, University of Canterbury.

Pearce, D. G. (1978c), Form and function in French resorts, *Annals of Tourism Research*, **5**(1), 142–56.

Pearce, D. G. (1979a), Towards a geography of tourism, *Annals of Tourism Research*, **6**(3), 245–72.

Pearce, D. G. (1979b), Geographical aspects of tourism in New Zealand, pp. 327–31 in *Proc. 10th NZ Geog. Conf. and 49th ANZAAS Congress (Geographical Sciences)*, NZ Geog. Soc., Auckland.

Pearce, D. G. (1981a), L'espace touristique de la grande ville: éléments de synthèse et application à Christchurch (Nouvelle-Zélande), *L'Espace Géographique*, **10**(3), 207–13.

Pearce, D. G. (1981b), Estimating visitor expenditure, a review and a New Zealand case study, *International Journal of Tourism Management*, **2**(4), 240–52.

Pearce, D. G. (1982a), Preparing a national tourist geography: the New Zealand example, pp. 136–49 in T. V. Singh, J. Kaur and D. P. Singh (eds) *Studies in Tourism, Wildlife Parks, Conservation*, Metropolitan Book Co., New Delhi.

Pearce, D. G. (1982b), *Westland National Park Economic Impact Study*, Dept of Lands and Survey/Dept of Geography, University of Canterbury, Christchurch.

Pearce, D. G. (1983a), Intra-regional traffic in the South Pacific, pp. 22–33 in 1983 *PATA Travel Research Conference Proceedings*, Pacific Area Travel Association, San Francisco.

Pearce, D. G. (1983b), The development and impact of large-scale tourism projects: Languedoc-Roussillon (France) and Cancun (Mexico) compared, pp. 57–71 in C. C. Kissling, N. J. Thrift, M. J. Taylor and C. Adrian (eds) *Papers, 7th Australian/NZ Regional Science Assn*, Canberra.

Pearce, D. G. (1984), International tourist flows: an integrated approach, with examples from French Polynesia, pp. 254–8 in *Proc. 12th NZ Geog. Conf.*, NZ Geog. Soc., Christchurch.

Pearce, D. G. (1987a), Spatial patterns of package tourism in Europe, *Annals of Tourism Research*, **14**(8), 183–201.

Pearce, D. G. (1987b), Mediterranean charters: a comparative geographic perspective, *Tourism Management*, **8**(4), 291–305.

Pearce, D. G. (1987c), *Tourism Today: a geographical analysis*, Longman, Harlow and Wiley, New York.

Pearce, D. G. (1987d), Motel location and choice in Christchurch, *New Zealand Geographer*, **43**(1), 10–17.

Pearce, D. G. (1988a), Tourist time–budgets, *Annals of Tourism Research*, **15**(1), 106–21.

Pearce, D. G. (1988b), The spatial structure of coastal tourism: a behavioural approach, *Tourism Recreation Research*, **23**(2), 11–14.

Pearce, D. G. (1989a), International and domestic tourism: interfaces and issues, *GeoJournal*, **19**(3), 257–62.

Pearce, D. G. (1989b), *Tourist Development*, Longman, Harlow, 2nd edn.

Pearce, D. G. (1990a), Tourism, the regions and restructuring in New Zealand, *Journal of Tourism Studies*, **1**(2), 33–42.

Pearce, D. G. (1990b) Tourist travel patterns in the South Pacific – analysis and implications, pp. 31–49 in C. C. Kissling (ed.) *Destination South Pacific: perspectives on island tourism*, Centre des Hautes Etudes Touristiques, Aix-en-Provence.

Pearce, D. G. (1991), Challenge and change in East European tourism: a Yugoslav example, pp. 223–40 in M. T. Sinclair and M. J. Stabler (eds) *The Tourism Industry: an international analysis*, CAB International, Wallingford.

Pearce, D. G. (1992a), Tourist travel patterns and regional impacts: issues and examples from New Zealand, pp. 145–60 in C. A. M. Fleischer-van Rooijen (ed.) *Spatial Implications of Tourism*, Geo Pers, Groningen.

Pearce, D. G. (1992b), *Tourist Organizations*, Longman, Harlow.

Pearce, D. G. (1992c), Tourism and the European Regional Development Fund: the first fourteen years, *Journal of Travel Research*, **30**(3), 44–51.

Pearce, D. G. (1993a), Domestic tourist travel patterns in New Zealand, *GeoJournal*, **29**(3), 225–32.

Pearce, D. G. (1993b), Comparative studies in tourism research, pp. 20–35 in D. G. Pearce and R. W. Butler (eds) *Tourism Research: critiques and challenges*, Routledge, London.

Pearce, D. G. (1994), Circuit tourism in Asia and the Pacific, *Proc. 17th NZ Geog. Soc. Conf.*, NZ Geog. Soc., Wellington.

Pearce, D. G. (1995), Planning for tourism in the 90s: an integrated, dynamic, multi-scale approach, pp. 229–44 in R. W. Butler and D. G. Pearce (eds) *Change in Tourism: people, places, processes*, Routledge, London.

Pearce, D. G. and Butler, R. W. (eds) (1993), *Tourism Research: critiques and challenges*, Routledge, London.

Pearce, D. G. and Elliott, J. M. C. (1983), The Trip Index, *Journal of Travel Research*, **22**(1), 6–9.

Pearce, D. G. and Grimmeau, J.-P. (1985), The spatial structure of tourist accommodation and hotel demand in Spain, *Geoforum*, **16**(1), 37–50.

Pearce, D. G. and Johnston, D. C. (1986), Travel within Tonga, *Journal of Travel Research*, **24**(3), 13–17.

Pearce, D. G. and Kirk, R. M. (1986), Carrying capacities for coastal tourism, *Industry and Environment*, **9**(1), 3–6.

Pearce, D. G. and Mings, R. C. (1984), Geography, tourism and recreation in the Antipodes, *GeoJournal*, **9**(1), 91–5.

Pearce, P. L. (1981), 'Environmental shock': a study of tourists' reactions to two tropical islands, *Journal of Applied Social Psychology*, **11**(3), 268–80.

Pearce, P. L. (1993), Fundamentals of tourist motivation, pp. 113–34 in D. G. Pearce and R. W. Butler (eds) *Tourism Research: critiques and challenges*, Routledge, London.

Perdue, R. R. and Gustke, L. D. (1985), Spatial patterns of leisure travel by trip purpose, *Annals of Tourism Research*, **12**(2), 167–80.

Piatier, A. (1956), *Sondages et Enquêtes au Service du Tourisme*, Geneva.

Piavaux, C.-M. (1977), *Valorisation de l'Environnement par le Tourisme dans le Luxembourg Belge. Propositions de specialisation par compartiments touristiques*, Fondation Universitaire Luxembourgeoise, Arlon, 1977.

Pigram, J. J. (1977), Beach resort morphology, *Habitat International*, **2**(5–6), 525–41.

Pinder, D. (1988), The Netherlands: tourist development in a crowded society, pp. 214–29 in A. M. Williams and G. Shaw (eds) *Tourism and Economic Development: Western European experiences*, Belhaven, London.

Plettner, H. J. (1979), *Geographical Aspects of Tourism in the Republic of Ireland*, Research Paper no. 9, Social Sciences Research Centre, University College, Galway.

Plog, S. C. (1973), Why destination areas rise and fall in popularity, *Cornell Hotel and Restaurant Administration Quarterly*, November, 13–16.

Plog, S. C. (1990), A carpenter's tools: an answer to Stephen L. J. Smith's review of psychocentrism/allocentrism, *Journal of Travel Research*, **28**(4), 43–5.

Plog, S. C. (1991a), *Leisure Travel: making it a growth market again*, Wiley, New York.

Plog, S. C. (1991b), A carpenter's tools re-visited: measuring allocentrism and psychocentrism properly . . . the first time, *Journal of Travel Research*, **29**(4), 51.

Poitier, F. and Cockerell, N. (1992), The European international short break market, *Travel and Tourism Analyst*, 5, 45–65.

Pollock, A. M., Tunner, A. and Crawford, G. S. (1975), *Visitors 74: a study of visitors to British Columbia in the summer of 1974*, B.C. Research, Vancouver.

Poncet, J. (1976), Le développement du tourisme en Bulgarie, *Annales de Geographie*, **468**,155–77.

Porteous, J. D. (1981), *The Modernisation of Easter Island*, Western Geographical Series vol. 19, University of Victoria, Victoria, Canada.

Potter, R. B. (1981), Tourism and development: the case of Barbados, West Indies, *Geography*, **68**, 46–50.

Potts, T. D. and Uysal, M. (1992), Tourism intensity as a function of accommodations, *Journal of Travel Research*, **31**(2), 18–21.

Pred, A. (1965), Industrialization, initial advantage, and American metropolitan growth, *Geographical Review*, **55**(2), 158–85.

Price, R. L. (1981), Tourist landscapes and tourist regions, a preliminary framework, Paper presented at the annual meeting of the AAAG, Los Angeles.

Priestley, G. K. (1986), El turismo y la transformación del territorio: un estudio de Tossa, Lloret de Mar y Blanes a traves de la fotografia aerea 1956–81, pp. 88–106 in *Jornades Tecniques Sobre Turisme i Mediambient Sant Feliu de Guixols–Costa Brava*, Barcelona.

Priestley, G. K. (1987), The role of golf as a tourist attraction: the case of Catalonia, Spain, pp. 288–302 in *Le Développement du Tourisme dans les Espaces Voisins des Grandes Zones de Fréquentation Touristique*, Office National du Tourisme Tunisien, Sousse.

Rajotte, F. (1975), The different travel patterns and spatial framework of recreation and tourism, pp. 43–52 in *Tourism as a Factor in National and Regional Development*, Dept of Geography, Trent University, Occasional Paper 4, Peterborough.

Rajotte, F. (1977), Evaluating the cultural and environmental impact of Pacific tourism, *Pacific Perspective*, **6**(1), 41–8.

Rasmussen, S. E. (1964), *Experiencing Architecture*, MIT Press, Cambridge, Mass.

Relph, E. (1976), *Place and Placelessness*, Pion, London.

Richez, G. and Richez-Battesti, J. (1982), Tourisme et mutations socio-economiques en Corse et a Mallorca, *Etudes Corses*, **10**, 18–19, 329–61.

Ritchie, J. B. R. (1993), Issues in price–value competitiveness of island tourism destinations, *World Travel and Tourism Review*, **3**, 299–305.

Robinson, G. W. S. (1953), The geographical region: form and function, *Scottish Geographical Magazine*, **69**(2), 51–7.

Robinson, H. (1976), *A Geography of Tourism*, Macdonald and Evans, London.

Rogers, A. W. (1977), Second homes in England and Wales: a spatial view, pp. 85–102 in J. T. Coppock (ed.) *Second Homes: curse or blessing?*, Pergamon, Oxford.

Rognant, L. (1990), *Un Géo-système Touristique National: l'Italie, essai systemique*, Centre des Hautes Etudes Touristiques, Aix-en-Provence.

Ronkainen, I. A. and Woodside, A. G. (1978), Cross-cultural analysis of market profiles of domestic and foreign travellers, *European Journal of Marketing*, **12**(8), 579–87.

Royer, L. E., McCool, S. F. and Hunt, J. D. (1974), The relative importance of tourism to state economies, *Journal of Travel Research*, **24**(1), 13–16.

RRC Associates (1991), *Reno–Sparks Convention and Visitors Authority 1991 Visitor Profile: Final Report*, Reno-Sparks Convention and Visitors Authority, Reno.

Ruppert, K. (1978), Mise au point sur une géographie générale des loisirs, *L'Espace Géographique*, **7**(3), 187–93.

Ryan, B. (1965), The dynamics of recreational development on the South Coast of New South Wales, *Australian Geographer*, **9**(6), 331–48.

Sabin, P. R. (1986), Finance from the public sector and EEC support, pp. 31–8 in *Planning for Tourism and Tourism in Developing Countries*, PRTC Educational Research Services, London.

Sarramea, J. (1978), L'origine géographique des touristes au cours d'une année, méthodes de recherches et exemple de Fréjus-Saint Raphael, *Méditerranée*, **33**(3), 67–73.

Sarramea, J. (1979), Origine géographique des touristes estivaux à Méribel, *Revue de Géographie Alpine*, **67**(1), 105–11.

Schmidhauser, H. P. (1975), Travel propensity and travel frequency, pp. 53–60 in A. J. Burkart and S. Medlik (eds) *The Management of Tourism*, Heinemann, London.

Schmidhauser, H. P. (1976), The Swiss travel market and its role within the main tourist generating countries of Europe, *Tourist Review*, **31**(4), 15–18.

Schmidhauser, H. P. (1977), *Le Marché touristique Suisse*, Institut du Tourisme de l'Ecole des Hautes Etudes Economiques et Sociales de Saint Gall, Saint Gall.

Schmidhauser, H. P. (1989), Tourist needs and motivations, pp. 569–72 in S. F. Witt and L. Moutinho (eds) *Tourism Marketing and Management Handbook*, Prentice Hall, New York.

Schnell, P. (1988), The Federal Republic of Germany: a growing international deficit, pp. 196–213 in A. M. Williams and G. Shaw (eds) *Tourism and Economic Development: Western European experiences*, Belhaven, London.

Secrétariat d'Etat au Tourisme (1977), Les flux interrégionaux de départ en vacances, *Statistiques du Tourisme*, 16.

Segui Llinas, M. (1991), Nature et tourisme: l'équilibre indispensable pour l'avenir de l'île de Majorque, *Méditerranée*, **72**(1), 15–20.

Service de la Recherche Socio-Economique (1977), *Le Touriste non-résident au Quebec 1975*, vol. III, Données de Base II, Ministere du Tourisme, de la Chasse et de la Péche Quebec.

Shand, R. T. (1980), Island smallness: some definitions and implications, pp. 3–20 in R. T. Shand (ed.) *The Island States of the Pacific and Indian Oceans: anatomy of development*, Development

Studies Monograph no. 23, Australian National University, Canberra.

Shirasaka, S. (1982), Foreign visitors flow in Japan, *Frankfurter Wirtschafts-und Social geographische Schriften*, Heft 41, 205–18.

Singapore Tourist Promotion Board (1992), *Survey of Overseas Visitors to Singapore*, Singapore Tourist Board, Singapore.

Singaravelou (1989), Le rôle du tourisme dans l'économie mauricienne, pp. 265–78 in *Iles et Tourisme en Milieux Tropical et Subtropical*, CRET, University of Bordeaux; CEGET–CRNS, Talence.

Smale, B. J. A. and Butler, R. W. (1985), Domestic tourism in Canada: regional and provincial patterns, *Ontario Geography*, **26**, 37–56.

Smith, R. A. (1988), The role of tourism in urban conservation: the case of Singapore, *Cities*, August, 245–59.

Smith, R. A. (1991), Beach resorts: a model of development evolution, *Landscape and Urban Planning*, **21**, 189–210.

Smith, R. A. (1992), Review of integrated beach resort development in Southeast Asia, *Land Use Policy*, 211–17.

Smith, R. H. T. (1970), Concepts and methods in commodity flow analysis, *Economic Geography*, **46**(2), 404–16.

Smith, S. L. J. (1983a), *Recreation Geography*, Longman, London.

Smith, S. L. J. (1983b), Restaurants and dining out: geography of a tourism business, *Annals of Tourism Research*, **10**(4), 515–49.

Smith, S. L. J. (1987), Regional analysis of tourism resources, *Annals of Tourism Research*, **14**(2), 254–73.

Smith, S. L. J. (1989), *Tourism Analysis: a handbook*, Longman, Harlow.

Smith, S. L. J. (1990a), A test of Plog's allocentric/psychocentric model: evidence from seven nations, *Journal of Travel Research*, **28**(4), 40–3.

Smith, S. L. J. (1990b), Another look at the carpenter's tools: a reply to Plog, *Journal of Travel Research*, **29**(2), 50–1.

Smith, V. L., Hetherington, A. and Brumbaugh, M. D. D. (1986), California's Highway 89: a regional tourism model, *Annals of Tourism Research*, **13**(3), 415–33.

Socias Fuster, M. (1986), La Oferta Turística de Baleares en 1986. Paper presented at meeting of the IGU's Commission on the Geography of Tourism and Leisure, Palma de Mallorca, August, 1986.

Soumagne, J. (1974), Cherbourg: la fonction d'escale maritime des touristes brittaniques, *Norois*, **82**, 223–39.

Spack, A. (1975), Aspects et problèmes touristiques en milieux urbains et periurbains: l'exemple de la ville de Metz et du pays messin, *Mosella*, **4**(1/2), 1–238.

Sprincova, S. (1968), Tourism as a regionalizing factor, pp. 197–210 in I. Koloman (ed.) *Function and Forming of Regions*, Slovak Pedagogical Publishers, Bratislava.

Stabler, M. J. (1991), Modelling the tourism industry: a new approach, pp. 15–43 in M. T. Sinclair and M. J. Stabler (eds) *The Tourism Industry: an international analysis*, CAB International, Wallingford.

Stang, F. (1979), Internationaler Tourismus in Indien, *Erdkunde*, **33**(1), 52–60.

Stansfield, C. A. (1964), A note on the urban–nonurban imbalance in American recreational research, *Tourist Review*, **19**(4), 196–200, **20**(1), 21–3.

Stansfield, C. A. (1969), Recreational land use patterns within an American seaside resort, *Tourist Review*, **24**(4), 128–36.

Stansfield, C. A. (1978), Atlantic City and the resort cycle: background to the legalization of gambling, *Annals of Tourism Research*, **5**(2), 238–51.

Stansfield, C. A. and Rickert, J. E. (1970), The Recreational Business District, *Journal of Leisure Research*, **2**(4), 213–25.

Statistics Canada (1987), *Touriscope: 1986 Domestic Travel*, Statistics Canada, Ottawa.

STB (1993), *British Tourism in Scotland*, 1992, Fact Sheet, Market Research Results RH27, Scottish Tourist Board, Edinburgh.

Strapp, J. D. (1988), The resort cycle and second homes, *Annals of Tourism Research*, **19**(2), 504–16.

Swizewski, C. and Oancea, D. I. (1978), La carte des types de tourisme de Roumanie, *Revue Roumaine de Géologie Géophysique et Géographie, Série Géographie*, **23**(2), 291–4.

Symanski, R. and Burley, N. (1973), *Tourist development in the Dominican Republic: an overview and an example*, Paper presented at the Conference of Latin American Geographers, Calgary (mimeo).

Takeuchi, K. (1984), Some remarks on the geography of tourism in Japan, *GeoJournal*, **9**(1), 85–90.

TDC (1991), *Annual Tourism Statistical Report 1990*, Tourist Development Corporation, Kuala Lumpur.

Texas Department of Commerce (1990), *1989 Texas Domestic Pleasure Travel Market Segments*, Texas Dept of Commerce, Austin.

Texas Department of Commerce (1993), *1992 Report of Travel to Texas*, Texas Dept of Commerce, Austin.

Thompson, P. T. (1971), *The Use of Mountain Recreational Resources: a Comparison of Recreation and Tourism in the Colorado Rockies and the Swiss Alps*, University of Colorado, Boulder.

Thurot, J. M. (1973), *Le Tourisme tropical Balnéaire:*

le modele caraibe et ses extensions, thesis, Centre d'Etudes du Tourisme, Aix-en-Provence.

Thurot, J. M. (1980), *Capacité de Charge et Production Touristique*, Etudes et Mémoires no. 43, Centre des Hautes Etudes Touristiques, Aix-en-Provence.

Tokuhisa, T. (1980), Tourism within, from and to Japan, *International Social Science Journal*, **32**(1), 128–50.

Tourism Council of the South Pacific (1989), *Vanuatu Visitor Survey 1988*, Tourism Council of the South Pacific, Suva.

Town and Country Planning Directorate (1984), *Tourism Reconnaissance Conclusions*, Ministry of Works and Development, Wellington.

Turner, L. and Ash, J. (1975), *The Golden Hordes: international tourism and the pleasure periphery*, Constable, London.

Turnock, D. (1977), Rumania and the geography of tourism, *Geoforum*, **8**(1), 51–6.

Um, S. and Crompton, J. L. (1990), Attitude determinants in tourism destination choice, *Annals of Tourism Research*, **17**(3), 432–48.

United Nations (1970), *Report of the Interregional Seminar on Physical Planning for Tourist Development*, Dubrovnik, Yugoslavia, 19 October–3 November 1970, ST/TAO/Ser. C/131, United Nations, New York.

US Travel Data Center (1989), *The 1988–1989 Economic Review of Travel in America*, Washington, DC, US Travel Data Center.

US Travel Data Center (1991), *The 1990–91 Economic Review of Travel in America*, Washington, DC, US Travel Data Center.

USTTA (1991), *In-flight Survey of International Air Travellers*, United States Travel and Tourism Administration, Washington.

Uysal, M. and McDonald, C. D. (1989), Visitor segmentation by Trip Index, *Journal of Travel Research*, **27**(3), 38–42.

Valenzuela, M. (1988), Spain: the phenomenon of mass tourism, pp. 39–57 in A. M. Williams and G. Shaw (eds) *Tourism and Economic Development: Western European experiences*, Belhaven, London.

Valenzuela Rubio, M. (1981), La incidencia de los grandes equipamientos recreativos en la configuración del espacio turístico litoral: la Costa de Malaga, Paper presented at the Coloquio Hispano-Francés Sobre Espacios Litorales, Madrid (mimeo).

van Doren, C. S. and Gustke, L. D. (1982), Spatial analysis of the U.S. lodging industry, 1963–1977, *Annals of Tourism Research*, **9**(4), 543–63.

Vanhove, N. (1980), Le littoral belge, *Hommes et Terres du Nord*, **4**, 52–62.

van Raaij, W. F. and Francken, D. A. (1984), Vacation decisions, activities, and satisfactions, *Annals of Tourism Research*, **11**(1), 101–12.

van Wagtendonk, J. W. (1980), Visitor use patterns in Yosemite National Park, *Journal of Travel Research*, **19**(2), 12–17.

Vera Rebollo, J. F. (1987), *Turismo y Urbanización en el Litoral Alicantino*, Instituto de Estudios Juan Gil-Albert, Alicante.

Vera Rebollo, J. F. (1991), La oferta complementaria en el turismo de sol y playa: una respuesta al agotamiento del modelo masivo en la Costa Blanca, pp. 91–9 in F. Fourneau and M. Marchena (eds) *Ordenación y Desarrollo del Turismo en España y en Francia*, Casa de Velázquez, Madrid.

Vila Fradera (1966), Trois cas d'actions de l'Etat pour la localisation touristique dans la nouvelle legislation espagnole, *Tourist Review*, **21**(4), 161–3.

Villegas Molina, F. (1975), Areas turisticas andaluzas, *Boletin de la Real Sociedad Geografica*, **1**, 309–22.

Vuoristo, K.-V. (1969), On the geographical features of tourism in Finland, *Fennia*, **99**(3), 1–48.

Wahlers, R. G. and Etzel, M. J. (1985), Vacation preference as a manifestation of optimal stimulation and lifestyle experience, *Journal of Leisure Research*, **17**(4), 283–295.

Wall, G. (1971), Car-owners and holiday activities, pp. 106–7 in P. Lavery (ed.) *Recreational Geography*, David and Charles, London.

Wall, G., Dudycha, D. and Hutchinson, J. (1985), Point pattern analyses of accommodation in Toronto, *Annals of Tourism Research*, **12**(4), 603–18.

Ward, M. (1975), Dependent development problems of economic planning in small developing countries, pp. 115–33 in P. Selwyn (ed.) *Development Policy in Small Countries*, Croom Helm, London.

Washer, R. M. (1977), *Holiday Homes on Banks Peninsula: an impact assessment*, MA thesis (unpublished), University of Canterbury, Christchurch.

Waters, S. R. (1992), *Travel Industry World Yearbook: The Big Picture 1992*, Child and Waters, New York.

Weaver, D. B. (1988), The evolution of a 'plantation' tourism landscape on the Caribbean island of Antigua, *Tijdschrift voor Economische en Sociale Geografie*, **79**(5), 319–31.

Weaver, D. B. (1992a), Gauging the magnitude of tourism in Caribbean destinations: the tourist population index and host/tourist ratio. Paper presented at the Islands of the World III Conference, Nassau, May (mimeo).

Weaver, D. B. (1992b), Tourism and the functional transformation of the Antiguan landscape, pp. 161–76 in C. A. M. Fleischer-van Rooijen (ed.) *Spatial Implications of Tourism*, Geo Pers, Groningen.

Weaver, D. B. (1993), Model of urban tourism for

small Caribbean islands, *Geographical Review*, **82**(3), 134–40.

Weightman, B. A. (1987), Third world tour landscapes, *Annals of Tourism Research*, **14**(2), 227–39.

White, P. and Woods, R. (1980), *The Geographical Impact of Migration*, Longman, London.

White, W. R. (1969), *Location of Motels in a Resort Area: an analysis of motel location in Barnstable County, Massachusetts*, MA thesis (unpublished), Clark University, Worcester.

Williams, A. M. and Shaw, G. (eds) (1988), *Tourism and Economic Development: West European experiences*, Belhaven, London.

Williams, A. V. and Zelinsky, W. (1970), On some patterns of international tourist flows, *Economic Geography*, **46**(4), 549–67.

Williams, P. W., Burke, J. F. and Dalton, M. J. (1979), The potential impact of gasoline futures on 1979 vacation travel strategies, *Journal of Travel Research*, **18**(1), 3–7.

Witt, C. A. and Wright, P. L. (1992), Tourist motivation: life after Maslow, pp. 33–55 in P. Johnson and B. Thomas (eds) *Choice and Demand in Tourism*, Mansell, London.

Wolfe, R. I. (1951), Summer cottagers in Ontario, *Economic Geography*, **27**(1), 10–32.

Wolfe, R. I. (1966), *Parameters of Recreational Travel in Ontario*, Downsview, Ontario.

Wolfe, R. I. (1970), Discussion of vacation homes, environmental preferences and spatial behaviour, *Journal of Leisure Research*, **2**(1), 85–7.

Wolman Associates (1978), *Yukon Tourism Development Strategy*, Dept of Tourism and Information, Yukon.

Wong, P. P. (1990), The geomorphological basis of beach resort sites – some Malaysian examples, *Ocean and Shoreline Management*, **13**, 127–47.

Woodside, A. G. and Etzel, M. J. (1980), Impact of physical and mental handicaps on vacation travel behaviour, *Journal of Travel Research*, **18**(3), 9–11.

Woodside, A. G. and Lysonski, S. (1989), A general model of traveller destination choice, *Journal of Travel Research*, **27**(4), 8–14.

Woodside, A. G. and Sherrell, D. (1977), Traveller evoked, inept and inert sets of vacation destinations, *Journal of Travel Research*, **16**(1), 14–18.

WTO (1978), *Methodological supplement to World Tourism Statistics*, World Tourism Organization, Madrid.

WTO (1981), *Technical handbook on the collection and presentation of domestic and international tourism statistics*, World Tourism Organization, Madrid.

WTO (1983a), *Domestic Tourism Statistics 1981–82*, World Tourism Organization, Madrid.

WTO (1983b), *Development of Leisure Time and the Right to Holidays*, World Tourism Organization, Madrid.

Yamamura, J. (1982), The course of development of tourism and recreation in Japan, *Frankfurter Wirtschafts-und Sozialgeographische Schriften*, **41**, 175–85.

Yokeno, N. (1968), La localisation de l'industrie touristique: application de l'analyse de Thunen-Weber, *Cahiers du Tourisme*, C-9, CHET, Aix-en-Provence.

Yokeno, N. (1974), The general equilibrium system of 'space-economics' for tourism, *Reports for the Japan Academic Society of Tourism*, **8**, 38–44.

Young, B. (1983), Touristization of traditional Maltese fishing–farming villages: a general model, *Tourism Management*, **4**(1), 38–44.

Young, G. (1973), *Tourism: blessing or blight?*, Penguin, Harmondsworth.

Yuan, S. and McDonald, C. (1990), Motivational determinants of international pleasure time, *Journal of Travel Research*, **29**(1), 42–4.

Zipf, G. K. (1946), The P_1P_2/D hypothesis: on intercity movement of persons, *American Sociological Review*, **11**, 677–86.

Author Index

Subject Index

Place Index

Brian
has
Dyslexia
A DOCTOR SPOT
CASE BOOK
C015678135

For Jasmine

First published in the UK in 2004 by Red Kite Books,
an imprint of Haldane Mason Ltd, PO Box 34196, London NW10 3YB
www.redkitebooks4kids.com

Reprinted 2009
Second edition published 2013

British Library CIP Data:
A catalogue record for this book is available from the British Library

ISBN 978-1-905339-86-0 2nd edition
(ISBN 978-1-902463-54-4 1st edition)

Printed in China
2 4 6 8 9 7 5 3 1

Medical consultant: **Dr Rob Hicks**

Please note:
The information presented in this book is intended as a support to professional advice and care. It is not a substitute for medical diagnosis or treatment. Always notify and consult your doctor if your child is ill.

Jenny Leigh

Illustrated by Woody Fox

Name: Brian Bear

Age: 7

Sex: Boy

Case Notes: Brian Bear's mother thought he was doing badly at school because he couldn't see the blackboard, but Brian's eyes were fine – there was another reason why he found reading and writing so hard.

Doctor R Spot

Brian had spent the day with his friend Humphrey. Humphrey's dad had cooked a yummy Sunday lunch, and the two friends had spent the afternoon riding their bikes along the track by the river. Brian was a bit wobbly on his bike. He just couldn't seem to balance, steer and pedal all at the same time. Humphrey was a really good bike rider and he raced ahead.

"Wait for me!" called Brian.

On his way home, Brian suddenly found a car driving straight towards him on the same side of the road! *Beep-beep!* The car swerved around Brian and screeched to a halt. Brian wobbled like mad and just managed to stop without falling off.

Mr Antelope jumped out of the car. "What do you think you're doing on the wrong side of the road?" he shouted. He was very cross!

"Sorry, Mr Antelope," said Brian. "I didn't think I *was* on the wrong side."

"Well, you should know your left from your right at your age, Brian!" snapped Mr Antelope angrily, and he drove off in his car, a bit too fast!

Mrs Bear was angry with Brian, too. "I had Mr Antelope on the 'phone for half an hour," she said. "Really, Brian, if I can't trust you to ride your bike safely, I won't let you ride it at all."

"Oh M-u-u-u-m!" wailed Brian. He loved his bike, and he was sure that Humphrey would find someone else to play with if he didn't have one.

"Lights out now," said Mrs Bear firmly. "You need to be up early for school tomorrow."

"School!" thought Brian as he lay in the dark, and he let out a big sigh. He found school work very hard.

The next morning, Mrs Bear couldn't get Brian out of bed. "My head hurts so much, I just want to lie still," he whispered.

"All right, dear," said Mrs Bear worriedly, "but I'm going to take you to see Doctor Spot, because this is the fourth bad headache you've had this term."

Later that day, Brian and Mrs Bear were in Doctor Spot's surgery, telling him all about Brian's headaches.

"Could his eyes be the problem?" asked Mrs Bear. "He does have a lot of trouble with his reading and writing — maybe he can't see the letters or numbers on the blackboard."

"Yes, it is a good idea to have his eyes tested," said Doctor Spot. "What's your favourite class at school, Brian?"

"I love art," said Brian, "and I love making up stories — as long as I don't have to write them down, or read them out."

"What happens when you do that?" asked Doctor Spot.

"Well, I get all in a muddle and they don't come out right," said Brian, sadly.

"Hmmm," said Doctor Spot and he turned to Mrs Bear. "I think it would be a good idea for you and Mr Bear to go and talk to Brian's teacher about this." Mrs Bear wasn't sure how the teacher could help Brian's headaches, but she said she would talk to the teacher anyway.

Mr and Mrs Bear went to the school for a meeting with Brian's teacher.

"I'm so glad you've come to see me," said Mrs Flamingo. "As I said in my letter, I am worried about Brian's progress at school."

"What letter?" asked Mr Bear.

"Why, I sent a letter home with Brian last week!" said Mrs Flamingo. "Brian is a very bright little bear," she went on. "He works very hard, but he finds reading and writing difficult. I would like him to meet someone who could tell us if Brian has a special learning problem."

"Oh, dear!" said Mrs Bear. "Is that what gives him his headaches?"

"Well, in a way it might be," said Mrs Flamingo.

Brian went to see Mr Parrot, who was a specialist teacher. Mr Parrot explained that they were going to do some tests to see if they could work out why Brian was finding some of his schoolwork so hard.

"Oh!" said Brian. "I'm not very good at tests."

"Why do you think that is, Brian?" asked Mr Parrot.

"My friend Mike says it's because I'm stupid," said Brian, glumly.

"Well, he doesn't sound like a very good friend," said Mr Parrot. "Besides, Miss Flamingo tells me you are very clever, and you won a prize in painting last week."

"Yes," said Brian slowly, "but everyone laughed at me because I had to write my name on the painting and it came out all wrong!" His eyes filled with tears.

"Brian, we're all better at some things than others. Is Mike good at art?" asked Mr Parrot.

"No way!" giggled Brian. "His animals always look like monsters! And Humphrey can't make up stories," he went on. "His ideas are all really boring!"

"Well, there you are," said Mr Parrot. "Mike finds drawing hard, Humphrey finds ideas hard, and you find words hard."

"But why do I find them hard?" asked Brian. "They are easy when I think them, but they go all wonky when I try to write them or read them."

"Some people's brains work in a different way to others," said Mr Parrot. "I think you may have a special difficulty with words called dyslexia."

"Can you make dyslexia go away?" asked Brian.

"No, I can't," said Mr Parrot, "but there are things we can do to help you."

Mr Parrot gave Brian lots of tests to do to see if he was right about Brian's dyslexia. There were letter and word tests, numbers and spelling, reading and writing, listening and memory. Then Mr Parrot asked Brian to look at a picture and make up a story about what was happening in it. By the time his parents came to collect him, poor Brian was quite worn out!

"Oh, dear!" said Mrs Bear when Mr Parrot told them that Brian did have dyslexia. "What do we do now?"

"There's lots you can do to help Brian, and lots that he can do to help himself," said Mr Parrot. "And I would like to see him once a week for some special teaching."

The next day, Mr Bear went out and bought a lovely set of wooden letters in bright colours. Each day, he and Brian did a little bit of work with them. At first Brian had to find the letter his father asked for. Then they both said the name of the letter out loud and the sound it made. Finally Mr Bear covered up the letter and Brian had to write it. Sometimes he got it

wrong and he was really cross with himself, but after a while he started to get more of them right.

Mrs Bear made some letter cards. Then she asked Brian to draw something on the other side of the card that started with that letter. He drew a hat on the back of the 'h' card, and a sun on the back of the 's', and he drew himself on the 'b' card! " 'B' can be for Brian or bear," he said, proudly. "This is fun, Mum!"

Humphrey and Brian rode their bikes along the track by the river.

"Why did your Mum let you have your bike back?" called Humphrey over his shoulder. "Have you learned your left from your right?"

"Not really," said Brian. "I still mix them up sometimes, but Mum decided that left and right don't matter so much — it is more important to know the wrong side of the road from the right side."

"So how do you know that?" asked Humphrey.

"Easy!" cried Brian. "Mum put a bell on my handlebar, and as long I ride with it on the same side as the edge of the road, I know I'm on my side!"

Brring brring! Brian rang his bell, and the two friends cycled off together.

"Dad! Mum!" shouted Brian as he arrived home from school. "Guess what? My team won the end-of-term quiz today!"

"Fantastic!" said his father, and he gave Brian a great big bear hug. "Did you get a prize?" Brian held up a large bag of sweets.

Brian smiled all through supper, and he smiled when he was in the bath. He was still smiling when Mrs Bear tucked him into bed.

"Who would have thought that a quiz would make you so happy?" she chuckled.

"It's because I'm not stupid after all, Mum," said Brian happily. "I've just got dyslexia!" And he fell fast asleep — smiling!

Parents' pages: Dyslexia

What are the indicators?

There are many indicators for dyslexia. If your child displays some of the behaviours listed below, it does not necessarily mean that they have dyslexia. If your child displays a cluster of the behaviours and does not progress, discuss your concerns with their school.

Pre-School

- Late speech development
- Persistent jumbled phrases
- Difficulty learning nursery rhymes
- Difficulty in dressing
- Likes being read to but shows no interest in letters or words
- Poor attention span
- Excessive tripping, bumping into things, and falling over
- Difficulty with catching, throwing, or kicking a ball
- Forgets names, colours, etc.
- Difficulty keeping a simple rhythm

Primary School

- Difficulty with reading and spelling
- Puts letters and figures the wrong way round (mirror writing)
- Leaves letters out of words or puts them in the wrong order
- Difficulty remembering tables, alphabet, formulas, etc.
- Sometimes confuses 'b' and 'd' and words such as 'no' and 'on'
- Still uses fingers or marks on paper to do simple sums
- Takes longer than average to do written work
- Difficulty with tying laces, tie, dressing
- Still confuses left and right
- Confuses days of the week, months of the year

What should I do?

If you are concerned about your child's development, discuss it with their teacher. The school will be able to advise you on appropriate assessment, and if necessary, specialist teaching help. If your child is dyslexic, it is important that you build good relationships with their teachers throughout their schooling. If the school is uncooperative, you may need to seek independent assessment and help. Have your child's eyesight and hearing checked to ensure these are not contributing to their difficulties.

Will my child always be dyslexic?

Much can be done to help dyslexic children, but dyslexia cannot be 'cured'. Dyslexic children are often labelled as 'stupid' or 'disruptive' and it is important that the condition is recognized as early as possible, and specialist help is sought. With the right support, many of the difficulties caused by dyslexia can be overcome.

Doctor Spot says:

- If you have a partner, make sure you are both involved in consultations with the school or any specialists
- Some dyslexic children work extremely hard to keep up at school. Don't pressure them to do extra reading and writing at home if they are tired
- When teaching children to do up buttons, always get them to start from the bottom where it is easier for them to see what they are doing
- Watch television with your child and discuss what you have seen
- Many dyslexic children have areas of high ability as well as learning difficulties. Encourage their strengths and praise them when they do well
- When teaching your child to tie shoelaces or a tie, stand behind them if you are both right or left-handed, but in front if you are opposite-handed. Get your child to describe each action
- Treat aggression and anti-social behaviour gently but firmly. Remember that all children behave like this sometimes – don't blame it all on their dyslexia!
- Don't let your child use dyslexia as an excuse, either. Reassure them that some things may take longer, but that they will get there in the end
- Your child may be teased at school by classmates. Watch out for stress signs such as bedwetting or introversion. Encourage your child to talk about their emotions
- A dyslexic child may take up a lot of time, but try to make sure this is not at the expense of other children in the family

For more advice and information on dyslexia, contact Dyslexia Action at www.dyslexiaaction.org.uk (training and resource centre: 01784 222300; email: training@dyslexiaaction.org.uk)

Titles in the Doctor Spot series:

Brian has Dyslexia ISBN: 978-1-905339-86-0
Brian is a very bright bear, but keeps getting his words muddled up. Doctor Spot helps him to understand his problem.

Charlie has Asthma ISBN: 978-1-905339-83-9
Charlie the Cheetah is always running out of breath. Doctor Spot tells him what's wrong and gives Charlie a special inhaler.

Emma has Measles ISBN: 978-1-905339-89-1
When Emma the Elephant catches measles, it looks as though she's going to miss her star part in the play.

Franklin has Conjunctivitis ISBN: 978-1-905339-85-3
Franklin can't open his eyes on a camping trip.

George has Meningitis ISBN: 978-1-905339-90-7
George the Gorilla is feeling very ill. His sister, Gloria, learned about meningitis at school, and her father calls Doctor Spot without delay.

Harriet has Tonsillitis ISBN: 978-1-905339-91-4
Harriet the Hippopotamus has a nasty case of tonsillitis.

Lawrence has Head Lice ISBN: 978-1-905339-88-4
Lawrence the Lion gets a shock when the barber finds nits in his mane. Doctor Spot tells Lawrence's class all about head lice.

Mike has Chicken-pox ISBN: 978-1-905339-84-6
Mike the Monkey comes out in spots and feels uncomfortably itchy.

Rachel has Eczema ISBN: 978-1-905339-87-7
Rachel the Rhino is sore and itchy and can't sleep at night. Doctor Spot prescribes ointments and dressings which soon make her feel better.

Zak has ADHD ISBN: 978-1-905339-82-2
Zak the Zebra is naughty, rude and unpopular, but all he wants is to make friends and be like the other children.